DNA Cloning 4

The Practical Approach Series

SERIES EDITORS

D. RICKWOOD
Department of Biology, University of Essex
Wivenhoe Park, Colchester, Essex CO4 3SQ, UK

B. D. HAMES
Department of Biochemistry and Molecular Biology
University of Leeds, Leeds LS2 9JT, UK

★ indicates new and forthcoming titles

Affinity Chromatography
Anaerobic Microbiology
Animal Cell Culture (2nd Edition)
Animal Virus Pathogenesis
Antibodies I and II
★ Basic Cell Culture
Behavioural Neuroscience
Biochemical Toxicology
★ Bioenergetics
Biological Data Analysis
Biological Membranes
Biomechanics—Materials
Biomechanics—Structures and Systems
Biosensors
★ Carbohydrate Analysis (2nd Edition)
Cell–Cell Interactions
★ The Cell Cycle
★ Cell Growth and Apoptosis
Cellular Calcium
Cellular Interactions in Development
Cellular Neurobiology
Clinical Immunology
Crystallization of Nucleic Acids and Proteins
★ Cytokines (2nd Edition)
The Cytoskeleton
Diagnostic Molecular Pathology I and II
Directed Mutagenesis
★ DNA Cloning 1: Core Techniques (2nd Edition)
★ DNA Cloning 2: Expression Systems (2nd Edition)
★ DNA Cloning 3: Complex Genomes (2nd Edition)
★ DNA Cloning 4: Mammalian Systems (2nd Edition)
Electron Microscopy in Biology
Electron Microscopy in Molecular Biology
Electrophysiology

Enzyme Assays
Essential Developmental Biology
Essential Molecular Biology I and II
Experimental Neuroanatomy
★ Extracellular Matrix
★ Flow Cytometry (2nd Edition)
Gas Chromatography
Gel Electrophoresis of Nucleic Acids (2nd Edition)
Gel Electrophoresis of Proteins (2nd Edition)
★ Gene Probes 1 and 2
Gene Targeting
Gene Transcription
Glycobiology
Growth Factors
Haemopoiesis
Histocompatibility Testing
★ HIV Volumes 1 and 2
Human Cytogenetics I and II (2nd Edition)
Human Genetic Disease Analysis
Immunocytochemistry
In Situ Hybridization
Iodinated Density Gradient Media
★ Ion Channels
Lipid Analysis
Lipid Modification of Proteins
Lipoprotein Analysis
Liposomes
Mammalian Cell Biotechnology
Medical Bacteriology
Medical Mycology
★ Medical Parasitology
★ Medical Virology
Microcomputers in Biology
Molecular Genetic Analysis of Populations
★ Molecular Genetics of Yeast
Molecular Imaging in Neuroscience
Molecular Neurobiology
Molecular Plant Pathology I and II
Molecular Virology
Monitoring Neuronal Activity
Mutagenicity Testing
★ Neural Cell Culture
Neural Transplantation
Neurochemistry
Neuronal Cell Lines
NMR of Biological Macromolecules
★ Non-isotopic Methods in Molecular Biology
Nucleic Acid Hybridization
Nucleic Acid and Protein Sequence Analysis
Oligonucleotides and Analogues
Oligonucleotide Synthesis
PCR 1
★ PCR 2
★ Peptide Antigens
Photosynthesis: Energy Transduction
★ Plant Cell Biology
★ Plant Cell Culture (2nd Edition)
Plant Molecular Biology

★ Plasmids (2nd Edition)
Pollination Ecology
Postimplantation Mammalian Embryos
Preparative Centrifugation
Prostaglandins and Related Substances
★ Protein Blotting
Protein Engineering
Protein Function
Protein Phosphorylation
Protein Purification Applications
Protein Purification Methods
Protein Sequencing
Protein Structure
Protein Targeting
Proteolytic Enzymes
★ Pulsed Field Gel Electrophoresis
Radioisotopes in Biology
Receptor Biochemistry
Receptor–Ligand Interactions
RNA Processing I and II
Signal Transduction
Solid Phase Peptide Synthesis
Transcription Factors
Transcription and Translation
Tumour Immunobiology
Virology
Yeast

DNA Cloning 4
Mammalian Systems
A Practical Approach

Edited by

D. M. GLOVER

Cell Cycle Genetics Research Group, CRC Laboratories, Department of Anatomy and Physiology, Medical Sciences Institute, The University, Dundee DD1 4HN, UK

and

B. D. HAMES

Department of Biochemistry and Molecular Biology, University of Leeds, Leeds LS2 9JT, UK

at
OXFORD UNIVERSITY PRESS
Oxford New York Tokyo

Walton Street, Oxford OX2 6DP
d New York
nd Bangkok Bombay
wn Dar es Salaam Delhi
Kong Istanbul Karachi
Madras Madrid Melbourne
Mexico City Nairobi Paris Singapore
Taipei Tokyo Toronto
and associated companies in
Berlin Ibadan

Oxford is a trade mark of Oxford University Press

Published in the United States
by Oxford University Press Inc., New York

A catalogue record for this book is available from the British Library

Library of Congress Cataloging-in-Publication Data
(Data available)

ISBN 0 19 963481 5 (Hbk)
ISBN 0 19 963480 7 (Pbk)

Typeset by Footnote Graphics, Warminster, Wilts
Printed in Great Britain by Information Press Ltd, Eynsham, Oxon.

Preface

It is a decade since the first editions of *DNA Cloning* were being prepared for the Practical Approach series. It is illuminating to look back at those volumes and reflect how the field has evolved over that period. We have tried to distil out of those earlier volumes such techniques as have withstood the test of time, and have asked many of the former contributors to update their chapters. We have also invited several new authors to write chapters in areas that have come to the forefront of this invaluable technology over recent years. The field is, however, far too large to cover comprehensively, and so we have had to be selective in the areas we have chosen. This has also led to each of the books being focused onto particular topics. This having been said, Volume 1 covers core techniques that are central to the cloning and analysis of DNA in most laboratories. Volume 2 on the other hand turns to the systems used for expressing cloned genes. Inevitably the descriptions of these techniques can be supplemented by reference to other cloning manuals such as *Molecular cloning* by Sambrook, Fritsch, and Maniatis (Cold Spring Harbor Laboratory, New York) as well as other books in the Practical Approach series. Volume 3 examines the analysis of complex genomes, an area in which there have been many important developments in recent years, both in the description of new vectors and in the strategic approaches to genome mapping. Again, companion volumes such as *Genome analysis: a practical approach* are also available. Finally in Volume 4 we look at DNA cloning and expression in mammalian systems; from cultured cells to the whole animal.

In recent years, heterologous gene expression in mammalian cells has become a key technique for studying many aspects of eukaryote cell and molecular biology. With this increasing use has come a plethora of vectors, reporter genes, transfection protocols, etc. which can represent a daunting prospect for the individual researcher. The first chapter of the present volume offers detailed background information and practical guidance on these issues leading to a rational choice of experimental approach to the individual expression project in hand. Chapter 2 deals with the use of vaccinia virus as a mammalian expression vector but the basic principles and approaches described are equally applicable to other poxviruses. Chapter 3 focuses on a class of selectable markers that can be used to obtain markedly increased efficiency of gene expression from integrating vectors by amplification. These are amongst the most efficient mammalian expression systems currently available. The two most widely used amplifiable marker genes are described in detail; dihydrofolate reductase (DHFR) and glutamine synthetase (GS). Retroviruses are also an attractive route for high-level expression given the existence of strong retroviral enhancers and the fact that the efficiency of infection of host cells with retroviruses is extremely high, permitting infection of almost 100% of the target cells in culture. The choice of retroviral vector, the generation of recombinants, and their expression in mammalian cells are described in Chapter 4. Embryonic stem (ES) cells are pluripotent stem cell lines established from mouse embryos. When introduced back into mouse embryos, they can contribute to all of the tissues of the resulting mouse and so any mutations introduced into the ES cell genome *in vitro* can be incorporated into the genome of the whole mouse. The potential for screening for rare genetic events in culture before introduction into embryos has led to increasing use of ES cells in gene targeting or gene trapping. Chapter 5 describes the genetic manipulation of ES cells in detail. Another set of extremely powerful techniques is mammalian transgenesis which has the major merit of permitting the study of expression within the whole organism. Most transgenic animals expressing transgenes do so in a manner that is appropriate to the regulatory elements present. Chapters 6 and 7 by Sarah Jane Waller, David Murphy, and their colleagues concentrate on the production of transgenic rodents by microinjection of fertilized one-cell eggs (the most efficient procedure available) and the genomic and expression analysis of transgenic animals respectively. Chapter 8 describes a defective Herpes simplex virus (HSV-1) vector system for expression analyses in cultured neurones and in neurones *in vivo*, that is, in the mammalian brain. This vector system has been used successfully to introduce functional gene products into multiple neuronal cell types and offers great potential for future work on brain function. Adenovirus vectors for high-efficiency transient expression of foreign genes in mammalian cells are the subject of the final chapter. The attraction of these vectors lies in the broad species and tissue trophism of adenoviruses coupled with the availability of simple techniques for the generation of recombinants and the high titres and high efficiency of gene transfer.

It is our hope that this book and its sister volumes will find their way on to bookshelves in laboratories and in so doing will become well-used, messy, and 'dog-eared'. It has been gratifying to see how widespread the use of the first edition has become. A similar success for the second edition would be rewarding to both the editors and the authors. Finally, and most importantly, we wish to thank all of the authors for their contributions and for their patience in bringing this project to fruition.

Dundee and Leeds David M. Glover
April 1995 B. David Hames

Contents

Contributors

CHRISTOPHER BEBBINGTON
Celltech Therapeutics Ltd, 216 Bath Road, Slough S11 4EN, UK.

ANTHONY M.C. BROWN
Department of Cell Biology and Anatomy, and Strang-Cornell Cancer Research Laboratory, Cornell University Medical College, New York, NY 10021, USA.

VANESSA CHISHOLM
Department of Cell Genetics, Genentech Inc., South San Francisco, CA 94080, USA.

KUM-FAI CHOOI
Neuropeptide Lab., Institute of Molecular & Cell Biology, National University of Singapore, 10 Kent Ridge Crescent, Singapore 0511.

JOSEPH P. DOUGHERTY
Department of Microbiology and Molecular Genetics, Robert Wood Johnson Medical School, University of Medicine and Dentistry of New Jersey, Piscataway, NJ 08854, USA.

JUDITH McNIFF FUNKHOUSER
Neuropeptide Lab., Institute of Molecular & Cell Biology, National University of Singapore, 10 Kent Ridge Crescent, Singapore 0511.

ALFRED I. GELLER
Childrens Hospital and Program in Neuroscience, Harvard Medical School, Boston, MA 02115, USA.

ROBERT D. GERARD
Molecular Cardiology Research Laboratories, NB11.102a, University of Texas Southwestern Medical Center, 6000 Harry Hines Blvd, Dallas, TX 75235–8573, USA.

DEAN HARTLEY
Division of Endocrinology, Childrens Hospital, 300 Longwood, Boston, MA 02115, USA.

MEI-YIN HO
Neuropeptide Lab., Institute of Molecular & Cell Biology, National University of Singapore, 10 Kent Ridge Crescent, Singapore 0511.

BEVERLY H. KOLLER
724 Burnett-Womack, CB# 7020, University of North Carolina at Chapel Hill, Chapel Hill, NC 27599, USA.

PHUNG LANG
Division of Endocrinology, Childrens Hospital, 300 Longwood, Boston, MA 02115, USA.

FILIP LIM
Division of Endocrinology, Childrens Hospital, 300 Longwood, Boston, MA 02115, USA.

MIKE MACKETT
CRC Christie Research Centre, Department of Molecular Biology, Christie Hospital, Manchester M20 9BX, UK.

ROBERT S. MEIDELL
Molecular Cardiology Research Laboratories, NB11.102a, University of Texas Southwestern Medical Center, 6000 Harry Hines Blvd, Dallas, TX 75235–8573, USA.

AMY MOHN
724 Burnett-Womack, CB# 7020, University of North Carolina at Chapel Hill, Chapel Hill, NC 27599, USA.

DAVID MURPHY
Neuropeptide Lab., Institute of Molecular & Cell Biology, National University of Singapore, 10 Kent Ridge Crescent, Singapore 0511.

SONG SONG
Division of Endocrinology, Childrens Hospital, 300 Longwood, Boston, MA 02115, USA.

PHILIP STARR
Division of Endocrinology, Childrens Hospital, 300 Longwood, Boston, MA 02115, USA.

SARAH JANE WALLER
Neuropeptide Lab., Institute of Molecular & Cell Biology, National University of Singapore, 10 Kent Ridge Crescent, Singapore 0511.

YAMING WANG
Division of Endocrinology, Childrens Hospital Medical Center, 300 Longwood, Boston, MA 02115, USA.

Abbreviations

ATCC	American Tissue Culture Collection
BSA	bovine serum albumin
BUdR	5-bromodeoxyuridine
CAT	chloramphenicol acetyl transferase
CHO	Chinese hamster ovary
CMV	cytomegalovirus
CPE	cytopathic effect
CTL	cytotoxic T lymphocytes
DEPC	diethylpyrocarbonate
DHFR	dihydrofolate reductase
DIC	differential interference contrast
DIF	differentiation-inhibition factor
DMEM	Dulbecco's modified Eagle's medium
DMSO	dimethyl sulfoxide
DTT	dithiothreitol
EBV	Epstein–Barr virus
EDTA	ethylenediaminetetraacetic acid
EEV	extracellular enveloped virus
EMCV	encephalomyocarditis virus
ES	embryonic stem
FACS	fluorescence activated cell sorting
FBS	fetal bovine serum
FCS	fetal calf serum
FSH	follicle stimulating hormone
gpt	guanine phosphoribosyl transferase (gene)
GRE	glucocorticoid response element
GS	glutamine synthetase
HAT	hypoxanthine, aminopterin, thymidine
HBS	Hepes-buffered saline
hCG	human chorionic gonadotrophin
HEK	human embryonic kidney
hisD	histidinol dehydrogenase
HIV	human immunodeficiency virus
hph	hygromycin-B-phosphotransferase (gene)
hprt	hypoxanthine phosphoribosyl transferase (gene)
HRP	horse-radish peroxidase
HS	horse serum

HSV	herpes simplex virus
IE	immediate early
INV	intracellular naked virus
IP	infectious particles
IPTG	β-D-thiogalactopyranoside
IRES	internal ribosome entry site
ISH	*in situ* hybridization
IVF	*in vitro* fertilization
lacZ	β-galactosidase (gene)
LIF	leukaemia-inhibition factor
LPS	lipopolysaccharide
LTR	long terminal repeat
MEM	minimal essential medium
MLV	murine leukaemia virus
MMTV	mouse mammary tumour virus
m.o.i.	multiplicity of infection
MPA	mycophenolic acid
MSX	methionine sulfoximine
MT	metallothionein
MTX	methotrexate
NEAA	non-essential amino acids
neo	neomycin phosphotransferase (gene)
ORF	open reading frame
PBS	phosphate-buffered saline
p.c.	post-coitum
PCR	polymerase chain reaction
PDL	poly-D-lysine
PEG	polyethylene glycol
p.f.u.	plaque-forming units
p.i.	post-infection
RSV	Rous sarcoma virus
rtPA	recombinant tissue plasminogen activator
RT-PCR	reverse transcriptase PCR
SDS	sodium dodecyl sulfate
SIN	self-inactivating
SNV	spleen necrosis virus
STO	embryonic fibroblast cell line
SV40	simian virus 40
TK	thymidine kinase
tPA	tissue plasminogen activator
ts	temperature sensitive
TU	transducing units
UTR	untranslated region
VA	viral associated

VP	virion protein
VSV	vesicular stomatitis virus
WR	Western Reserve strain
wt	wild-type
X-Gal	5-bromo-4-chloro-3-indolyl-β-D-galactopyranoside
YAC	yeast artificial chromosome

1

High efficiency gene transfer into mammalian cells

VANESSA CHISHOLM

1. Introduction

Over the past decade, recombinant or heterologous gene expression in mammalian cells has emerged as an essential tool for studying multiple aspects of eukaryotic biology. The development of highly efficient vector-based gene transfection methods has provided researchers with almost unlimited possibilities for the use of engineered protein products derived from mammalian cell culture. What has become apparent is that highly individualized scientific questions can be routinely addressed using a relatively distinct battery of these types of expression protocols. The choice of technique and how it can be effectively used depends primarily on the specific requirements of the experiment. An extensive series of expression vector designs have been developed. Some are more suited to introducing and analysing gene products transiently while others are better for the stable and continued expression of heterologous DNA in cells. Vectors have been constructed to exploit certain cell types. Likewise, transfection methods can be categorized into those more effective for short-term expression analysis and those more suited to select stably expressing cell lines. The identification of recombinant products has evolved from time-consuming radioactive assays to more straightforward immunoassays and luminescent techniques. What may appear to be a most formidable field of players to choose from is best approached by becoming familiar with the basic characteristics essential for the efficient transfer and subsequent expression of genes in mammalian cells. In general terms these include:

- the key mammalian vector components and how they function
- the transfection method and/or system that will best suit the experimental goals
- outside parameters that affect transfection efficiencies
- simple techniques available for expression analysis that could be used to ascertain the transcriptional potency of an expression construct or the presence of recombinant protein

- additional aspects that may enhance expression in particular systems or cell types

The primary focus of this chapter is to provide individual investigators with framework techniques and information on which to build their own specific experimental approach to a given transfection/expression problem. Although not a compendium of all the possible techniques available to date, the protocols in this chapter were chosen primarily for their simplicity and high probabilities of success. Where possible, emphasis has been given to those methods which do not require large amounts of special equipment or radioactivity.

2. Vector design

The basic mammalian vector contains at the very least six key components, four of which are directly involved in promoting efficient expression of a foreign sequence, whether it be cDNA, genomic DNA, or even a bacterial antibiotic resistance gene. The salient design features of these vectors include:

- backbone 'shuttle' sequences for providing replication and maintenance functions in bacteria
- an appropriate translational initiation sequence
- an intron
- a mammalian polyadenylation signal
- a mammalian-derived promoter/enhancer capable of driving transcription in the cells of choice
- a marker gene for co-selection in mammalian cells (optional; see section 5)

These general features do, of course, vary from plasmid to plasmid. Some of the more widely used elements and their combinations from each category are defined below.

2.1 Mammalian 'shuttle' vectors

'Shuttle' vectors by definition contain components which allow the movement of plasmids from one organism to another. To maintain the ability to manipulate and propagate mammalian expression vectors efficiently, bacterial replication and selection sequences are routinely included as the backbone of the plasmid. The basic framework often includes the pBR322 replication origin, ColE1 for plasmid propagation, and the bacterial β-lactamase gene which confers ampicillin resistance for dominant selection in *E. coli*. Other genes encoding bacterial selectable markers which may be present include those conferring tetracycline resistance and chloramphenicol resistance, and a variety of suppressor tRNA genes for selection by suppression of nonsense alleles present in defined genes within the bacterial host. It is worth the effort to become familiar with these selection systems since many vectors require very specific bacterial hosts in which to propagate. Many of the companies which supply vectors and molecular biology reagents include this information in their catalogues (i.e. New England BioLabs®). The pUC vector sequences have become increasingly popular because they provide an additional element, the f1 or M13 origin of replication, as a means of generating single-stranded DNA templates for sequencing and *in vitro* mutagenesis. Also included, again for ease of molecular manipulation, are promoter sequences for T3 or T7 bacteriophage DNA-dependent RNA polymerase, to allow the generation of RNA transcripts from the DNA insert.

These framework sequences, although not directly involved in the functionality of the mammalian transcription unit, are of utmost importance in maintaining the overall integrity of the vector. One should take care that their activities are not affected during design of expression constructs. Most standard vectors include multiple cloning sites (a series of restriction endonuclease sites, usually occurring only once in the vector) for ease of insertion of the test gene without disrupting the framework sequences. The framework sequences can also have an overall effect, albeit indirectly, on expression in transfected cells. The classic example of this is the presence of 'poison sequences' found in pBR322 which affected the replication rates of SV40 DNA/bacterial chimeras in simian cells (1). Vectors which contained these sequences were shown to be markedly depressed in viral replication functions. It is therefore wise to be aware of differences in framework sequences when comparing the expression efficiencies of different vectors.

2.2 Inclusion of the 'Kozak' consensus translational initiation sequence

The conditions for optimal translational efficiency of expressed mammalian genes have been studied extensively (2). As a result, it has become almost routine practice to include the 'Kozak' defined translational initiation sequence (CCA/GCC*ATG*) before the initiating ATG of cDNA gene expression constructs. The optimal base at the +4 position is a G (standard numbering begins +1 at the A of the initiating ATG), but depending on the amino acid at that position in the protein product, the presence of a G may or may not be possible.

The secondary structure of the 5′ untranslated region (UTR) of many genes can potentially inhibit expression in a recombinant system. In addition, mammalian cDNAs often have multiple ATG codons within this region, only one of which can act as the appropriate translational initiation signal. Expression constructs should therefore be carefully designed and the inclusion of any 5′ UTR sequences tested for their effects on expression levels. In general, however, it is usually better to delete these sequences from the final expression vector.

2.3 Introns

Probably the most controversial and mechanistically complex sequences associated with overall expression levels for commonly used vectors are introns. Splicing of primary mRNA transcripts is a common event in mammalian cells. Highly conserved sequence signals direct the cleavage of a 5′ intron/exon border (the splice donor), ligation of the 5′ end of the intron to a 2′ OH of an adenylate residue within the intron sequences (the branch point), cleavage at the 3′ intron/exon border (the splice acceptor), and subsequent ligation of the two exon ends. Although examples can be found where gene expression does not appear to be dependent on splicing, many studies have indicated that the addition of an intron to the 5′ untranslated leader in a vector can significantly improve expression (3).

2.4 Polyadenylation signals

Although not fully understood, polyadenylation of mature mRNA is known to be critical for message stability. Many standard expression vectors include the SV40 early polyadenylation signal so as to negate any need to include sequences beyond the stop codon when transferring the test cDNA gene into the vector (4).

2.5 Mammalian promoter/enhancers

The promoter consists of those sequences where RNA polymerase II transcription is initiated. However, other sites facilitating initiation by RNA polymerase II, such as CAAT and TATAA boxes, are also regarded as promoter elements. Promoters are spatially oriented to direct transcription in a defined direction. Enhancer elements, on the other hand, are composed of sequences which augment transcription from a promoter in a manner which is at least partially independent of orientation and distance. The presence of enhancer elements is essential to achieve the highest levels of transcriptional activation.

The most generic and consequently the most versatile expression assemblies have originated from viral systems. Commonly used vectors, especially many of those commercially available, take advantage of the high transcriptional activities of either the SV40 early enhancer/promoter (5), the Rous sarcoma virus long terminal repeat (RSV-LTR) (6), or the human cytomegalovirus major immediate early enhancer/promoter (hCMVE/P) (7) to drive gene expression. The choice of which combination to use depends mainly on the cell type. For example, the hCMV, active in human cell lines generally, has been shown to be very active in adenovirus-transformed human lines like HEK-293 (see section 3.2). In this case, certain proteins from the adenovirus E1 region were shown to *trans*-activate the hCMV IE1 early promoter but in turn repress the function of both the SV40 and RSV enhancer sequences (8).

Regulated transcription has become an intense area of interest for those involved in heterologous gene expression particularly in cases where the protein product is toxic or when regulation of expression of the foreign gene is desirable. Aside from the many classic 'inducible' promoters which exist (e.g. the metallothionein (MT), or heat shock promoters), present attention is focused on engineered systems where the interacting components are co-introduced into cells to establish function. One subset of this type of system requires an exogenously added chemical or hormone to induce expression. Among the best studied of these are the steroid hormone regulated promoters. The glucocorticoid-inducible mouse mammary tumour virus (MMTV) LTR relies on the presence of a ligand regulated steroid receptor protein to bind and activate transcription from glucocorticoid response elements (GREs) in the promoter (9). A similar hormone response system based on the *Drosophila* ecdysone receptor has been extensively characterized in HEK-293 cells (10). The usefulness of these systems is increased by the ability to produce chimeric receptors by mixing different DNA binding and *trans*-activation regions from a variety of sources to customize expression.

Bacterial regulated promoters have also been engineered for use in mammalian systems. Based on the *lac* operon of *E. coli*, the '*lac* switch' (now available from Stratagene) relies on a modified *lac* repressor protein, targeted to the cell nucleus, to repress the expression of genes transcriptionally driven from promoters whose functions are regulated by *lac* operator regions (11). The expression of these genes is then induced by the addition of β-D-thiogalactopyranoside (IPTG). Another genetically regulated system of this type is based on the Tn10-encoded tetracycline resistance operon. Here, the *tet* repressor was fused with the C-terminal activating domain of the herpes simplex virus (HSV) virion protein 16 (VP16), known to be essential for the transcription of immediate early viral genes. This *trans*-activating chimeric protein can stimulate promoter function from a hCMV promoter immediate early region fused with *tet* operator sequences (12).

Other *trans*-activation systems have been useful in converting cell lines which lack these factors to ones where dependent promoters become highly active, sometimes beyond levels of expression previously attained with standard promoter/enhancers. Examples include host cell/vector systems based on the use of adenovirus transformed cells (e.g. HEK-293) and a multi-component promoter element capable of being highly activated by the E1A tumour antigen (13), and a 'cascade'-like system where a hormone regulated herpes VP16 protein transcriptionally activates IE promoter elements (14). Stable gene expression levels were demonstrated with this latter system to be superior to standard CHO cell levels derived from vectors driven by SV40 early, RSV-LTR, and mouse metallothionein promoter (mMT) promoter/enhancer elements.

Many of these systems have their own advantages and disadvantages for use in a given experimental scenario. It is clear, however, that in the broad sense, they reflect the state of the art in transfection technology with overall advantages in both transient and stable gene expression studies.

3. Providing good substrates for transfection; DNA and cells

As a preface to the actual transfection methodology, it is important to consider the preparation of the key components, the DNA and the cells. Section 3.1 describes the provision of high quality DNA as template for transfer and section 3.2 covers basic cell culture parameters for ensuring efficient transfection.

3.1 Preparation of DNA for transfection

Successful gene transfer experiments rely on the use of high quality reagents, in particular the quality of the DNA. Caesium chloride (CsCl) centrifugation has been the method of choice for DNA preparation for many years, the 'gold standard' for preparing pure DNA template for transfections. Recently an alternative methodology, QIAGEN® anion-exchange chromatography, has been demonstrated to provide DNA that may be more effective in transient expression; less DNA prepared by this method was shown to be required for maximum expression when compared to CsCl prepared DNA in an assay using NIH 3T3 cells (15). The QIAGEN® method of DNA purification also requires considerably less time and equipment than CsCl banding of DNA. Furthermore, many of the toxic reagents usually associated with DNA purification (ethidium bromide and organics such as phenol and butanol) are used minimally or not at all. As well as the QIAGEN® columns themselves, all of the solutions required for cell lysis (alkaline lysis method), column equilibration, and DNA elution, as well as protocols for optimal bacterial culture growth, and final DNA column purification, are available from QIAGEN® Inc. For pUC framework vectors, we simply grow 50 ml cultures in LB broth supplemented with the appropriate antibiotic, overnight at 37°C (approx. 16–20 h) inoculated from a reasonably fresh (⩽ one-week-old) master plate of the *E. coli* transformant of interest. Using the QIAGEN®-tip 500 protocol (*Protocol 1*) enough DNA (200–400 μg/preparation) is obtained for use in multiple transient or stable transfections. Many preparations of this type can be carried out at one time, making direct comparison of multiple vector constructions very efficient. *Protocol 1* highlights the simplicity of this technique for purifying plasmid DNA.

Alternatively, DNA can still be prepared using CsCl centrifugation coupled with any number of cell lysis methods. A procedure of this type is provided in *Protocol 2*. DNA preparations obtained by this method should be monitored as to their degree of supercoiling since expression in a transient assay can be affected by the presence of nicked and relaxed vector. Closed, circular, supercoiled plasmid is the superior DNA template.

Protocol 1. Preparation of plasmid DNA by alkaline lysis/QIAGEN® anion-exchange chromatography (maxi preparation for up to 500 μg of DNA)[a]

Equipment and reagents

- 50 ml polyallomer centrifuge tubes (e.g. Nalgene Oak Ridge centrifuge tubes 3119–0050)
- 30 ml Corex centrifuge tube or equivalent
- Sorvall centrifuge with SS-34 rotor and HB-4 or HB-6 rotor or equivalents
- QIAGEN®-tip 500 anion-exchange columns (supplied by QIAGEN Inc.) (store the columns at room temperature)
- Appropriate culture volume of *E. coli* transformant[b]
- Buffer P1: 100 μg/ml RNase A, 50 mM Tris–HCl, 20 mM EDTA pH 8.0 (store this buffer at 4°C)
- Buffer P2: 0.2 M NaOH, 1% SDS (store at room temperature; at lower temperatures the SDS will precipitate out)
- Buffer P3: 3 M potassium acetate pH 5.5 (store at 4°C)
- Buffer QBT: 750 mM NaCl, 15% ethanol, 0.15% Triton X-100, 50 mM MOPS pH 7.0 (store at room temperature)
- Buffer QC: 1 M NaCl, 15% ethanol, 50 mM MOPS pH 7.0 (store at room temperature)
- Buffer QF: 1.25 M NaCl, 15% ethanol, 50 mM Tris–HCl pH 8.5 (store at room temperature)
- TE buffer: 10 mM Tris–HCl, 1 mM EDTA pH 8.0
- STE buffer: 10 mM Tris–HCl, 1 mM EDTA, 100 mM NaCl pH 8.0
- 70% ethanol (store this solution at −20°C and use it cold)
- Isopropanol

Method

1. Pellet the transformed bacterial cells (use a 50 ml polyallomer centrifuge tube) by centrifugation at 6000 *g* for 12 min at 4°C.
2. Resuspend the bacterial pellet in 10 ml of buffer P1. Resuspend the cells completely and be sure there are no clumps of cells remaining.
3. Add 10 ml of buffer P2 to the solution from step 2 and mix by gently inverting the tube four to six times. Do not vortex.
4. Incubate the mixture at room temperature for no more than 5 min.
5. Add 10 ml of chilled buffer P3 and mix immediately by gently inverting the tube four to six times. Do not vortex. The solution should become cloudy and very viscous.
6. Leave this solution on ice for 20 min.
7. Mix the samples gently by inverting the tube once or twice immediately prior to centrifugation.
8. Centrifuge the tubes at 4°C for 30 min at 30 000 *g* in a Sorvall SS-34 rotor or equivalent.
9. While centrifuging the cell extracts, equilibrate the QIAGEN®-tip 500 column by applying 10 ml of buffer QBT and allowing the column to empty by gravity flow.[c]
10. Decant the supernatant from the pelleted material. If the supernatant

Protocol 1. *Continued*

is clear, it can be directly loaded on to the pre-equilibrated column from step 9. If the supernatant is not clear, a second, shorter centrifugation (15 min at 30 000 *g* at 4°C) should be done prior to loading the column from step 9. This step will prevent applying particulate material to the column which will clog it and reduce the gravity flow rate.

11. After the extract has passed through the column, wash the column twice with 30 ml of buffer QC. This step removes any contaminants (i.e. RNA) from the bound plasmid DNA.
12. Elute the bound plasmid DNA from the column with 15 ml of buffer QF. Collect the eluate in a 30 ml Corex tube or equivalent.
13. Precipitate the plasmid DNA by adding 0.7 vol. of isopropanol that has been pre-equilibrated to room temperature to minimize salt precipitation.
14. Collect the precipitated plasmid DNA immediately by centrifugation at $\geqslant$ 15 000 *g* at 4°C for 30 min in a Sorvall HB-4 or HB-6 rotor or equivalent. The pelleted plasmid DNA is glassy and translucent in appearance, and care should be taken when pouring off the isopropanol after this centrifugation.
15. Wash the pelleted plasmid DNA briefly with 15 ml of cold 70% ethanol. As an option, the DNA can be washed with room temperature ethanol if it is to be used under conditions where high salt concentrations will interfere. It is usually necessary to briefly re-pellet the DNA by centrifugation ($\geqslant$ 15 000 *g* for 15 min at 4°C) prior to decanting the ethanol wash.
16. Air dry the DNA pellet for 10–15 min and then resuspend it in a small volume of TE buffer.[d] Store the plasmid DNA at −20°C.

[a] This protocol is a condensed version of the information found in the QIAGEN® Plasmid Handbook and is presented here by courtesy of QIAGEN® Inc.
[b] The QIAGEN® Plasmid Handbook gives extensive information on preparing suitable bacterial cultures for QIAGEN® column DNA purification. This handbook also includes extensive background information for using and troubleshooting the QIAGEN® columns.
[c] QIAGEN® columns empty by gravity flow. The flow of the buffer stops when the meniscus reaches the upper frit in the column. This prevents drying out and thus the columns do not need to be attended at all times.
[d] Depending on the culture volume and amounts of DNA expected, the volume of TE used for dissolving the plasmid DNA can be adjusted to give reasonable working concentrations. The DNA can be quantitated by determining the absorbance of the DNA solution at 260 nm. An A_{260} of 1.0 is equivalent to 50 μg/ml of double-stranded DNA.

Protocol 2. Preparation of plasmid DNA by alkaline lysis[a] and CsCl banding

Equipment and reagents

- Ultracentrifuge and rotor (Beckman VTi 80 rotor or equivalent)
- 5 ml Beckman polyallomer Quick Seal® ultracentrifuge tubes and sealer
- 50 ml polyallomer centrifuge tubes (as in *Protocol 1*)
- Hand-held medium wave UV transilluminator
- Safety face-shield for working with UV light
- 3–5 ml syringe plus 18 gauge needles
- Luria Bertani (LB) broth: 10% tryptone, 5% yeast extract, 10% NaCl pH 7.5
- 2 × YT broth: 16% tryptone, 10% yeast extract, 5% NaCl pH 7.0
- STE buffer: 100 mM NaCl, 10 mM EDTA, 10 mM Tris–HCl pH 8.0
- TE buffer: 10 mM EDTA, 10 mM Tris–HCl pH 8.0
- Solution I: 50 mM glucose, 10 mM EDTA, 25 mM Tris–HCl, pH 8.0
- Solution II: 0.5 M NaOH, 1% SDS (make fresh from 10 × stocks prior to use)
- 10 mg/ml lysozyme: chicken egg white lysozyme (Sigma L-6876) in solution I
- 1 g/ml CsCl in TE buffer (e.g. from Gibco-BRL)
- Ethidium bromide (5 mg/ml in water) (**NB**: this reagent is highly carcinogenic so handle it using appropriate safety precautions; it is also light-sensitive so store in a light-proof container)
- Water-saturated *n*-butanol
- Ethanol (100% and 70%)
- 3 M potassium acetate pH 4.8
- 3 M sodium acetate pH 7.0
- Isopropanol

Method

1. Inoculate a 5 ml culture of the plasmid-bearing *E. coli* strain in LB broth containing the appropriate antibiotic(s) for plasmid selection. Grow at 37°C overnight.
2. Use the 5 ml culture to inoculate 500 ml 2 × YT broth containing the appropriate antibiotic(s) in a 2 litre flask (1/100 dilution of the overnight culture). Grow at 37°C for exactly 12 h.
3. Pellet the cells by centrifugation at 6000 *g* for 12 min at 4°C.
4. Resuspend each pellet in 15 ml of STE buffer and transfer to a 50 ml polyallomer tube.
5. Pellet the cells by centrifugation at 6000 *g* for 12 min at 4°C.
6. Resuspend each pellet in 4 ml of solution I. Vortex the pellet to make sure there are no clumps and then add 1 ml of 10 mg/ml lysozyme in solution I to each tube (2 mg/ml final concentration). Vortex well after addition and vortex again after 5 min. Incubate at room temperature for a total of 10 min.
7. Add 10 ml of solution II to each tube. Mix by gently inverting two or three times and keep on ice for 10 min. The solution should become clear and very viscous. *Caution*; this step denatures all the DNA and hence vigorous shaking will result in the breakage of large DNA molecules, especially chromosomal DNA.
8. Add 7.5 ml of 3 M potassium acetate pH 4.8, to each tube, mix by inversion, and keep on ice for 10 min.[b]

Protocol 2. *Continued*

9. Centrifuge the lysates at 34 500 *g* for 45 min at 4°C.
10. Decant the supernatants and transfer these to new 50 ml polyallomer tubes (or 30 ml Corex tubes). Add 0.6 vol. of isopropanol (14 ml) and incubate at room temperature for 10 min to precipitate the nucleic acid.
11. Collect the precipitate by centrifugation at 23 000 *g* for 30 min at room temperature. The pellet should be translucent in appearance.
12. Decant the supernatant carefully and stand the tubes upside-down to drain, being careful not to allow the pellet to dislodge. Alternatively, one can hold the tubes at a slight downward angle and carefully wipe the isopropanol from the sides of the tube using a clean tissue paper. Wash the pellet with 10 ml of ice-cold 70% ethanol. If the pellet dislodges completely, re-centrifuge at 16 500 *g* for 10 min.
13. Add 2.5 ml of TE buffer to the pellet. Shake gently for 30 min or so since the DNA will go into solution only gradually.
14. Weigh out exactly 3.5 g of CsCl into a 15 ml conical tube. Add the dissolved DNA solution to the tube. Rinse the original DNA-containing tube with 1.0 ml of TE buffer and add this to the CsCl solution, bringing the final concentration of CsCl to 1 g/ml.
15. Transfer the CsCl/DNA mixture[c] to a 5 ml Beckman Quick Seal tube (or equivalent) and add 395 μl of 5 mg/ml ethidium bromide. Overlay with 1 g/ml CsCl in TE buffer to fill the tube. Seal and centrifuge for 8–12 h in a VTi80 ultracentrifuge rotor (or equivalent) at 65 000 r.p.m. at 20°C.
16. After centrifugation, monitor the DNA banding pattern in the tube using a hand-held medium wave UV transilluminator (wear a face-shield for safety). Puncture the tube below the DNA bands using a syringe (3–5 ml) fitted with a 18 gauge needle, venting the top of the tube by piercing with another 18 gauge needle. Collect the lower of the two bands (the plasmid DNA band). The top, usually more diffuse band is chromosomal DNA. The plasmid band can be re-banded (as in steps 14–16), to ensure complete separation of chromosomal DNA and RNA from the plasmid.
17. Dilute the plasmid DNA band (approx. 1 ml) with 9 vol. of water-saturated *n*-butanol. Shake well and let this stand for 1 min. Discard the upper, red (butanol) phase. Repeat this extraction step at least four times.
18. Dilute the extracted DNA solution to 4 ml with TE buffer. Add 450 μl of 3 M sodium acetate pH 7.0. Mix well and then precipitate the DNA by adding 9.5 ml of 100% ethanol. Mix and store overnight at −20°C.

19. Pellet the DNA by centrifugation at 23 000 *g* for 30 min at 4°C. Decant the supernatant and rinse the pellet with 5 ml cold 70% ethanol. Reprecipitate the DNA (steps 18 and 19) three times.
20. Dissolve the final air dried pellet in 500 μl of TE buffer and determine the concentration of the DNA.[d] It is useful for all subsequent manipulations to adjust the concentration of these stock DNAs to 1 mg/ml. Store at −20°C.

[a] See ref. 16.
[b] This step neutralizes the mixture. Under these conditions, plasmid DNA should renature instantly and remain in solution whereas larger macromolecules should stay denatured. The viscosity of the solution will decrease and a white precipitate will form.
[c] Density should be between 1.55 and 1.59 g/ml.
[d] Measure the A_{260} (see *Protocol 1*, step 15 and footnote [d]).

3.2 Cell culture

The importance of the condition of the cells used in a transfection experiment can not be stressed enough. Since there is no single way to maintain optimally all the types of cell lines one might use for transfection experiments, investigators are urged to establish sound culturing techniques for their individual requirements empirically. Listed below are some of the most important parameters to consider.

(i) Basic media. Different types of media are commercially available for the optimum growth of mammalian cells. It is worth determining the appropriate medium and supplements for the cell line(s) being used. For Chinese hamster ovary (CHO), SV40-transformed African green monkey kidney (COS), and human embryonic kidney (HEK)-293 cells grown attached to dishes in monolayer, Dulbecco's modified Eagle medium: nutrient mixture F12 (DMEM/F-12) supplied by Gibco-BRL works well. Extra glutamine (2 mM final concentration) is also added due to its general instability and its absolute necessity as a growth supplement. Penicillin (100 U/ml) and streptomycin (100 μg/ml) can also be added to prevent any bacterial contamination. There is general concern, however, as to the problems which routine maintenance in the presence of antibiotics may cause, particularly in masking mycoplasma contaminations. A prudent investigator will test cell lines routinely for mycoplasma, by far the most insidious contaminant for any tissue culture laboratory. Mycoplasma test kits are available from a number of suppliers (e.g. Gibco-BRL 189–5672AV).

(ii) Serum. For the types of lines mentioned above, the concentration of fetal bovine serum (FBS) most often used to support cell growth is 10% (v/v). The serum is routinely heat treated at 55°C for 30 min to inactivate complement and other potential contaminants. Differences between serum batches have been observed to affect cellular viability and other culture parameters, so it is wise to be aware of the source and actual batch of the serum used in order to track possible future problems. Some selection media require dialysed serum (see section 5.1.5). In these cases, 100 ml of serum is dialysed against 2 litres of 0.9% NaCl (or PBS) at 4°C for two days. This process is repeated three times and the serum is then filter sterilized before use.

(iii) pH. The pH of the medium is normally maintained through the bicarbonate content of the medium and the adjustable carbon dioxide (CO_2) atmosphere of the incubator. Since many of the transfection methods described here, particularly the calcium phosphate procedure, are fine-tuned to narrow pH windows, one should always be aware of the pH buffering capacity of the medium in use. Normally 2.0–3.7 g/litre of sodium bicarbonate is sufficient to maintain the pH of a medium between pH 7.2 and pH 7.4 in a 5–10% CO_2 incubator. Sometimes, 10 mM Hepes buffer pH 7.2, can also be added to help maintain the appropriate pH range. It is useful to monitor the buffering capacity of a particular medium by setting-up a test culture overnight in the medium and incubator to be used for the transfections.

(iv) Special media. Serum-free conditions are often desirable during the transfection protocol or post-transfection, especially in experiments where the protein product accumulates in the medium. The presence of serum can interfere with the biochemistry of the DNA/transfectant matrix and subsequent protein characterization techniques, as well as having the potential to cause product proteolysis. Several good media for use with low serum content or no serum have been developed and are commercially available, for example OPTI-MEM® from Gibco-BRL (expressly suggested for use in the lipofection transfection protocol, see section 4.2) and CHO-S-SFM (for serum-free suspension cultures). One should be aware that the optimization of a serum-free medium for cell culture growth is complex and is not simply a case of preparing standard growth medium without serum. These media are primarily used for cells growing in suspension. They have also been used successfully at different steps in a number of the transfection protocols presented here where certain reagents, sensitive to serum components, are introduced to cell monolayers for long incubation periods. In the case where the secretion of a recombinant protein is being monitored over several days post-transfection (monolayer culture), the use of serum-free medium alleviates the potential problems caused by the use of serum (see above) without sacrificing culture viability. It is important for the investigator to determine the optimum medium of this type empirically for the particular set of experimental conditions which exist.

3.3 Cell propagation (monolayer cultures)

There are a number of cell maintenance parameters which are inherent to successful transfection experiments:

- the number of times the cells have been passaged
- the density of the culture used to seed the transfection experiment
- the extent of confluency in the transfection dishes

As contact-inhibited cells reach confluency, that is grow to completely cover the surface of the tissue culture dish, they will stop growing. Cells which have stopped growing take longer to return to growth or logarithmic phase than those that are maintained there and subcultured. Depending on the cell line, cells subcultured at this stage for a transfection experiment may not return to a healthy growth phase before the transfection is actually performed. An overly dense culture can contain many non-viable cells, thus lowering the total number of viable, healthy cells plated for a transfection experiment. To avoid these problems, it is best to determine the optimum propagation routine for the particular cell line being used. Adherent cell lines like CHO, COS, and HEK-293 are reasonably easy to maintain. The following are guide-lines for providing effective transfection cultures with these types of more robust cell lines (see also ref. 17).

(i) Routine maintenance. Dishes of cells used to inoculate transfections ('seed cultures') should be in the range of about 50–80% confluency. For the above mentioned lines, seed cultures like this can be maintained by routine subculturing (1/10 splits) every three to five days. Generally this type of continual subculturing should not be done more than 15 times (i.e. over a period of about eight weeks) before a fresh cell line is established from frozen stock. One may notice a fall off in transfection frequency and expression levels as the age of the culture increases. This inevitably signals the need for fresh cells.

(ii) Trypsinization. Solutions of 0.05% trypsin, 0.53 mM EDTA in phosphate-buffered saline (PBS) are normally used to remove cells from monolayers for further manipulation. This solution can be prepared in the laboratory but is also available pre-made (e.g. from Gibco-BRL). Over-trypsinizing the cells can have a serious effect on cell viability and plating efficiencies, especially in serum-reduced media. As a general routine, 1 ml trypsin is added to a PBS washed monolayer of cells in a 100 mm dish. The cells are washed with PBS because serum inhibits trypsin activity. The cells are incubated at either room temperature or 37°C depending on the adherent character of the cell line. The more adherent the cells are, the higher the incubation temperature. Nevertheless, one should always be careful and not exceed temperatures of 40°C due to detrimental 'heat shock' effects that can occur. The cells should be monitored under the inverted microscope at 1–2 min intervals during the trypsinization. As soon as the cells become rounded, singular, and easily dislodged from the dish, trypsinization should be stopped by adding either serum-containing medium (ninefold volume) or trypsin inhibitor for cell lines cultured in low serum (< 2%) medium. CHO cells grown in 5–10% serum are more resistant to trypsinization, taking > 5 min at 37°C to detach. HEK-293 cells, however, may be detached in less than 1 min under similar conditions and are usually treated at room temperature.

(iii) Plating densities. Many investigators have a 'feel' for the density of the cells that have been plated. Counting the number of cells, however, results in a more accurate, and hence more reproducible, inoculation for any transfection protocol. This can be done using a Coulter counter if available or a haemocytometer (see *Protocol 3A*). A good general cell density for the transfection protocols presented here would be to seed $1–4 \times 10^4$ cells/cm^2 at 24 h pre-transfection. The optimal seeding density varies with cell line and transfection method and would need to be determined experimentally. It is also well worth checking the cell viability of the transfection seed cultures as described in *Protocol 3B*.

Protocol 3. Cell counting and viability staining

Equipment and reagents

- PBS: 4.3 mM $Na_2HPO_4.7H_2O$, 1.4 mM KH_2PO_4 pH 7.2, 137 mM NaCl, 2.7 mM KCl
- 0.05% trypsin, 0.53 mM EDTA in PBS (also available from Gibco-BRL)
- 70% ethanol
- 0.4% (w/v) trypan blue stain in PBS (also available from Gibco-BRL)
- Neubauer haemocytometer and coverslip
- Inverted microscope for tissue culture

A. *Cell counting*

1. Clean the surface of the haemocytometer and coverslip with 70% ethanol.
2. Wet the edge of the coverslip slightly and press it down over the counting area.
3. Wash and trypsinize the cells (see section 3.3) and dilute them if necessary. Transfer 15–30 μl of the cell suspension to the edge of the haemocytometer counting area and allow the cells to be drawn under the coverslip. Avoid under- or overfilling the chamber since this will affect counting accuracy.
4. Using a × 10 objective lens, focus on the grid lines in counting chamber. Orient the slide to visualize the entire central area of the grid (1 mm^2 area bounded by three lines with 25 smaller areas each with 16 smaller areas in them). Count all of the cells within this 1 mm^2 area.
5. Count the 1 mm^2 areas (each with only 25 subdivisions) on the top, bottom, left, and right of the central area. Average the counts for all five 1 mm^2 areas.
6. Calculate the cell density as: cells/ml = cells/1 mm^2 area $\times 10^4 \times$ dilution factor. The dilution factor (if any) is that used to prepare the cell sample counted. Cell densities need to be at least 10^5 cells/ml for this counting method to be reasonably accurate.

B. *Trypan blue staining for cell viability*

1. Prepare trypsinized cells suspensions at approximately 5×10^5 cells/ml in PBS (see section 3.3).
2. Add 0.5 ml of 0.4% (w/v) trypan blue solution to 0.5 ml of cell suspension and incubate at room temperature for 10 min.
3. Transfer the stained cells to the haemocytometer chamber as described in part A, step 3.
4. Separately count both the stained (dead) and unstained (viable) cells as in Part A, steps 4 and 5.
5. Calculate the percentage of viable cells by dividing the number of unstained cells by the total (stained and unstained) number of cells counted.

4. Transfection procedures and host/vector systems

The questions as to what cell type and vector to use, how the vector should be introduced into cells, how expression will be analysed, and for what duration, are all highly interrelated and the answer to one question may be strictly dependent on the answer to another. Presented below are a series of useful host/vector options and transfection protocols designed for optimal transient and stable expression.

4.1 Transient transfection

The methods of introduction of DNA into mammalian cells have been routinely characterized by the overall duration of product expression. In transient transfection, vector sequences are not integrated into the cellular chromatin but are maintained for short periods of time extrachromosomally, between 12–80 hours. Protein production is usually rapid but limited in quantity and highly dependent upon the transfection efficiency. Transient transfections are best suited to multiparameter experiments where, for example, the strengths of several promoter/enhancer constructs are to be tested or mutant proteins characterized, for expression cloning of genes, and for the characterization of post-translational processing events. All of the protocols presented here can be used for the transient transfection of cells. The first two methods described here, based on the use of DEAE–dextran (*Protocol 4*) and calcium phosphate (*Protocol 5*), are discussed with regard to two extensively optimized host/vector systems used for transient expression analysis. Lipofection, a relatively new transfection technique, is included in this section as well (see section 4.2). It is fast becoming the method of choice for both transient and stable analysis of gene expression.

4.1.1 Plasmid replication-based transient transfection

The classic host/vector systems for this type of expression protocol are the SV40-transformed kidney cell lines COS-1 or COS-7 (ATCC CRL1650, CRL1651) transfected with vectors containing SV40 origins of replication (18). Other systems of this type are also being developed for efficient use in transient transfection assay, for example, systems based on the polyoma replication origin used in SV40–polyoma transformed NIH 3T3 mouse cells, MOP-8 (ATCC CRL1709) (19), and more recently in CHO cells (20). Originally designed as a means to support the replication of defective viruses with deletions in the early region genes, these host/vector combinations have evolved into extremely effective, non-infectious, plasmid-based expression systems. The cell lines described are capable of supporting the replication of very high copy numbers of introduced vector albeit for limited periods of time. These enormous amounts of replication provide the basis for high gene expression levels but also tend to be toxic to the cells. This characteristic also appears to be the primary element in the general inability to select stably transfected cell lines using these systems. Variations include establishing cell lines with conditionally expressed or regulated T antigen sequences to potentially circumvent the detrimental effects of overreplication (21); growth of the cells under non-permissive conditions yields non-functional T antigen and subsequently no vector replication. Upon shift to permissive conditions, the antigen promotes replication and, in essence, an induction of expression for genes present on transcription units linked to the viral replication origins.

Probably the most commonly used method for the introduction of replication-based vectors into COS cells is by DEAE–dextran transfection. One of the original transfection vehicles, DEAE–dextran is still a very popular and easy method for introducing DNA transiently into both adherent and suspension cell lines (22). The basic method has evolved with many permutations, usually including one or more ‘enhancing’ variations (section 4.1.3). *Protocol 4* includes several of these variations for optimal transfection efficiency including chloroquine treatment (23) and DMSO shock (24). It works very well for COS cells as well as for CHO cells. However, the DEAE–dextran method is the only one presented in this section that is not suitable for selecting stable expression cell lines.

Convenient vectors for transient expression include pcDNA1 which is commercially available from Invitrogen (see also ref. 25). This vector has both the polyoma and SV40 origins of replication as well as the hCMV enhancer/promoter sequences for efficient expression in COS cells. A number of vector derivatives utilizing different enhancer/promoter combinations are also available.

The literature contains many modifications for transfection of cells using DEAE–dextran. This myriad of options only emphasizes that a successful transfection protocol will require the user to establish the optimal parameters for each particular set of experiments. The key elements to good transfection using *Protocol 4* are cell density, DNA concentration, and DEAE–dextran concentration. *Figure 1A–C* shows representative data using this technique to transfect CHO and COS cells with recombinant tissue plasminogen activator (rtPA) expression plasmids under a variety of conditions. Using 12-well dishes, one can vary each of the critical parameters in a defined manner in order to establish the optimal set of conditions for a particular cell line and vector.

Protocol 4. Transfection of mammalian cells using DEAE–dextran

Equipment and reagents

- 1 X TBS buffer: 25 mM Tris–HCl pH 7.4, 137 mM NaCl, 5 mM KCl, 0.6 mM NaH_2PO_4, 0.7 mM $CaCl_2$, 0.5 mM $MgCl_2$. This buffer is normally made by mixing an appropriate volume of a 10 × concentrated solution of the first four components (stock stored at −20°C) to an appropriate volume of a 100 × concentrated solution of the last two components (stock stored at −20°C). The latter is added dropwise to allow all the salts to dissolve). Prepare fresh 1 × TBS prior to each transfection.
- 10 mg/ml DEAE–dextran (Sigma D-9885, average M_r 500 000) in 1 × TBS (filter sterilize and store at 4°C)
- Sterile microcentrifuge tubes
- Serum-free medium (see section 3.2)
- Host cells: 12-well transfection dishes seeded at 2×10^5–8×10^5 cells/well[a] (see section 3.2)
- 10% dimethyl sulfoxide (DMSO, Sigma D-5879) in water[b] (filter sterilize and store at 4°C)
- 10 mg/ml chloroquine (Sigma C-6628) in water. Dilute to 100 μM in serum-free media (see section 3.2) plus 10 mM Hepes pH 7.2[c] for transfections. Chloroquine is light-sensitive, and hence the 10 mg/ml stock solution should be stored at 4°C in the dark, wrapped in foil.
- Plasmid DNA (adjusted to a concentration of 1 μg/μl for ease of handling throughout the transfection protocol)

Method

NOTE: All manipulations should be performed under sterile conditions in a laminar flow tissue culture hood to limit the possibility of contamination. All reagents should be brought to ambient temperature prior to use except media which should be at 37°C.

1. In sterilized microcentrifuge tubes, mix plasmid DNA (0.5–4 μg/well) with 1 × TBS not exceeding a final volume of 7.5 μl/well.[d]
2. Add 15 μl of 10 mg/ml DEAE–dextran solution per well to each DNA mixture above by allowing drops of the DEAE–dextran to slowly slide down the sides of the microcentrifuge tubes.[e] Gently flick the tubes to mix the contents after the addition of DEAE–dextran.
3. For each well to be transfected, add 0.5 ml of 100 μM chloroquine in serum-free medium to a sterile microcentrifuge tube.
4. Add the DNA/DEAE–dextran mixture (from step 2) to the chloroquine in serum-free medium (from step 3). Flick the tubes or invert gently to mix. Allow the mixture to incubate at room temperature in a tissue culture hood while the cells are prepared (see step 5 below).

Protocol 4. *Continued*

5. Remove the cell cultures from the incubator. Aspirate the growth medium and wash the cells twice with pre-warmed, serum-free medium[f] to dilute out any serum from the growth medium.
6. Add 0.5 ml of the appropriate mixture from step 4 to each well of the 12-well transfection dish.
7. Incubate for 2.5 h at 37°C in a 5% CO_2 incubator. Depending on the cell line, cells may appear to be dying off. CHO cells come through these steps in reasonable condition.
8. At the end of the incubation period, aspirate the transfection mixtures from the cells.
9. Shock the cells by adding 0.5 ml of 10% DMSO to each well for 1 min at room temperature. This can be conveniently done with one 12-well dish at a time. At the end of 1 min dilute the DMSO quickly by adding 2–3 ml of serum-free medium to each well. Aspirate the diluted DMSO and replace with 1.0 ml of fresh growth medium (standard media for cell maintenance plus serum). Return the dishes to the 37°C incubator.
10. The next day some cells will have died, the extent of cell death depending primarily on the sensitivity of the particular cell line to the DEAE–dextran concentration used or the DMSO shock. The CHO-DUKX-B11 line survives this protocol reasonably well as do COS cells. If using complete medium at this step, replace it with serum-free medium after 24 h in cases where secreted protein products will be monitored.[g] This medium replacement is not necessary in assays of intracellularly expressed products.
11. Analyse the medium for the expressed protein between 48 h and 120 h (secreted product) after the onset of serum-free conditions in step 10. Intracellular expression of certain reporter genes can be monitored as early as 12 h post-transfection.[h]

[a] This protocol can be scaled up to 100 mm dishes. All solution volumes are adjusted based on a total volume of 5 ml/dish. The total amount of plasmid DNA for optimum transfection (DNA saturation) at this scale ranges from 1 μg/ml to 7 μg/ml and would need to be determined empirically.
[b] 10% DMSO can also be prepared in 1 × TBS or PBS.
[c] Penicillin and/or streptomycin can be added to help prevent bacterial contamination.
[d] Reporter plasmids are also often included to monitor transfection efficiency.
[e] The final concentration of DEAE–dextran in this protocol is 300 μg/ml of medium. This can be adjusted to other transfection parameters as suitable (see *Figure 1*).
[f] Although cells are washed with serum-free medium in step 5, 1 × TBS buffer can be used instead. PBS buffer, a common tissue culture reagent, should not be used to wash the cells for DEAE–dextran transfection, since high concentrations of phosphate as well as the presence of serum can affect the efficiency of this method.
[g] Cells can be subcultured at this point if required or desirable.
[h] The temporal expression and/or production profile of different vector constructs and protein products will need to be empirically determined for each set of transfections

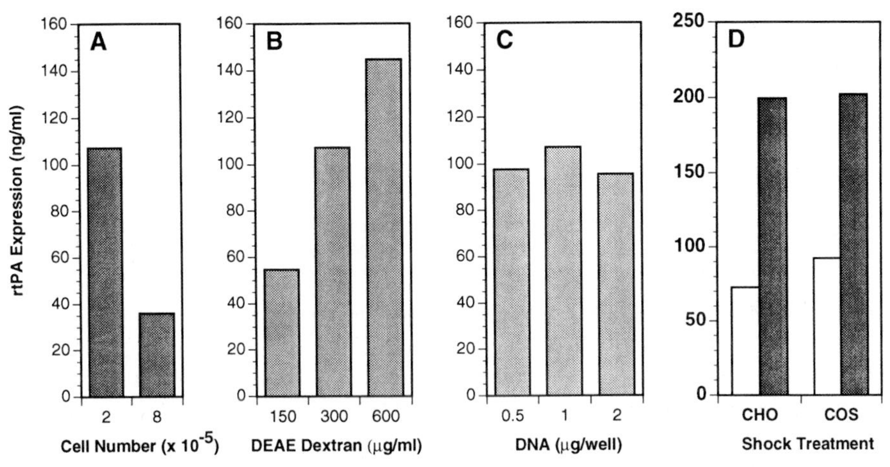

Figure 1. Varying parameters for DEAE–dextran transfection. Each panel shows the effect on transfection/expression when (A) cell number, (B) DEAE–dextran concentration, (C) DNA concentration, or (D) shock method—15% glycerol (white bar) versus 10% DMSO (grey bar) were varied. In this case, the CHO-DUXB 11 derivative cells used for transfection showed distinct differences in recombinant tPA (rtPA) expression (the reporter gene for transfection efficiency) when cell number, DEAE–dextran, and shock method (see also COS cells) were varied. Changes in DNA concentration, however, for this particular vector (SV40 early enhancer/promoter) had little or no effect on expression. Recombinant tPA expression was assayed three days post-transfection via ELISA (performed by Genentech Inc. Immunoassay services).

4.1.2 Non-replicative transient expression

In this category, one of the most versatile systems developed to date has been the use of HEK-293 cells for transient gene expression (26). This human cell line is adenovirus-transformed making it a good candidate line for transfection with hCMV transcriptional control sequences. The excellent transient expression obtained is based on the exceptional stability of transfected DNA in this cell line and not (as presented in the previous section) dependent on a replicative vector function. Since special vector elements are not necessary for the successful use of this system, many more types of vector and expression units can be tested. However, note that some enhancers, for instance the early SV40 enhancer, are transcriptionally repressed by certain adenovirus-specific proteins and so would not be good candidate vectors for use with these particular cells. A method for the optimal transient transfection of HEK-293 cells, developed by Cori Gorman, using calcium phosphate is presented in *Protocol 5*. As with the DEAE–dextran method, this protocol includes modifications and conditions optimized for its efficient use with the (HEK-293)/vector (hCMV regulatory region) system. The method takes advantage of the formation of calcium phosphate–DNA complexes which adsorb to the cell surface and are subsequently endocytosed (27).

Protocol 5. Calcium phosphate transient transfection of HEK-293 cells

Equipment and reagents

- Host cells (5–10 × 10^5/100 mm dish) seeded the night before use
- PBS
- 2 × Hepes-buffered saline (HBS) pH 7.12: 280 mM NaCl, 50 mM Hepes, 1.5 mM Na_2HPO_4. Adjust pH to exactly 7.12 with 1 M NaOH at room temperature. The pH of this solution is critical, so care should be used when making this solution. Filter sterilize and store as 10 ml aliquots at 4°C)
- 2.5 M $CaCl_2$ (filter sterilize and store at 4°C)
- 0.1 × TE buffer: 1 mM Tris–HCl pH 8.0, 0.1 mM EDTA
- 15% glycerol in PBS (filter sterilize and store at 4°C)
- Plasmid DNA: precipitate the plasmid DNA with ethanol (*Protocol 2*, steps 18–19) and allow the pellet to dry in a tissue culture hood to ensure sterility. Resuspend the DNA (at least 10 μg) in 450 μl of 0.1 × TE buffer per transfection dish[a].
- Falcon snap-cap tissue culture tubes
- Adenovirus viral associated (VA) RNA plasmid DNA (optional)[a]
- Serum-free medium (see section 3.2)

Method

1. Re-feed the cells with 10 ml complete medium 3 h prior to transfection. Incubate at 37°C.
2. Set-up two tubes for each transfection (Falcon snap-cap tissue culture tubes work well at this step). To tube A (per transfection dish[b]), add 450 μl of resuspended DNA solution (at least 10 μg of the plasmid DNA and at least 1 μg of Ad VA DNA[a] (optional) in 0.1 × TE). To this add 50 μl of 2.5 M $CaCl_2$ per transfection dish. Gently flick the tubes to mix the contents. To tube B (per transfection dish[b]), add 500 μl of 2 × HBS pH 7.12.
3. Add the DNA/$CaCl_2$ mixture in tube A to tube B carefully, dropwise. Pipette the mixture slowly up and down several times with a 5 ml plastic pipette.[c] A fine DNA/$CaCl_2$ precipitate should be visible in the tube and the solution often has a slightly blue opalescent colour. Incubate at room temperature for at least 20 min to allow the precipitate to form completely.
4. Add 1.0 ml of the DNA/$CaPO_4$ precipitate per tissue culture dish. Distribute the mixture evenly over all the cells by gently swirling the dish while adding the precipitate.
5. Incubate the cells with the precipitate for 3 h[d] at 37°C. Once during this incubation period it is useful to observe the precipitate on the cells microscopically. As a rule, the finer the precipitate, the more effective the transfection.
6. Aspirate the precipitate from the cells and shock them by adding 1.0 ml of 15% glycerol in PBS at room temperature to each plate. The duration of the shock is 30 sec so it is best to proceed with shocking only one or, at most two, plates at a time.

7. After 30 sec, aspirate the glycerol from each plate, and wash the cells once with 5 ml PBS.
8. Add 10 ml of fresh growth medium (plus serum) to each plate and incubate at 37°C under 5% CO_2.
9. When monitoring the production of a secreted product, replace the growth medium with serum-free medium after a 24 h recovery period (see *Protocol 4*, steps 10 and 11). Intracellular expression does not require serum-free conditions.[e]
10. Usually after 48 h, the medium can be assayed for secreted protein products. As is often the case with this method, media samples can be collected after 48 h under serum-free conditions and the cells re-fed with fresh serum-free growth medium to obtain further production samples. Cell viability and the overall health of the culture should be monitored during this post-transfection production period.

[a] Use of the adenovirus viral associated (VA) RNA plasmid DNA is optional (see section 4.1.4).
[b] A single DNA/$CaPO_4$ precipitate can be made up for multiple transfections.
[c] Alternately, gently blow a slow stream of air bubbles through the solution using a sterile Pasteur pipette.
[d] Precipitates can be incubated with cells for up to 18 h.
[e] The temporal expression and/or production profile of different vector constructs and protein products will need to be empirically determined for each set of transfections.

As stated earlier, there have been many modifications to the basic calcium phosphate transfection technique. One modification (*Protocol 6*), developed by Chen and Okayama (28) who have done extensive work in optimizing $CaPO_4$ transfection technology in general, utilizes a $CaCl_2$/*N,N*-bis[2-hydroxyethyl]-2-aminoethanesulfonic acid (BES)-buffered saline solution to deliver the DNA to the cells. Use of the modified buffer, adjusted to a slightly more acidic pH, allows the DNA–$CaPO_4$ complex to form slowly in the culture dish (i.e. *in situ* rather than in a tube) during the 16 h incubation at 37°C. This method, also good for transient transfection, has been shown to be a very efficient method for generating stably transfected cell lines as well (see section 5).

Protocol 6. Calcium phosphate transfection using BES buffer

Equipment and reagents

- Host cells (as in *Protocol 5*)
- 2 × BES buffer: 280 mM NaCl, 1.5 mM Na_2HPO_4, 50 mM BES pH 6.95. Adjust the pH by adding 1 M NaOH at room temperature (the pH of this buffer is absolutely critical to the success of this procedure. Filter sterilize and store in aliquots at −20°C).
- 2.5 M $CaCl_2$ (see *Protocol 5*)
- Falcon snap-cap tissue culture tubes
- Plasmid DNA (5–50 μg DNA in 450 μl water per transfection dish)
- Serum-free medium (see section 3.2)

Protocol 6. *Continued*

Method

1. Re-feed the cells as described in *Protocol 5*, step 1.
2. Add 50 μl of 2.5 M $CaCl_2$ to 450 μl of plasmid DNA (5–50 μg) in water.
3. Add 500 μl of 2 × BES buffer to the DNA/$CaCl_2$ solution and incubate at room temperature for 10 min.
4. Add the mixture (1 ml) directly to the cells as in *Protocol 5*, step 4, but incubate for 16–24 h in a 3% CO_2 atmosphere. Reduced atmospheric CO_2 is needed to allow formation of the DNA/$CaPO_4$ precipitate.
5. After incubation, aspirate the medium containing the precipitate.
6. Wash the cells once with growth medium or PBS and re-feed with fresh growth medium. In this protocol, no shock is done. Post-transfection cells are returned to 5% CO_2 as in *Protocol 5*, step 8 and monitored for expression as in *Protocol 5*, steps 9 and 10.

Protocols 5 and 6 are highly efficient at transfecting many fibroblast and endothelial cell lines, e.g. CHO, mouse L, HeLa, and HEK-293. The key parameters to good transfection with both of these methods are:

- the concentration of the DNA in the precipitate (approx. 10 μg/ml for *Protocol 5* and 20–30 μg/ml for *Protocol 6*)
- the quality of the precipitate
- the condition of the DNA (linear DNA is inactive)
- the pH

Careful note must be made in $CaPO_4$ transfection protocols of the very narrow pH range for media and buffers for successful transfection, especially the 2 × HBS and the 2 × BES buffers. As mentioned in section 3.2, it is important to adjust the CO_2 tension of the incubator properly to maintain the proper buffering capacities. With regard to DNA optimization, although the techniques themselves have been developed for optimal DNA/precipitate formation, the total concentration of expression vector required to reach a saturation level for expression may differ from vector to vector and be well below the standard concentrations given above. The use of carrier DNA to adjust the apparent concentration of presented plasmid is a controversial area and very experiment-dependent. If carrier DNA is used at all, it should be carefully selected and the results interpreted wisely.

4.1.3 Chemical 'enhancers' of transient transfection

A number of modifications to the above methods rely on empirically tested chemical treatments which enhance the efficiency of transfection or expression. The use of chloroquine (23) and cell shock by DMSO or glycerol (24, 29) are not inherent to the originally defined protocols but potentially provide conditions to improve experimental effectiveness. Although the protocols presented above have some of these modifications in place, they should be calibrated for optimal performance in any given experimental situation.

(a) Chloroquine, which prevents endosomal acidification, has been used both during DNA exposure as in the protocols given above and as a post-transfection treatment. It is somewhat toxic to the cells so exposure time is often limited to 3–4 h.

(b) Glycerol and DMSO, used to osmotically shock cells, can be tested for optimal concentration (between 10% and 30%), and for optimal duration of exposure (from 30 sec to 3 min) to minimize toxicity while maximizing expression levels. The data presented in *Figure 1D* highlights transfection expression differences which can occur using either 15% glycerol or 10% DMSO to shock CHO and COS cells incubated with DNA/DEAE–dextran complexes.

(c) Butyrate is another enhancer which has been used. The treatment of transiently transfected cells with butyrate can potentially provide an increase in the number of cells which express the transfected DNA (30). Butyrate has also been shown to affect the transcriptional efficiency of certain viral control regions, among these SV40 and polyoma. A solution of sodium butyrate (e.g. Sigma B5887) (0.5 M in water, pH adjusted to 7.0, filter sterilized, and stored frozen at −20°C) is diluted into complete medium (1–10 mM working concentration) and added to the cells after the glycerol shock (e.g. *Protocol 5*, step 8). It is removed 24 h later, the cells are washed, and re-fed with fresh medium. Butyrate treatment has also been shown to be effective post-electroporation (31).

4.1.4 The adenovirus viral associated (VA) RNA gene

Apparently unique to transient systems is the potential enhancement of expression derived from including the adenovirus viral associated (VA) RNA gene in transfections (32). It has been proposed that the appearance of dsRNA, a potential product of transcription from both strands of the transfected DNA template, may cause translational arrest via activation of a kinase (DAI) which, by phosphorylation, inhibits the function of the translational initiation factor eIF_2. Inclusion of the VA RNA gene in transfections allows for the accumulation of a specific RNA known to inhibit the kinase (DAI), thus potentiating translation. Some vectors and host systems are unresponsive to the presence of the VA RNA gene, and so it needs to be empirically tested in each experimental case to determine its ability to enhance gene expression. It has been successfully implemented with both *Protocols 5* and *6* to attain optimal results. Normally 1 μg/ml of Ad (VA) RNA plasmid DNA can be co-transfected with the plasmid DNA of interest in *Protocol 5*. Several vectors have been developed which contain these sequences for use in transient analysis (33).

4.1.5 Monitoring transfection efficiencies

Transfection efficiencies, the percentage of cells which have taken up and are expressing the DNA presented, do vary from dish to dish within a single transfection experiment as well as across experiments. This inherent variability can be monitored and data subsequently normalized by the inclusion in each transfection of a constant amount of internal control plasmid. Ideally the expression of the product from this control vector should not be affected by any of the experimental conditions being examined. Likewise, expression from this vector should not affect or be affected by the vector constructs being tested (e.g. competition for limited transcription factors). An appropriate control vector must be empirically determined for each experimental design. Thus the absolute expression level of this vector will be a direct indication of the overall efficiency of plasmid uptake by the cells from dish to dish. These levels can then be used to normalize those obtained with experimental vectors; vectors whose expression levels are potentially affected. Most investigators repeat transient experiments at least twice in order to verify their observations. It is also a procedural advantage in most cases to use a secreted standard in assays where secreted protein levels are being monitored. Where intracellular expression is assayed, the use of a reporter protein whose activity is compatible with the cell extract protocol used for the experimental vectors is likewise preferable.

4.2 Gene delivery via lipofection

The most recent development in transfection methodology, and probably the simplest technically, has been the use of cationic lipid-mediated gene delivery or lipofection. The rationale for its use stems from the recognition that virtually all cell membranes have a net negative surface charge with which a positively charged lipid vesicle would be expected to fuse and deliver its contents (34). In this technique, commercially prepared cationic liposomes consisting of a synthetic lipid bearing a quaternary amine are blended with a neutral phospholipid. When mixed with DNA, they react spontaneously with the phosphate backbone and, at a critical liposome density, result in liposome/polynucleotide complexes which are taken up efficiently by cells. The transfection efficiency is related to the efficiency of formation of these complexes. The methodology has become widespread due to its convenience and effectiveness in a wide variety of cell types. There are various commercially prepared lipofection reagents available, e.g. Lipofectin™, LipofectAMINE™, and Transfectace™ (Gibco-BRL), Transfectam® (Promega), and DOTAP (Boehringer Mannheim). The following protocol uses the LipofectAMINE™ (Gibco-BRL) reagent. The lipofection protocol has been extensively characterized and compared for a number of the reagents listed above (35) and LipofectAMINE™ in this case, was found to be superior. The formation of the liposome–DNA complexes (*Protocol 7*, step 2) is rapid and the nature and stability of the complex is sensitive to both the ratio of DNA to lipid as well as the volume in which they are combined. The method is appropriate for most cell types grown in 60 mm dishes.

Protocol 7. Lipofection transfection

Reagents

- Host cells: 60 mm dishes seeded with 5 × 10^5 cells and incubated overnight at 37°C. Cells should reach a confluency of 50–60% [a]
- Cationic lipid reagent: LipofectAMINE™ (Gibco-BRL 18324–012)
- 150 mM NaCl, 20 mM Hepes pH 7.5
- Serum-free medium,[b] e.g. OPTI-MEM® Reduced Serum Medium (Gibco-BRL 31985–021)
- Plasmid DNA (1–5 μg)

Method

1. Prepare DNA–liposome complexes as follows:
   (a) Dilute 1–5 μl of DNA to 100 μl in 150 mM NaCl, 20 mM Hepes pH 7.5, or serum-free medium.[b]
   (b) Dilute 2–20 μl (2–20 μg) of cationic liposomes (LipofectAMINE™) to 100 μl in 150 mM NaCl, 20 mM Hepes pH 7.5, or serum-free medium.[b]
   (c) Add the DNA solution to the liposome solution and mix gently. Allow to stand at room temperature for 10–30 min.
   (d) Add 800 μl of serum-free medium[b] to the DNA–liposome complexes just before adding to the cells.
2. Just prior to DNA–liposome addition, wash the cells once with serum-free medium. Layer 1 ml of DNA–liposome complexes on to the cells.
3. Incubate the cells for 1–24 h at 37°C (typically 5–6 h works best for most cell types[c]) in the appropriate CO_2 concentration.[d]
4. At the end of this incubation period, aspirate the medium containing the complexes and replace with 5 ml of normal growth medium (with serum).
5. Incubate the cells for 24–48 h at 37°C for transient expression (see also *Protocol 4*, steps 10 and 11) or subculture into a suitable selection medium for stable expression.

[a] The transfection efficiency may vary with the density of the cell culture. Cationic liposomes tend to be somewhat cytotoxic, a characteristic which requires the cell density to be carefully optimized for each line.
[b] Serum has been shown to have a negative effect on transfection by lipofection. OPTIMEM® medium is used to provide good cell viability under the potentially long incubation periods described in step 3, and in this case is used without a serum supplement (see section 3.2).
[c] The incubation time will affect transfection efficiency and must be optimized for each cell type. If the cells are found to be very sensitive to the DNA–liposome complexes, small amounts of serum (< 2%) can be added to reduce the toxicity. The addition of serum will require adjustment or re-optimization of the DNA to lipid ratio (step 1) needed to produce complexes that result in maximal transfection efficiency.
[d] For efficient transfection, the optimal percentage of carbon dioxide in the incubator has to be determined for the particular serum-free medium used in the protocol.

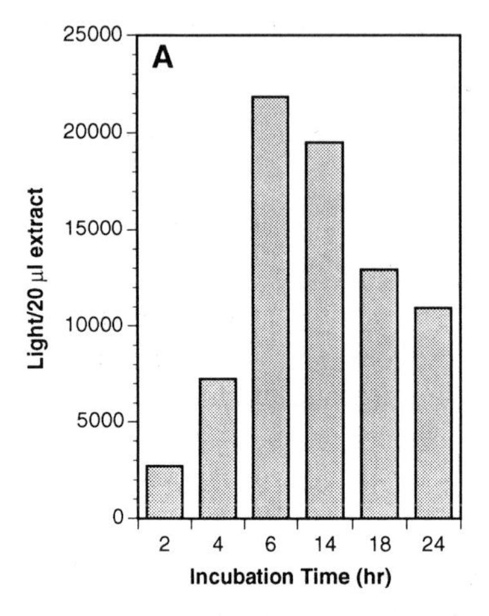

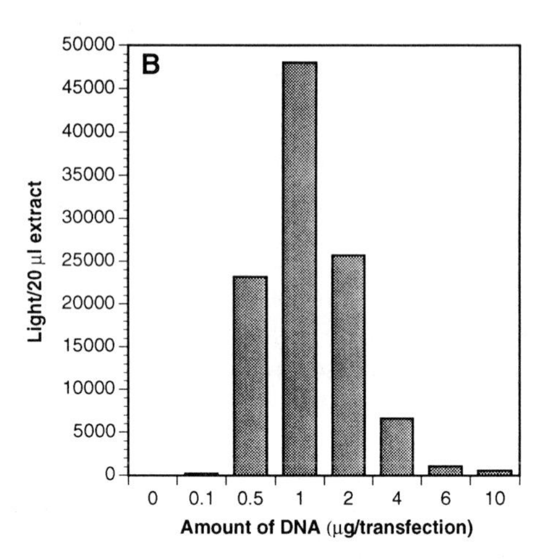

Figure 2. Optimization of transfection using liposomes. The data depicted here illustrates the importance of experimentally determining both (A) the incubation time on transfection efficiency and (B) the influence of optimal DNA to lipid ratio. A plasmid containing the firefly luciferase gene under the control of the hCMV IE gene enhancer/promoter region was transfected into HEK-293 cells using commercially available liposomes (LipofectAMINE™). In (A), only the incubation time was varied with constant DNA and liposome values and in (B), the DNA concentration was varied with the liposome concentration remaining constant. After 48 h, whole cell extracts were prepared and assayed for luciferase activity. The ordinate value shown, light /20 µl of extract, is a measure of integrated light units or the integrated light output between 1 and 16 sec. The optimal efficiency of transfection, as measured by luciferase expression, was observed when 1 µg of plasmid was mixed with 4 µg of LipofectAMINE™ reagent and the optimal incubation time appeared to fall between 6–14 h. These two parameters require optimization for each plasmid and liposome preparation.

The data shown in *Figure 2* illustrate the importance of determining both the optimal DNA to lipid ratio and the effect of incubation time on transfection efficiency for each experiment. In the example shown in *Figure 2*, HEK-293 cells were transiently transfected with vectors containing the firefly luciferase gene transcriptionally driven by the hCMV promoter/enhancer (see section 4.3).

4.3 The use of reporter genes

A number of reporter genes, whose expression levels are used as indicators of transfection or transcriptional efficiency, have been developed for use in transient transfection systems (36). Probably the most popular is chloramphenicol acetyl transferase (CAT).

4.3.1 Assays for chloramphenicol acetyl transferase

Methods for the assay of CAT have evolved over the years to the point where investigators have quite a choice. Those used most often rely on radioactive substrates:

- the original assay (for detailed protocol see ref. 37) using [^{14}C]chloramphenicol and thin-layer chromatographic analysis of its acylated derivatives
- a variety of organic extraction methods which use [^{14}C]chloramphenicol, [^{3}H]chloramphenicol, or sodium acetate (38, 39)

Many vectors for use of the CAT reporter gene have been developed (40). Promega supplies a series of basic pCAT reporter vectors as well as kits for assaying CAT activity by both thin-layer and liquid scintillation counting. In addition, Amersham supplies a highly sensitive and rapid CAT assay kit called *Quan-T*-CAT. The method presented in *Protocol 8* is a slightly modified version of that described by Nordeen *et al.* (39). It is extremely sensitive and utilizes probably the least expensive radioactive substrate available, [^{3}H]sodium acetate. This protocol also includes a generally acceptable procedure for preparing cell extracts by freeze–thawing.

Protocol 8. Assay of chloramphenicol acetyl transferase

Reagents

- Transfected cells (from one of the earlier protocols, *Protocols 4–7*)
- Mock transfected cells (as negative control)
- PBS
- Cell harvest buffer: 40 mM Tris–HCl pH 7.5, 150 mM NaCl, 1 mM EDTA
- Freeze–thaw buffer: 250 mM Tris–HCl pH 7.8
- *S*-acetyl coenzyme A synthetase (Sigma A-5269)
- 5 mM acetyl coenzyme A in water (Ac-CoA: Sigma C-3019) (this solution can be stored at −20°C for no longer than two weeks)
- 0.5 M chloramphenicol in ethanol
- [^{3}H]sodium acetate (2–5 Ci/mmol; New England Nuclear NET-003H NEN)
- 10 mM sodium acetate
- 4 × assay buffer: 0.4 M Tris–HCl pH 7.8, 24 mM $MgCl_2$, 0.3 M KCl
- 100 mM ATP (stored at −20°C)
- Chloramphenicol acetyl transferase (Sigma C-2900. This can be used as a control in cases where cell extracts may contain substances which inhibit CAT activity (e.g. some plant extracts)
- Benzene

A. *Preparation of cell extract*

Prepare extracts from both the transfected cells and non-transfected (or mock transfected cells) as a negative control.

1. At 48 h after transfection, wash the cells three times with PBS.
2. Incubate the cells with 1 ml (60 mm dish) or 1.3 ml (100 mm dish) of cell harvest buffer for 5 min at room temperature.
3. Scrape the cells from the tissue culture dish and transfer them to a microcentrifuge tube. Some cell lines will easily dislodge from monolayer cultures using gentle pipetting whereas others will require the use of a rubber policeman to physically remove the cells.
4. Pellet the cells by centrifugation for 2 min at 4°C in a microcentrifuge.

Protocol 8. *Continued*

5. Aspirate the supernatant and resuspend the cells in freeze–thaw buffer (100 μl/60 mm dish or 300 μl/100 mm dish).
6. Cycle the extracts three times from a dry ice–ethanol bath (5 min) to a 37°C bath (5 min). Place the extract on ice when completed.
7. Clarify the cell extract by centrifugation as in step 4. Recover the extract, taking care to avoid cell debris. The extract can be stored frozen at −20°C until assayed.

B. *CAT assay*

1. Make up the appropriate volume of reaction cocktail (200 μl/extract sample) using the following recipe (volumes are for five assays):
   - H_2O 530 μl
   - 4 × assay buffer 330 μl
   - 5 mM Ac-CoA 107 μl
   - 100 mM ATP 40 μl
   - Acetyl CoA synthetase 4 μl
   - 10 mM sodium acetate 51 μl
   - [^{3}H]sodium acetate 1.65 μl
2. Remove a 200 μl aliquot of the reaction cocktail for each extract sample to be assayed and place in an assay tube. These will serve as the chloramphenicol ('substrate-minus' controls.
3. Add 0.5 M chloramphenicol to the rest of the reaction cocktail (2.7 μl per five assays).
4. Add 200 μl of the reaction cocktail from step 3 to another assay tube for each extract sample to be assayed. Pre-incubate all tubes (steps 2 and 3) at 37°C for 5–10 min.
5. Add extract to each pair of tubes (with and without substrate) to initiate the assay.[a] Initially use 25–50 μl of extract per reaction.[b]
6. At various times during the assay (usually between 45 min and 1 h), remove a 50 μl aliquot (usually duplicate aliquots per time point) of the reaction mixture and extract with 1.0 ml of cold benzene to stop the reaction. Microcentrifuge tubes containing 1.0 ml aliquots of benzene can be set-up beforehand and kept on crushed ice or in a 60°C water-bath to prevent them from freezing. Vortex for 10 sec and then separate the phases by microcentrifugation for 30 sec.[c]
7. Remove the top 800 μl of the organic phase, which will contain the ^{3}H-acetylated chloramphenicol, and place this into a scintillation vial in the radioactive hood. Be careful not to remove any interface or aqueous phase since this 'carryover' will give erratic results.

8. Allow benzene to evaporate from the samples at room temperature in a vented hood (usually 4 h–overnight).
9. Add 3 ml of scintillation fluid to each sample and determine radioactivity using a liquid scintillation counter.

[a] In addition to the substrate-minus controls (see step 2) which are used to determine the background activity for each extract assayed, mock transfected cell extracts should be 'spiked' (0.02 U is adequate per reaction) with commercially available bacterial CAT enzyme to test for the presence of any inhibitor of CAT activity.
[b] The optimal amounts of cell extract to add will need to be determined empirically for each extract due to differences in promoter activity and cell expression levels.
[c] Multiple time points per reaction for both substrate plus and minus reactions should be taken to determine the linear range of CAT activity for the assay. The chloramphenicol control values (extract samples with no chloramphenicol substrate added) are subtracted from each appropriate experimental value. This normalizes all samples for endogenous levels of activity present which may vary from cell line to cell line or with the manner of extract preparation. The ratio of the control (chloramphenicol minus) and experimental values should remain constant through the linear range of the assay. Only assay values falling within the linear range can be compared.

The data presented in *Figure 3* reflect the sensitivity of *Protocol 8* for assaying CAT.

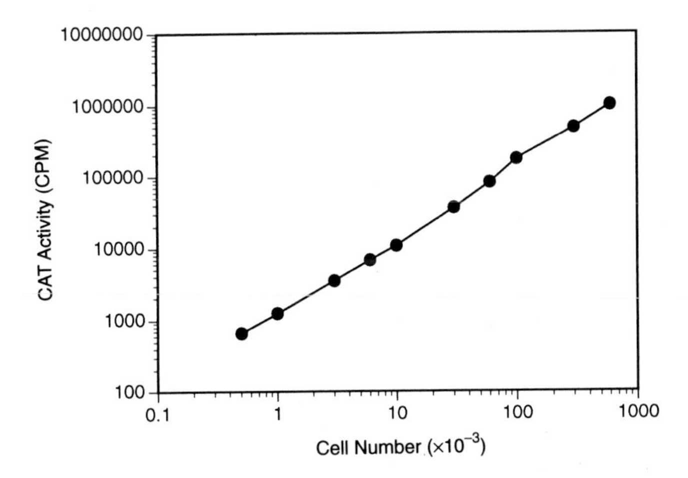

Figure 3. Standard curve of CAT activity determined against increasing cell number. This CAT activity curve was generated using *Protocol 5*. Stably expressing CAT positive PC-3 cell clones (prostate carcinoma cell line) were generated by transfection of the CAT reporter gene under the control of the hCMV enhancer/promoter. Increasing numbers of these cells were assayed for CAT activity. The log of cell number is plotted against the log of CAT activity as measured in c.p.m.

Kits are now also commercially available (i.e. Gibco-BRL) for the non-isotopic immunoassay of a variety of reporter genes, CAT being one of them. These kits employ a sandwich ELISA technique to capture, recognize, and spectrophotometrically quantitate the reporter protein. The advantages of such systems are the complete elimination of radioactivity, good sensitivity, and simplicity since the kit provides almost all of the necessary reagents. Multiple samples can be manipulated at one time, usually difficult to achieve when using the radioactive assay. The major disadvantage is the need for a microwell plate/strip reader. In addition, immunoassays such as these often take the investigator a period of time to become technically proficient with. Nevertheless, these methods provide a viable non-radioactive alternative for CAT determination. Developments in immunoassay techniques (and their commercial availability) have also facilitated the assay of a number of secreted molecules such as growth hormone (hGH or bGH). These types of assays require only conditioned medium samples and not cell extracts.

4.3.2 Assay for luciferase

Simple protocols for the use of luciferase activity as an expression reporter gene have been developed (41), a method which is rapidly overtaking CAT as the assay of choice due to its sensitivity. This firefly enzyme catalyses an ATP-dependent oxidation of the luciferin substrate, a bioluminescent reaction. Since light is the actual product that is measured, the assay requires the use of non-standard laboratory equipment, namely an appropriate luminometer (e.g. Turner TD-20e), to quantitate activity. Cells transfected with luciferase reporter vectors can be prepared for assay by the cell extraction freeze–thaw method described in *Protocol 8*. Kits containing all the reagents to assay luciferase activity are supplied by several companies (e.g. Promega). In outline, the assay is carried out as follows:

(a) Mix a fresh reaction cocktail consisting of 25 mM glycylglycine pH 7.8, 5 mM ATP pH 7.5 (added fresh), and 15 mM $MgSO_4$.

(b) Prepare the substrate solution (1 mM D(−)luciferin in 25 mM glycylglycine) and dilute to 200 μM in the same buffer for assay. Keep this solution in the dark.

(c) Pipette 10 μl cell extract and 350 μl reaction cocktail into a luminometer vial and mix well.

(d) Load the vial into the luminometer and inject 100 μl substrate into the vial.

(e) Read the light emission at room temperature.

As with any assay of this sort, positive controls (addition of defined amount of luciferase) and negative controls (no substrate) need to be included. One can vary the amount of cell extract used in the assay depending on the activity present. Standard protein assay reagents can be used to quantitate the total protein content of the extracts. The luciferase enzyme activity appears to decrease rapidly and so storage of extracts at −20°C or below is suggested for only a few weeks. An example of the use of the luciferase assay is shown in *Figure 2*.

4.4 Antibody staining

Provided antibodies are available, a very quick means of determining transfection efficiency or stable product expression is by staining the cells with an antibody directed against the expressed antigen. These types of protocols are not quantitative but they can give the investigator a good qualitative assessment of transfection or general expression levels. Success with these types of immunomethods will depend, as with any immuno-based protocols, on the quality and specificity of the antibody(s) used. A very simple immunoperoxidase-based method for cell staining is presented in *Protocol 9* (37). This method is a good first step in determining transfection efficiencies. The effective titre and specificity of the primary antibodies must be determined empirically.

Protocol 9. Cell staining using peroxidase-coupled detection[a]

Reagents

- Transfected cells in 60 mm dishes (see *Protocols 4–7*)
- PBS
- Methanol/acetone (1:1 v/v) (make fresh prior to use)
- Secondary antibody: this is horse-radish peroxidase (HRP)-conjugated anti-primary antibody (usually goat anti-mouse IgG-HRP or anti-rabbit IgG-HRP) (e.g. commercially available from Dako Corporation or Pierce Immunotechnology). Dilute the antibody in 10% FBS/PBS and titrate over a range of 1:50 to 1:1000 to determine the best dilution for use; a dilution of 1:150 is usually a good starting value.
- 10% fetal bovine serum in PBS (FBS/PBS)
- Primary antibody diluted in 10% FBS/PBS (the optimal concentration of this antibody must be determined empirically by testing dilutions over a range of 1:100 to 1:5000 but a 1:500 dilution is a useful starting value)
- Colour reagent: 100 ml of PBS, 10 μl hydrogen peroxide (H_2O_2), 1 ml of a saturated *ortho*-diansidine solution. Dissolve the *ortho*-diansidine (e.g. Sigma) in absolute ethanol until the substrate begins to precipitate out of solution; after allowing the precipitate to settle, remove the clear solution and store it covered at 4°C.

Method

(Carry out all steps at room temperature unless otherwise indicated.)

1. Wash the transfected cells twice with PBS. Sometimes cells dislodge very easily at this step so care should be taken to avoid this and hence maintain the monolayer.
2. Fix the cells for 2 min with 4 ml of freshly made methanol/acetone (1:1).
3. Wash the cells twice with PBS.
4. Add 4 ml of 10% FBS/PBS and incubate for 5 min. This step blocks sites that would result in non-specific binding of antibodies to the dish.
5. Add 4 ml of the primary antibody diluted appropriately in 10% FBS/PBS (1:500 dilution initially). Incubate with gentle shaking for at least 2 h.
6. Remove the primary antibody and wash the cells five times with PBS

Protocol 9. *Continued*

within a span of 5 min. This can be conveniently done by gently squirting PBS from a wash bottle and aspirating the wash solution.

7. Add 4 ml of the secondary peroxidase-conjugated antibody[b] (diluted 1:150 in 10% FBS/PBS). Incubate with gentle shaking for at least 2h.[c] Save 0.5 ml of the secondary antibody dilution to test the colour reaction by adding 0.5 ml of colour reagent to it. The reaction should be immediate and produce a dark brownish-blue colour.
8. Repeat step 6.
9. Add 4 ml of colour reagent to each dish. Allow the colour reaction to develop for 30–45 min.
10. Stop colour development by aspirating the colour reagent and washing the cells three times with water. Leave 3 ml of water on the cells when finished. The colour should remain reasonably stable for about a week at 4°C.

[a] Ref. 37 provides excellent photographs of transfected cells stained via this method.
[b] A fluorescent labelled antibody can be used for the secondary antibody but will require a fluorescent microscope for visualization.
[c] Incubation at this step can be done overnight at 4°C.

5. Stable transfection

Stably transfected cell lines produce the product proteins on a continuous, long-term basis (months to years). Selection is normally carried out clonally. In most cases, good expression requires the amplification of the integrated sequences.

One of the characteristics of stable transfection is the random nature of integration events. Expression levels in stable integrative transfections are highly influenced by the chromatin structure at or near the site of integration as well as the number of copies of the exogenous DNA which ultimately become integrated. As a result, the transcriptional activity of the transfected gene in different clones generated from identical vector constructions often shows extreme variability due to both position effects and multiple integrations. Genomic targeting vectors have been developed which address these constraints and potentially provide a level of predictability and consistency to stable gene expression, allowing for an overall enhancement in the efficiency of transfection. Among these are several vector integration systems based on targeting the exogenous DNA to predetermined sites within the genome with site-specific recombinases. Again, the development of non-mammalian component systems such as the Cre recombinase of coliphage P1 (42) and the FLP protein of *Saccharomyces cerevisiae* (43) have facilitated the evolution of site-directed recombination in mammalian cells.

Over the past years, popular methods to maintain stably expressing cell lines have been established using high copy number, episomally replicated vectors. These systems are free from chromatin position effects and are not dependent on gene amplification, but do depend on efficient plasmid replication and stability for good overall expression levels. Unlike the SV40 ori/COS system, replication of the vector is controlled, allowing cell viability to be maintained. One popular choice of system, based on viral replication components, is the Epstein–Barr virus (EBV) vectors (44). Good expression depends on the presence of the 1.8 kb origin of replication region of EBV (ori P), for replication in cell lines either expressing the viral EBNAI nuclear antigen or its expression via co-transfection. The vectors will replicate in a wide variety of cell lines, but not in rodent cells. EBV vectors and cell lines designed for use with this transfection system are available from Invitrogen.

5.1 Selectable markers in mammalian cells

The classic approach for obtaining the desired transformants, termed co-transformation, allows one to select for the genomic incorporation of one gene construct by co-transfecting the host cells with a second construction carrying one of the many available phenotypically 'selectable' genes. Many vectors in use today have been developed so as to contain both the marker gene and the desired expression units on the same construction. These selectable markers essentially fall into two categories:

- genes of non-mammalian origin which establish drug resistances in cell culture
- selectable markers dependent on certain mammalian cell genotypes, e.g. DHFR selection

A few commonly used selectable marker genes and their requirements are presented below (others have been described in ref. 37). In most cases, especially where the selections are of bacterial origin, the vectors are constructed with these genes under the control of a mammalian promoter. Emphasis is given to simple selection systems without regard to potential uses in gene amplification, an area covered in Chapter 3.

5.1.1 Neomycin (G418) resistance

Inclusion of the bacterial aminoglycoside phosphotransferase gene (*aph*) on mammalian vectors allows for the detoxification of the protein synthesis inhibiting drug G418 (45). Many vectors for use with this selection are commercially available on a variety of different vector designs (e.g. Invitrogen mammalian selectable expression vectors pcDNA1/Neo etc.). This is probably the most commonly used drug selection in mammalian cell culture, although the effective dose range of G418 varies widely from cell line to cell line. It is necessary to titrate the amount of drug used in a selection according to both its effective concentration and cell number. The effective concentration (or potency) of each batch of G418 sulfate (Geneticin®, Gibco-BRL, 860–1811) varies and yet often in the literature a concentration of G418 is cited without regard to these differences. Taking the time to do a simple 'kill curve' using perhaps two cell densities (0.5 and 1.0 × 10^6 cells/100 mm dish) and a series of drug dosages (50, 100, 200, 400, 600, 800, and 1000 μg G418/ml) will allow one to determine an effective concentration to use with any particular cell line. The dose required for selection is the lowest concentration that results in 100% cell death. The presence or absence of serum in the medium will alter the effective concentration of G418 for selection, so this variable should be examined also.

G418 is normally made up in PBS at a concentration of 40 mg active material/ml, filtered sterilized, and stored at −20°C in aliquots. Each vial of G418 has a different potency indicated on its label. For example, a label may indicate a potency of 694 μg/mg or 694 μg of active material per milligram total weight. This value changes from lot to lot of G418. Making the stock solution based on active material and not total weight ensures that a consistent and reproducible amount of G418 is used each time medium is supplemented with this drug. Dilutions of this stock solution are made into complete growth medium in which G418 is stable for about two months. Cells should be re-fed every four or five days under this selection. An important point to note here is that recent data have suggested that the *aph* gene can exhibit a negative *cis* effect on adjacent promoters when incorporated into the genome (46).

5.1.2 Hygromycin-B resistance

The hygromycin-B-phosphotransferase gene (*hph*), isolated from an *E. coli* plasmid, detoxifies the drug hygromycin-B, a potent inhibitor of protein synthesis via phosphorylation (47). Several hygromycin resistance selectable vectors are commercially available for use (e.g. the hygromycin selectable episomal series from Invitrogen). This selection is very similar to that for neomycin resistance (see section 5.1.1) in that a titration curve should first be performed to determine the most effective concentration (10–1000 μg/ml) of this drug, again at two different cell densities. A hygromycin-B (Boehringer Mannheim 843–555) stock solution, 32 mg/ml, is made in water, filtered sterilized, and stored at −20°C. In growth medium, hygromycin-B is not stable for more than two weeks when stored at 4°C. During selection, cells need to be re-fed every two to three days. The generally poor stability of this drug at 37°C allows for more growth of non-resistant cells if the selection is not consistently maintained and so the selection must be monitored carefully.

5.1.3 New dominant drug markers

In recent years, the need has arisen for more selection systems to be developed, particularly in experiments where multiple expression units are to be introduced independently and analysed. The puromycin *N*-acetyl transferase gene (*pac*) from *Streptomyces alboniger* confers resistance to the antibiotic puromycin. The *ble* gene from *Streptoalloteichus hindustanus* encodes a high affinity binding protein for the glycopeptide antibiotics bleomycin and phleomycin. Both genes have been used successfully as positive selectable markers in mammalian cell culture (48, 49). The concentration of drug required in the *pac* selection is very low (2.5–10 μg/ml range) making the selection protocol reasonably economical. Cloned into a standard pSV2 vector, this gene was used to transfect a number of commonly used cell lines successfully, including BHK21 and mouse L cells, resulting in puromycin resistance. The concentration of drug required in the *ble* selection is also very low (10–20 μg/ml range). The only potential drawback to using these two new selection systems is the limited availability at present of suitable vector constructs bearing the relevant genes. Unlike *Neo*^{R}, a selection almost universally included on all vector designs, these markers are not as yet in widespread use.

5.1.4 Tryptophan synthetase and histidinol dehydrogenase selections

The *trpB* gene of *E. coli* and the *hisD* gene derived from *Salmonella typhimurium* have also been developed for use as positive selections in mammalian cell culture (50). The *trpB* gene transfected into cells allows for survival of clones in tryptophan-minus medium in the presence of exogenously supplied indole. Since this selection requires a very special medium formulation, it may not be the most convenient to use. The product of the *hisD* gene, histidinol dehydrogenase, protects cells from the toxic effects of high concentrations of histidinol which is known to inhibit histidyl-tRNA synthase. The commercial availability of *hisD* selectable vectors (e.g. Invitrogen), along with a supply of inexpensive substrate present this as a potentially more convenient and useful selection scheme. Cells should be titrated for cytotoxicity of histidinol (Sigma H-5542) in complete growth medium. Generally this marker can confer resistance to 2.5 mM histidinol in medium containing 10% serum. One item worthy of note, as seen in CHO cells, is the possibility of generating histinidol resistance by the amplification of the endogenous synthetase gene. These types of events often occur as the concentration of drug is gradually increased rather than via large initial concentrations.

5.1.5 Dihydrofolate reductase (DHFR) selection

This is the classic selectable marker for co-transfection in CHO cells and is used extensively for gene expression studies in these cells. Dihydrofolate reductase catalyses the conversion of folate to tetrahydrofolate which is required for a number of *de novo* and salvage pathways for both amino acid and nucleoside synthesis. Two selection routines are possible:

(a) Using the CHO-DUKX-B11 (dhfr^{-}) cell line (51) or a derivative, vectors carrying the mouse *dhfr* cDNA can be selected efficiently in nucleoside-free media (DMEM/F-12; Gibco-BRL—special ordered without nucleosides) supplemented with dialysed fetal calf serum (see section 3.2).

(b) A mutant form of the *dhfr* gene can be used as a dominant selection in wild-type ($dhfr^+$) cell lines (52). This selection is based on the use of methotrexate, a drug which inhibits dihydrofolate reductase and so is cytotoxic when cells are grown in nucleoside-free media. The DHFR enzyme encoded by the mutant gene is more resistant to inhibition by methotrexate than the wild-type enzyme and so methotrexate concentrations (50–500 nM) can be titrated to select for transfection events. One needs to be extremely careful, however, to isolate true transfected clones from a potentially high background of resistant clones due to amplification of the cellular *dhfr* gene.

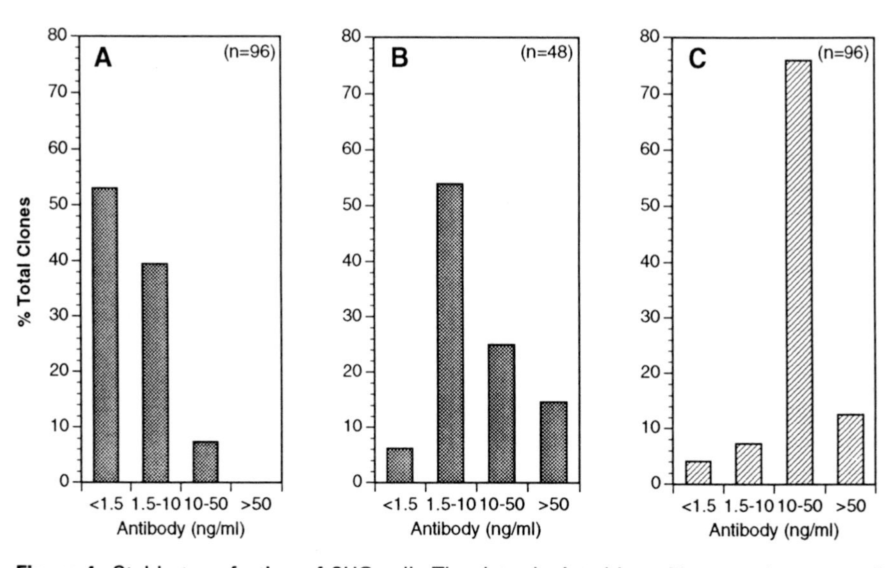

Figure 4. Stable transfection of CHO cells The data depicted here illustrate the range of clonal expression of recombinant antibody observed in CHO cells stably transfected by co-transfection. The ordinate plots the precentage of the cell population which express antibody after transfection and the abscissa plots the degree of expression (ng/ml antibody secreted into the medium). In each panel, n refers to the number of independent clones examined from 96-well dishes. (A) 1 μg of vector containing the selectable marker (DHFR) was co-transfected with 20 μg each of vectors containing heavy and light chain expression units. Supercoiled DNA was introduced into cells by calcium phosphate transfection as described in *Protocol 5*. (B) Vectors designed to include both the selectable marker gene and either the heavy or light chain expression unit on the same plasmid (e.g. DHFR/ light chain and hygromycin/ heavy chain) were transfected as supercoiled vector (10 μg : 10 μg) by the calcium phosphate method. (C) Vectors designed as in (B) but in this case the plasmids were linearized and introduced into the cells by electroporation. In each case medium from 70–90% confluent cultures was assayed after five days of product accumulation. Assays for antibody production provided by Genentech Inc. Immunoassay services.

5.2 High efficiency method for stable transfection

As outlined in section 5.1, a number of selectable genes have been developed for use in stable transfection. These genes are routinely introduced with the experimental DNA either on a separate vector at an empirically determined ratio (the standard co-transformation method), or on the same plasmid within a separate transcription unit. Very efficient transfection of the latter type vectors can be obtained using electroporation of linearized plasmids. *Figure 4* gives a broad overview of stable selections carried out using both a single vector and multiple vector approach to selection as well as the use of either $CaPO_4$ or electroporation as a transfection method. The latter method (*Protocol 10*) uses the application of short, high voltage pulses to induce pore formation in the plasma membrane, subsequently allowing DNA to enter the cell (53). Although specialized equipment is required, this method is technically simple to perform. The efficiency of uptake of DNA appears to be influenced by many factors, among them the nature of the reagents (e.g. DNA, buffers, and cell line). However, of particular importance is the applied electric field and its duration. As with many of the other transfection methods described in this chapter, each parameter should be optimized for best results. The method is included in this section since it is very effective in generating stably transfected cells. However, it is also a very efficient procedure to transiently transfect cell lines as well. All of the other transfection techniques presented in this chapter, except the DEAE–dextran method (*Protocol 4*), are also useful for establishing stably expressing cell lines.

Protocol 10. Electroporation transfection

Equipment and reagents

- Host cells (approx. 10–15 monolayer cultures in 100 mm dishes)
- Sterile microcentrifuge tubes (optional)
- Electroporation apparatus (e.g. CellPorator Electroporation System I; Gibco-BRL)
- Electroporation cuvettes (0.4 cm electrode gap)
- Buffer A: this is DMEM/F12 medium (Gibco-BRL) without serum but (optionally) containing penicillin and streptomycin; alternatively use 20 mM Hepes pH 7.05, 137 mM NaCl, 5 mM KCl, 0.7 mM Na_2HPO_4, and 6 mM dextrose. Adjust the pH carefully using NaOH just prior to using this solution.
- Serum-free medium (see section 3.2)
- Complete growth medium
- Plasmid DNA. (The DNA used for stable transfection with this method is usually first linearized with an appropriate restriction endonuclease. For a routine stable transfection into CHO cells, linearize 5–50 μg DNA, phenol/chloroform extract it to remove the restriction enzyme, and precipitate with ethanol. Electrophorese a small aliquot in an agarose gel to confirm linearity. Dry the DNA in the tissue culture hood and resuspend it in 50 μl of buffer A just before transfection.)

Method

1. Trypsinize the cells (see section 3.3) and collect them by centrifugation.
2. Wash the cells twice with serum-free medium or buffer A.

Protocol 10. *Continued*

3. Resuspend 0.5–1.0 × 10^7 cells in 750 μl of buffer A in a sterile microcentrifuge tube.
4. Add the plasmid DNA to the cells. Close the tube cap and gently invert the tube to mix.[a] Allow the DNA/cell mixture to incubate in a tissue culture hood at room temperature for 30 min. Gently invert the tube five to six times during this incubation period to keep the cells from settling out.
5. Set-up the electroporation apparatus according to the manufacturer's directions.
6. If DNA/cell mixture is in a microcentrifuge tube, transfer it to an electroporation cuvette using a sterile Pasteur pipette. Make sure the cells are suspended by gently inverting the cuvette several times. Place the cuvette in the electroporation unit.
7. Set the electroporation parameters. For the CHO-DUKX-B11 line and derivatives, a voltage of 350–400 V and a capacitance of 330 μF are effective conditions.[b]
8. Discharge the voltage to the sample and allow it to rest within the unit for 2–5 min before removing it.
9. Place the cuvette on ice for 10 min.[c]
10. Remove the cells using a sterile 1.0 ml pipette and plate them in complete growth medium in a 60 mm tissue culture dish. Incubate 37°C for 24 h.
11. After this incubation monitor the viability of the cells by trypan blue exclusion (*Protocol 3*).
12. For selection of stable transfectants, plate the cells (after the initial 24 h incubation at step 10) into the appropriate selection medium (see section 5.1) at approximately 10^6 cells/10 cm^2 dish (the exact plating densities need to be optimized for the selection system and the cell line used). Incubate at 37°C.

[a] The cells can also be incubated with the DNA directly in the sterile electroporation chamber.
[b] These parameters need to be optimized for each cell line.
[c] The overall effect of this incubation on transfection efficiency appears to be very experiment-dependent. Its value therefore needs to be determined empirically.

The key to the effective use of the electroporation method (*Protocol 10*) for stable transfection is probably vector choice. Transfection with a pre-linearized vector at a designated site (well outside of the expression units) allows for the selection of stable cell lines whose expression of the selectable marker ensures a 90–100% chance of co-expression of the linked experimental gene. In methods where vectors are not normally linearized before use (see earlier in this chapter), breakage and integration of the supercoiled vector is random. Thus, in cells where breakage is within the coding region of the gene of interest, expression either fails to occur or produces a premature termination product. Electroporation also provides integrated DNA that is more structurally defined than in the $CaPO_4$ procedure, where large concatemers are often ligated and integrated into the chromatin, particularly in the presence of carrier DNA, even before amplification takes place. Another reason for introducing DNA for stable expression via electroporation (or lipofection, see *Protocol 8*) is the increasing concern over the potential generation of mutations in the transfected DNA using $CaPO_4$-mediated gene transfer. This is of particular importance where recombinant proteins are purified from large scale (> 50 litres) continuous cell cultures. The risks of generating unpredictable changes in the transfected DNA, in these cases, can be less than cost-effective.

Acknowledgements

I wish to acknowledge primarily my mentor, Cori Gorman, whose excellent article on this subject (see ref. 37) remained the focal point of this update and my Genentech colleagues who contributed both data (Brian Lucas, Ken Carey) and support (Susan Frey, Craig Crowley, Rich DeMarco, and Paul Godowski). I would also like to acknowledge the many researchers whose work on transfection technology and reporter systems have contributed immeasurably to our present understanding of gene expression. I also thank my family, George and Heather, who inevitably donate time to these projects as well.

References

1. Lusky, M. and Botchan, M. (1981). *Nature*, **293**, 79.
2. Kozak, M. (1986). *Cell*, **44**, 283.
3. Huang, M. and Gorman, C. M. (1990). *Nucleic Acids Res.*, **18**, 937.
4. Kaufman, R. J. and Sharp, P. A. (1982). *Mol. Cell. Biol.*, **2**, 1304.
5. Berg, P. (1981). *Science*, **213**, 296.
6. Gorman, C. M., Moffat, L. F., and Howard, B. H. (1982). *Mol. Cell. Biol.*, **2**, 1044.
7. Foeking, M. K. and Hofstetter, H. (1986). *Gene*, **45**, 101.
8. Gorman, C. M., Gies, D., McCray, G., and Huang, M. (1989). *Virology*, **171**, 377.
9. Israel, D. I. and Kaufman, R. J. (1989). *Nucleic Acids Res.*, **17**, 4589.
10. Christopherson, K. S., Mark, M. R., Bajaj, V., and Godowski, P. J. (1992). *Proc. Natl Acad. Sci. USA*, **89**, 6314.
11. Fieck, A., Wyborski, D. L., and Short, J. M. (1992). *Nucleic Acids Res.*, **20**, 1785.
12. Bujard, H. and Grossen, M. (1992). *Proc. Natl Acad. Sci. USA*, **89**, 5547.
13. Berg, D. T., McClure, D. B., and Grinnell, B. W. (1993). *Biotechniques*, **14**, 972.

14. Hippenmeyer, P. and Highkin, M. (1993). *BioTechnology*, **11**, 1037.
15. Ehlert, F., Bierbaum, P., and Schorr, J. (1993). *Biotechniques*, **14**, 546.
16. Birnboim, H. C. (1983). In *Methods in enzymology* (R. Wu, L. Grossman, K. Moldave eds), Vol. 100, pp. 243–255. Academic Press, Inc., New York.
17. Freshney, R. I. (1987). *Culture of animal cells, a manual of basic technique*. Alan R. Liss, Inc., New York.
18. Mellon, P., Parker, V., Gluzman, Y., and Maniatis, T. (1981). *Cell*, **27**, 279.
19. Muller, W. J., Naujokas, M. A., and Hassell, J. A. (1984). *Mol. Cell. Biol.*, **4**, 2406.
20. Heffernan, M. and Dennis, J. W. (1990). *Nucleic Acids Res.*, **19**, 85.
21. Gerard, R. D. and Gluzman, Y. (1985). *Mol. Cell. Biol.*, **5**, 3231.
22. McCutchan, J. H. and Pagano, J. S. (1968). *J. Natl Cancer Inst.*, **41**, 351.
23. Luthman, H. and Magnusson, G. (1983). *Nucleic Acids Res.*, **11**, 1295.
24. Lopato, M. A., Cleveland, D. W., and Sollner-Webb, B. (1984). *Nucleic Acids Res.*, **12**, 5707.
25. Seed, B. and Aruffo, A. (1987). *Proc. Natl Acad. Sci. USA*, **84**, 3365.
26. Gorman, C. M., Gies, D. R., and McCray, G. (1990). *DNA Prot. Eng. Tech.*, **2**, 3.
27. Graham, F. L. and van der Eb, A. J. (1973). *Virology*, **52**, 456.
28. Chen, C. and Okayama, H. (1987). *Mol. Cell. Biol.*, **7**, 2745.
29. Sussman, D. J. and Milman, G. (1984). *Mol. Cell. Biol.*, **4**, 1641.
30. Gorman, C. M. and Howard, B. H. (1983). *Nucleic Acids Res.*, **11**, 7631.
31. Goldstein, S., Fordin, C. M., and Howard, B. H. (1989). *Nucleic Acids Res.*, **17**, 3959.
32. Akusjari, G., Svensson, C., and Nyard, O. (1987). *Mol. Cell. Biol.*, **7**, 549.
33. Kaufman, R. (1985). *Proc. Natl Acad. Sci. USA*, **82**, 689.
34. Felgner, J., Martin, M., Tsai, Y., and Felgner, P. L. (1993). *J. Tiss. Cult. Meth.*, **15**, 63.
35. Hawley-Nelson, P., Ciccarone, V., Gebeyehu, G., Jesse, J., and Felgner, P. L. (1993). *FOCUS*, **15**, 73.
36. Alam, J. and Cook, J. L. (1990). *Anal. Biochem.*, **188**, 245.
37. Gorman, C. (1985). In *DNA cloning: a practical approach*, Vol. II (D. M. Glover ed.), pp. 143–90. IRL Press, Oxford.
38. Seed, B. and Sheen, J.-Y. (1988). *Gene*, **67**, 271.
39. Nordeen, S. K., Green, P. P. I., and Fowlkes, D. M. (1987). *DNA*, **6**, 173.
40. Luckow, B. and Schutz, G. (1987). *Nucleic Acids Res.*, **15**, 5490.
41. de Wet, J. R., Wood, K. V., DeLuca, M., Helinski, D. R., and Subramani, I. (1987). *Mol. Cell. Biol.*, **7**, 725.
42. Fukushige, S. and Sauer, B. (1992). *Proc. Natl Acad. Sci. USA*, **89**, 7905.
43. O'Gorman, S., Fox, D. T., and Wahl, G. M. (1991). *Science*, **251**, 1351.
44. Yates, J. L., Warren, N., and Sudgen, B. (1985). *Nature*, **313**, 812.
45. Colbere-Garapin, F., Horodniceanu, F., Kaurilskyu, P., and Garapin, A.-C. (1981). *J. Mol. Biol.*, **150**, 1.
46. Artlet, P., Grannemann, R., Stocking, C., Friel, J., Bartsch, J., and Hauser, H. (1991). *Gene*, **99**, 249.
47. Blochlinger, K. and Diggelmann, H. (1984). *Mol. Cell. Biol.*, **4**, 2929.
48. Vara, J. A., Pontela, A., Ortin, J., and Jimenez, A. (1986). *Nucleic Acids Res.*, **14**, 4617.

49. Mulsant, P., Gatignol, A., Daleno, M., and Tiraby, G. (1988). *Somat. Cell Cell Mol. Genet.*, **14**, 243.
50. Hartman, S. and Mulligan, R. (1988). *Proc. Natl Acad. Sci. USA*, **85**, 8047.
51. Urlaub, G. and Chasin, L. A. (1980). *Proc. Natl Acad. Sci. USA*, **77**, 4216.
52. Simonsen, C. and Levinson, A. (1983). *Proc. Natl Acad. Sci. USA*, **80**, 2495.
53. Andreason, G. L. (1993). *J. Tiss. Cult. Meth.*, **15**, 56.

2

Construction and characterization of vaccinia virus recombinants

MIKE MACKETT

1. Introduction

The use of cowpox to vaccinate against smallpox was popularized by Jenner at the end of the 18th century and ultimately led to the eradication of smallpox. Although cowpox was used as a vaccine for a short period, the closely related poxvirus vaccinia has been in use for this purpose since the 1850s. Cowpox and vaccinia virus have played an important role in the history of human medicine since they represent the first widely used human vaccine and the only vaccine whose use has resulted in the eradication of a human disease. Surprisingly, interest in poxviruses in general, and in vaccinia virus in particular, increased substantially after the disappearance of smallpox. This interest has arisen largely from the development of vaccinia virus as an expression vector with the potential use of vaccinia virus recombinants as live vaccines against pathogens other than the smallpox virus.

The first edition of this book (1) described the basic theory and protocols central to generating vaccinia virus recombinants. Although in principle many of the protocols remain largely unchanged, the process has been simplified and is now more flexible.

2. Biology of vaccinia

Although this chapter deals with vaccinia virus and the generation of recombinants, it should be noted that the basic principles can be applied to other poxviruses, such as fowlpox and canarypox. Both of these avipoxviruses have been used recently as vectors for foreign genes and canarypox has been used in Phase 1 human trials as a vaccine (2). The protocols for generating recombinants in these viruses differ only in the conditions required for growth of avipoxviruses.

Excellent reviews have been written on poxvirus molecular biology (3) and pathogenesis (4), and on the enzymology of vaccinia virus transcription (5), DNA replication (6), virus glycoproteins, and immune evasion (7). A further useful and very thorough source of information on the construction, characterization, and use of vaccinia recombinants is ref. 8.

2.1 Virus structure and replication

Vaccinia virus has a complex architecture with an oval or 'brick-shaped' structure approximately 250 × 350 nm in size, making it at the limits of detection by light microscopy. Over 100 proteins are contained in the particle which has an outer lipid envelope and a biconcave-shaped core, containing the DNA genome and many virus-coded enzymes, flanked by structures termed lateral bodies.

Two major forms of infectious virus exist. One form, intracellular naked virus (INV), is found in the cytoplasm of virus-infected cells and represents the great majority of infectious progeny; 99% of the total progeny of the WR strain of wild-type vaccinia virus is cell-associated. The second form, termed extracellular enveloped virus (EEV), is released from infected cells and possesses an additional lipid envelope which is acquired from the Golgi membrane of the host cell. Several virus-coded glycoproteins are present only in this envelope (9) and give EEV its distinct immunological and biological properties. In particular, EEV production is required for plaque formation and for the efficient spread of virus *in vitro* and *in vivo* (9).

The virus genome is 67% A+T and is a large double-stranded DNA molecule of 180–200 kb, the precise size depending on the virus strain. The termini of the two DNA strands are joined by a single-stranded DNA hairpin loop. The central two-thirds of the genome show high conservation between orthopoxviruses and contain genes essential for virus replication while the remaining third is not required for replication *in vitro* (10). Variation between vaccinia strains are generally limited to the termini and can be the result of duplication, deletion, or transposition. The complete sequence of the 180 kb genome of the Copenhagen strain (11) and much of the WR strain (e.g. ref. 12) has been reported as, incidently, has much of the sequence of variola (13).

Poxviruses replicate in the cytoplasm of the infected cell and as a consequence code for a large number of enzymes which allow replication and transcription to occur independently of the cell nucleus. Many of these enzymes are packaged within the virion core, but some are present only in the infected cell. The key virion enzyme is a multisubunit DNA-dependent RNA polymerase which recognizes unique motifs contained in poxvirus gene promoters. Other enzymes, such as DNA polymerase, a topoisomerase, a DNA ligase, and several enzymes involved in nucleotide triphosphate precursor synthesis, also aid poxvirus replication in the cytoplasm.

2.2 Gene expression

Immediately after infection, the virus-encoded DNA-dependent RNA polymerase transcribes the early class of genes. Some of these genes encode enzymes needed for DNA synthesis, and others encode proteins that mediate further uncoating of the particle to permit DNA replication. At the onset of DNA replication, transcription of most early genes stops and an intermediate class of genes is activated. At least three intermediate proteins are transcription factors which are required for late transcription (14). Late genes are distinct from intermediate genes in that they are absolutely dependent upon prior DNA synthesis. They encode late virus structural proteins and some enzymes and early transcription factors which are packaged into maturing particles. A fourth class of gene is expressed throughout infection due to the presence of early and late promoters upstream of the gene.

3. Construction of vaccinia virus recombinants

The major technical problems involved in the construction of vaccinia virus recombinants are the insertion of DNA into the virus, the efficient expression of the foreign gene, and selection of the recombinant virus. Unlike smaller DNA viruses, the size of the virus genome made the construction of recombinant genomes *in vitro* particularly difficult and, since isolated virus DNA is non-infectious, it was difficult to produce infectious recombinant viruses from recombinant genomes constructed *in vitro*. Although this has now been achieved (15), the approach described below is still the method of choice. Recombinants are generated by a two-step procedure, as illustrated in *Figure 1*; a plasmid vector is constructed in which the foreign gene is linked to a vaccinia virus promoter, and then the gene is transferred into the virus genome by homologous recombination in virus-infected cells transfected with the recombinant plasmid.

3.1 Plasmid vectors

A plethora of plasmid insertion vectors for transfer of foreign genes into vaccinia has been described (e.g. *Table 1*). Careful consideration of the vector to be used is required as this will determine:

- the time, level, and duration of gene expression
- the site of insertion within the vaccinia genome
- the method required for selection

The simplest insertion vectors have structures illustrated by pGS20 and pMM4 in *Figure 2*. A vaccinia promoter, in the case of pGS20 the 7.5K promoter and in the case of pMM4 the TK promoter, is positioned immediately upstream of several unique restriction endonuclease sites in which foreign genes can be inserted. The 7.5K promoter has two elements giving expression both before and after DNA replication, whereas the TK promoter is weaker than the 7.5K promoter and operates only before DNA replication. The promoter and cloning sites are flanked by DNA from a non-essential

Table 1. Plasmid insertion vectors for transfer of genes to vaccinia virus

Plasmid	Promoter	ATG[a]	Special feature(s)	Insertion site	Selection[b]	Reference
pMM4	TK	–	–	TK[c]	BUdR	16
pGS20	7.5K	–	–	TK	BUdR	16
pSC11	7.5K	–	–	TK	X-Gal in overlay	17
pMJ601	11K	–	–	TK	X-Gal in overlay	18
p2001	CPX	–	High level expression	TK	BUdR	26
pBCB01–3,06	HF	–	–	TK	TK	27
pTKgpt-F1–3s	11K	+		TK	MPA	23
pTM1	T7.10	+	High level (transient) expression	TK	BUdR	28
pTM3	T7.10	+	High level (transient) expression	TK	MPA	28
pPR34,35	4b-op hybrid	–	Inducible expression	TK	BUdR	29
pgpt-ATA18–2	11K(mod)	–	Good levels of expression, large polylinker	TK	MPA	30
pHES1–3	H6	+	Large polylinker, easy screen for recombinants	TK	Host range	31

[a] + or – indicates the presence or absence of an ATG codon downstream from the vaccinia virus promoter to which foreign DNA sequences can be fused.
[b] The abbreviations referred to here are as follows: BUdR, 5-bromodeoxyuridine; TK, thymidine kinase; MPA, mycophenolic acid.
[c] TK promoter of vaccinia virus.

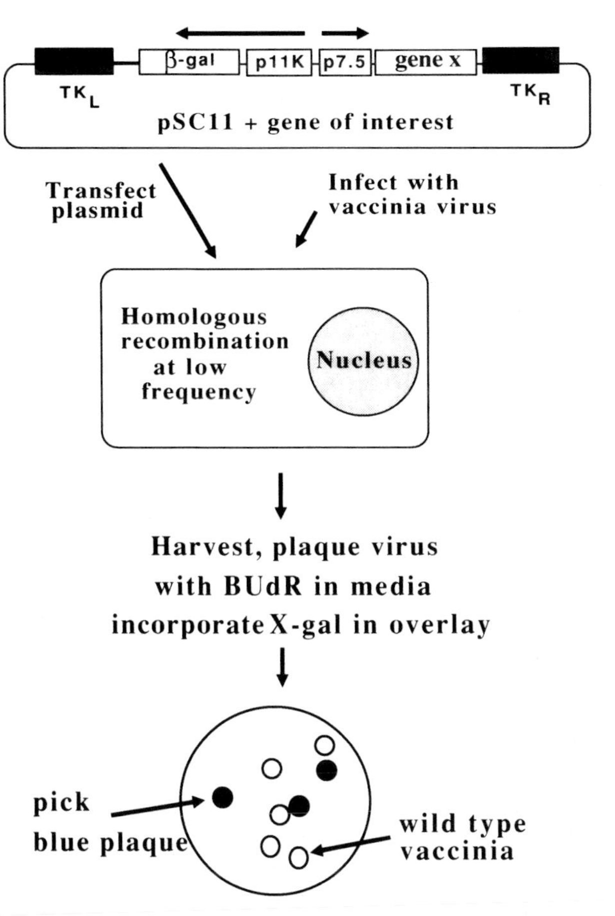

Figure 1. Schematic representation of the generation and selection of vaccinia recombinants. A plasmid vector, in this case pSC11, containing the gene of interest (gene x) linked to a vaccinia promoter (p7.5) and a second promoter (p11K) driving expression of β-galactosidase (β-gal), is transfected into cells infected with vaccinia virus. TK_L and TK_R represent the left-hand and right-hand regions of the vaccinia TK gene, respectively. Recombination between the vaccinia virus DNA flanking the gene and the corresponding DNA from the virus genome inserts β-galactosidase and the gene of interest into the vaccinia genome. This is replicated and packaged into infectious progeny virus. Recombinant virus can then be detected by plaquing the virus in the presence of BUdR with X-Gal in the agarose overlay. Wild-type TK positive virus incorporates BUdR into replicating DNA causing miscoding and as a consequence the virus will not grow. Recombinants are TK negative and will grow in TK negative cell lines in the presence of BUdR. Blue plaques are formed by virus that expresses β-galactosidase and the gene of interest. (If co-expression of *E. coli gpt* has been used for selecting recombinants, all viruses that plaque in the presence of mycophenolic acid will express the *gpt* gene and the gene of interest.)

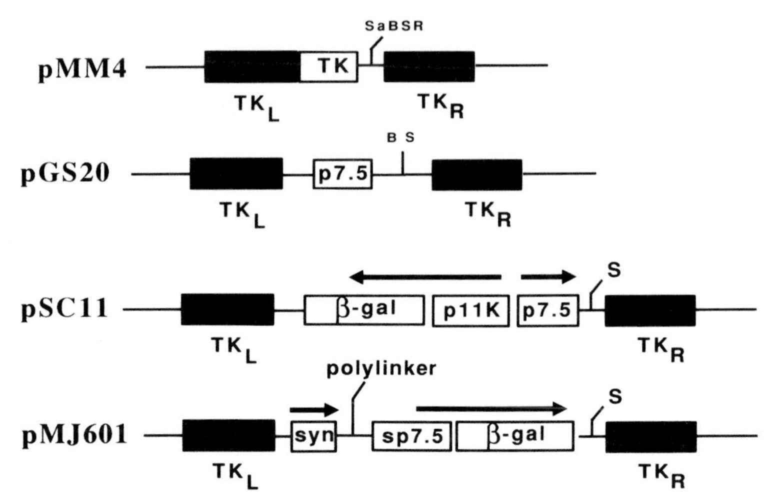

Figure 2. Vectors for insertion of foreign genes into vaccinia virus. pMM4 (16), pGS20 (16), pSC11 (17), and pMJ601 (18) are all insertion vectors that, after homologous recombination, will locate the foreign gene in the thymidine kinase (TK) locus of the virus. Shaded boxes indicate the left-hand (L) or right-hand (R) region of the TK gene. Arrows indicate the direction of transcription of the various units. p11K is a promoter for a late structural gene (21), p7.5K is a promoter that is active before and after DNA replication (16) as is the synthetic truncated version of the promoter sp7.5 present in pMJ601, syn is a synthetic promoter constructed from oligonucleotides which directs high levels of foreign gene expression. β-gal is the *E. coli* β-galactosidase gene allowing recombinant virus to be detected on the basis that it co-expresses the enzyme which in the presence of the X-Gal will give a blue plaque. S, *Sma*I; B, *Bam*HI; R, *Eco*RI; Sa, *Sal*I, are sites for the restriction endonucleases indicated.

locus in the virus genome, in this case the thymidine kinase gene (*tk*), which directs insertion of the foreign gene to the *tk*. The *tk* locus has been used extensively, since interruption of this gene facilitates selection of recombinants (see section 3.4 and *Protocol 7*).

Vectors such as pMM4 (16), pGS20 (16), pSC11 (17), and pMJ601 (18) (*Figure 1* and *Table 1*), do not have an ATG codon positioned downstream from the vaccinia virus promoter to which foreign sequences can be fused, and are therefore suitable for expression of complete open reading frames (ORFs) which contain their own translational initiation codon. Other vectors have an ATG codon positioned upstream from the restriction sites used for insertion which are themselves arranged in all three reading frames and are followed by translational termination codons in all reading frames to enable expression of any protein-coding sequence (19, 20).

When constructing a transfer vector, the promoter of the foreign gene should be removed so that the protein-coding sequence can be placed as close as possible to the transcriptional start site of the vaccinia virus promoter. Additional ATG codons upstream from the ORF should also be removed.

3.2 Choice of promoter

In general, vaccinia promoters for major late structural genes are the strongest natural promoters, for example the 11K (21) or 4b (22) promoters. When the 11K promoter was used to drive β-galactosidase expression in a vaccinia recombinant, yields were in the order of 1–2 mg of protein per litre of infected cells (23). Even higher yields can be achieved by transfer of the prokaryotic T7 RNA polymerase gene to vaccinia and expressing the foreign gene from a T7 promoter (see section 8). Although giving levels 10–20-fold lower than the 11K promoter, the most frequently used promoter is from the 7.5K gene which is active throughout the virus replication cycle. (For a comprehensive list of the promoters used see ref. 8.) *Table 1* is not intended as an exhaustive list but should aid in the choice of an appropriate vector.

In some situations it is important to express the foreign gene either early or late during infection. Vaccinia virus terminates early transcription approximately 50 nucleotides downstream of the sequence TTTTTNT. Thus, if expression in vaccinia of a gene containing this motif is required, it is necessary to express the gene from a late promoter (24) or remove the termination signals to keep the amino acid coding potential (25). Conversely, there are some situations in which early expression is important. For instance, during late vaccinia virus infection the presentation of some peptide epitopes to class I restricted cytotoxic T lymphocytes (CTLs) is blocked, whereas the same epitopes may be presented early during infection (32, 33). If recombinants are constructed in order to analyse CTLs, it is therefore sensible, and may be essential, to express the antigen from an early promoter.

3.3 Insertion site

For most applications the site of insertion is less important than the kinetics and level of expression. Approximately one-third of the virus genome is dispensable *in vitro* (10) and, in theory, foreign genes could be inserted into any one of these (more than 55) genes. However, insertion into genes that are non-essential *in vitro* may have a marked effect on the virulence of the virus *in vivo*. This may be important where recombinants are used to generate antibody or cytotoxic T cell responses or as vaccines. The most convenient and widely used site is *tk*, since insertion here readily allows isolation of recombinants (see section 3.4). However, many other sites have also been used (for an extensive list see ref. 8). If the insertion site is not important for other reasons, however, foreign genes are better inserted into the central region of the genome (e.g. *tk*), since it is more stable than the variable termini.

3.4 Methods for selecting recombinants

Recombinants are formed by homologous recombination within cells infected with vaccinia virus and transfected with a plasmid vector (*Figure 1*). This method, described in detail later (*Protocols 5* and *6*), produces recombinants at a frequency of approximately 0.1% and hence methods to distinguish or select parental from recombinant viruses are required.

As mentioned in section 3.3, the *tk* gene is the most extensively used site for recombination since insertion of the foreign gene within the *tk* locus interrupts the gene and recombinants are unable to produce active *tk* enzyme. Wild-type virus with a functional thymidine kinase will incorporate 5-bromodeoxyuridine (BUdR) into replicating DNA causing lethal mutations. Consequently recombinant virus may be selected from parental wild-type virus by its ability to replicate in a *tk* negative cell line in the presence of BUdR (34). Other methods for distinguishing recombinant and parental virus involve co-expression of β-galactosidase (section 6.4.2, *Protocol 7*) with a visual selection for blue plaques when X-Gal is added to the overlay (17, *Figure 1*), DNA hybridization (*Protocol 8*), and PCR (*Protocol 9*). Dominant selection can be achieved using the *E. coli* guanine phosphoribosyl transferase gene (refs 23, 35, and *Protocol 11*), either incorporating the *gpt* gene into the recombinant virus or allowing for transient expression of the gene. Wild-type virus will not grow in the presence of mycophenolic acid (MPA), an inhibitor of purine nucleotide metabolism. Acquisition of the *gpt* gene allows a recombinant to grow. Thus, if the gene of interest under control of a vaccinia promoter is tagged to the *gpt* gene under a different promoter, virus resistant to the action of MPA should also express the gene under study.

Other methods for the selection of recombinants have also been reported. These include co-insertion of the gene of interest with the neomycin resistance gene and selection with G418 (36), selection based on host range restriction (37) where the parent virus will not grow on a particular cell line and the plasmid used to generate the recombinant restores the parent virus growth properties, and plaque morphology (38) where the parent virus gives a small plaque phenotype which is restored by the transfected plasmid. Although these are efficient methods for generating recombinants, only the most extensively used methods are described in practical detail in this chapter (section 6.2, *Protocols 7–12*).

4. Initial considerations for working with vaccinia

The minimum requirements to work with vaccinia or vaccinia recombinants in the UK are those associated with working with any Category 2 human pathogen such as herpes simplex virus or Epstein–Barr virus. Individual researchers should contact their Biological Safety Officer for specific details. Researchers in other countries should be aware of and comply with national and local guide-lines relevant to such work.

4.1 Safety

Although used in the past as a vaccine, vaccinia virus is a human pathogen and can cause serious disease with life-threatening sequelae. Fortunately, serious complications are exceedingly rare but individuals who are immunosuppressed have difficulty in dealing with the virus and in some cases deaths have been reported (39). A booklet from the Health and Safety Executive (40) is the most important publication in the UK in the area of assessing risks from vaccinia. Vaccination for work with vaccinia is no longer recommended in the UK unless the manipulation of the virus is considered likely to enhance its virulence, very large amounts of virus are being used, or the work involves animals. In the latter situation, a case-by-case evaluation of the need for vaccination should be undertaken. Vaccination is strictly contra-indicated in potential vaccinees with eczema or any form of immunological deficiency.

In the UK, the regulations concerning genetic manipulation can be found in the relevant HSE booklet (41). A centre where work with vaccinia recombinants is to be carried out must also be registered with the HSE. New applications to generate recombinants can take up to 90 days to process and are subject to a fee before approval to carry out the work is given.

4.2 Virus stocks

Many different vaccinia strains have been used to make recombinants. The most commonly used wild-type virus for this purpose is the WR (Western Reserve) strain, but Copenhagen, Lister, and several attenuated strains have all been used as well. The Wyeth vaccine strain (New York City Board of Health) has also been used, but does not grow as well as the WR and Copenhagen strains. However it may be preferred as a vector on safety grounds. Both WR and Wyeth are available from the American Tissue Culture Collection (ATCC catalogue number VR119). Virus stocks are generally stable at −20°C or −70°C for many years. For long-term storage, repeated freeze–thawing of stocks should be avoided and, if at all possible, stocks kept at −70°C.

4.3 Cell lines and culture

Standard cell culture techniques are all that is required to propagate vaccinia virus and generate recombinants. This virus is one of the simplest to grow and will replicate in a wide variety of cell lines. At high multiplicities of infection, replication takes less than 24 hours. The virus remains cell-associated rather than budding into the media. The total amount of virus yield varies significantly between cell lines and even the same virus grown in the same cell line from two different sources will give different yields. The same stock titred in monkey kidney CV-1 cells or rabbit kidney RK13 cells will give a three- to fivefold difference in the number of plaques produced (RK13 > CV-1). However, yields of 50–100 plaque-forming units (p.f.u.) per cell are not uncommon.

One of the advantages of using vaccinia as a vector rather than other viruses is the wide host range of the virus. There are very few cells, even

Table 2. Cell lines for growth of vaccinia virus

Cell line	ECACC code [a]	ATCC code [b]
BSC-1 African green moneky epithelial line	85011422	CCL 26
Vero African green monkey kidney line	84113001	CCL 81
HeLa S3 cervical carcinoma spinner cells	87110901	CCL 2.2
HTK−143 human osteosarcoma	91112502	CRL 8303
CV-1 African green monkey kidney fibroblast line	87032605	CCL 70
Hep 2 human larynx carcinoma cells	86030501	CCL 231
RK13 rabbit kidney	88062427	CCL 37

[a] ECACC is the European Collection of Animal Cell Cultures.
[b] ATCC is the American Tissue Culture Collection.

primary lines, where the virus will not replicate at all, and all of the cell lines in *Table 2* will give good yields of virus. Infection with recombinants have been carried out in a wide array of cell lines including polarized lines such as MDCK and Vero cells, primary rat atrial fibroblasts, B cells, T cells, fibroblasts of various origins, epithelial cells. Transient expression of recombinant vectors can occur even in *Xenopus* oocytes after microinjection. Assays on the recombinant product can thus be carried out in the context of the cell that is the natural site of expression of the gene, or the most convenient cell line.

5. Assay, growth, and purification of vaccinia virus

5.1. Plaque assay for vaccinia virus infectivity

An assay for vaccinia virus based on virus infectivity is described in *Protocol 1*.

Protocol 1. Assay of vaccinia virus (assay for infectivity)

Equipment and reagents

- Source of vaccinia virus for assay; usually infected cells
- 10 mM Tris–HCl pH 9.0 (or 5 mM phosphate–citrate buffer pH 7.4)
- 2.5 mg/ml trypsin [a] (Difco 1:250 product code 0152-13-1 or 2 or 3)
- Virus dilution medium; Hanks balanced salt solution (Gibco-BRL) containing 0.1% BSA, *or* phosphate-buffered saline (PBS) containing 0.1% BSA, *or* PBS containing 1% fetal calf serum (FCS)
- Monolayer cultures of a cell line chosen for infectivity assay (6 cm tissue culture dishes or 1.5 cm^2 monolayers in 24-well trays) [b]
- Phosphate-buffered saline (PBS): 170 mM NaCl, 3.4 mM KCl, 10 mM Na_2HPO_4, 1.8 mM KH_2PO_4 pH 7.2
- Appropriate cell culture media
- 0.2% crystal violet (in 20% ethanol)
- Water-bath sonicator, e.g. Sonicor SC220 (Orme Scientific)

Method

1. Harvest the infected cells by scraping them into the growth medium followed by pelleting at low speed in a bench-top centrifuge. Resuspend the infected cells in 10 mM Tris–HCl pH 9.0 (or 5 mM phosphate–citrate buffer pH 7.4) and release the virus by cell lysis with two cycles of freezing followed by thawing.
2. Disperse any aggregates of cell debris and virus by gentle sonication (e.g. 30 sec in a water-bath sonicator).
3. Add 0.1 vol. of 2.5 mg/ml trypsin and incubate at 37°C for 30 min.[c]
4. Make a tenfold serial dilution[d] of the virus preparation in Hanks balanced salt solution containing 0.1% BSA. Alternatively PBS containing 0.1% BSA or PBS containing 1% FCS can be used as diluent.
5. Remove the medium from a monolayer culture of the appropriate cell line chosen for the assay that has just reached confluence. Wash once with PBS and then drain the monolayer.
6. Add the virus dilution to the cell monolayer in a small volume (0.5 ml for a monolayer in a 6 cm diameter tissue culture dish or 0.15 ml for cells grown as 1.5 cm^2 monolayers in 24-well trays) and incubate at 37°C for 1–2 h.
7. Remove the virus inoculum, add the appropriate growth medium (composition depends on cell type) containing 5% FCS. Incubate at 37°C for 36–48 h.
8. Remove the medium and stain the monolayer with 0.2% crystal violet; 1 ml is sufficient to stain a 6 cm dish in 2–3 min at room temperature.
9. After staining, observe areas of cell degeneration and detachment which appear as 'holes' in the monolayer (plaques). Count these plaques. Use monolayers with approximately 100 plaques to calculate the virus titre of the stock. A knowledge of the dilution used to produce a certain number of plaques allows one to calculate the infectivity of the stock (it is assumed that a single infectious virus particle is responsible for each plaque). This infectivity is usually expressed as plaque-forming units/ml (p.f.u./ml).[e]

[a] It is not vital to use a good grade of trypsin.
[b] CV-1, BSC1, Vero, or RK13 are all suitable hosts for plaque assay (ATCC or ECCAC). Murine or BHK cells give smaller plaques but may be used if necessary.
[c] Virus yields can be severely affected if too much trypsin is used. *This step can be left out*, but in doing so the titre of virus will be decreased two- to fivefold.
[d] To minimize errors the dilutions should be done in duplicate and each dilution titrated on several monolayers.
[e] Note that the titre of a virus stock may vary as much as fivefold when plaqued on different cell lines.

5.2 Growth and purification of vaccinia virus

Vaccinia virus is prepared from a single plaque as described in *Protocol 2* and can then be used to infect a small monolayer of cells to yield a small stock of the virus (*Protocol 3*).

Protocol 2. Plaque purification[a]

Reagents

- All reagents listed in *Protocol 1* except 0.2% crystal violet
- Eagle's minimal essential medium (MEM; from Gibco-BRL 041 01575H or M)
- 1% sterile neutral red (Difco) stock solution
- Agarose overlay: MEM containing 1% Noble agar[b] (Difco) and 5% FCS
- 1% agarose containing 0.01% neutral red (prepared using 1% neutral red stock)
- Sterile Pasteur pipettes

Method

1. Carry out *Protocol 1*, steps 1–6.
2. Remove the virus inoculum and cover the monolayer with agarose overlay (4 ml for a 6 cm diameter tissue culture plate). If selective medium is required, this can be incorporated into the agarose overlay. Incubate at 37°C for 36–48 h.
3. Stain the cells by pouring 1% agarose containing 0.01% neutral red on to the agarose overlay. Incubate at 37°C for several hours to visualize virus plaques which appear as clear areas in the monolayer since only living cells take up the neutral red.
4. Pick well isolated plaques using a Pasteur pipette with a small bulb attached to it. Put the Pasteur pipette into the agarose directly over a plaque so that the pipette is in contact with the monolayer. Remove the agarose plug using suction. This removes the majority of the infected monolayer in the region of the plaque.
5. Transfer the plug and virus-containing monolayer to 0.5 ml of MEM.
6. Release the virus from the monolayer with two or three cycles of freeze–thawing. This can be achieved conveniently by alternate use of dry ice and a water-bath. Between 1×10^2 and 5×10^4 p.f.u. of virus can normally be recovered from a single virus plaque.

[a] To ensure that virus stocks are grown from the progeny of a single recombinant it is necessary carry out at least two rounds of plaque purification. Virus released from step 6 is further plaque purified by taking it through *Protocol 1*, steps 4–6 and *Protocol 2*, steps 4–6.

[b] On some cell lines with selective media we have found that 1% low gelling temperature agarose gives better results instead of 1% Noble agar.

Protocol 3. Preparation of small stocks of virus

Reagents

- Virus purified from a plaque (from *Protocol 2*)
- Growth medium containing 5% FCS (the medium required depends on the cell type being used and whether selection is being applied)
- Cell monolayer (25 cm^2) that has just reached confluence
- Hypotonic buffer: this may be *either* 10 mM Tris–HCl, 1 mM EDTA pH 7.0 *or* 5 mM citrate–phosphate buffer pH 7.4

Method

1. Use half of the virus recovered from a plaque to infect the small monolayer of cells which have just reached confluence.
2. Add the appropriate medium (depending on the cell type and whether selection is required) containing 5% FCS. Incubate at 37 °C until cells in the monolayer are visibly affected (48–72 h). This viral cytopathic effect can either be focal in the form of plaques or all the cells may be infected depending on the yields of virus from the plaque.
3. Scrape the cells into the medium and recover them by pelleting at low speed in a bench-top centrifuge.
4. Resuspend the cells in 2 ml of hypotonic buffer.
5. Release the virus by three cycles of freeze–thawing.
6. Store at −70 °C (if a −70 °C freezer is not available a −20 °C freezer is adequate for storage periods of up to a year).

The small virus stock prepared using *Protocol 3* can be used as required to generate larger stocks (*Protocol 4*). After thawing from −70 °C, the cell debris is often aggregated and if desired can be dispersed by mild sonication before use in *Protocol 4*. Note that either a crude virus stock or purified virus stock may be made. Crude virus stocks can be used for most purposes involving analysis of the biochemical properties of a recombinant gene product. For immunization of animals and use of recombinants in immunological assays we use purified stocks.

Protocol 4. Large scale preparation of virus

Equipment and reagents

For crude virus stocks:

- Small virus stock (from *Protocol 3*)
- SMEM (Gibco-BRL 041 01380H or M)
- MEM (Gibco-BRL 041 01575H or M)
- MEM containing 5% FCS
- 2.5 mg/ml trypsin
- Eight 150 cm^2 tissue culture bottles
- HeLa S3 spinner cells (ATCC)[a]
- PBS containing 0.1% BSA

Protocol 4. *Continued*

For purified virus stocks, the reagents listed above are required plus the following:

- 2–10 litres of HeLa S3 spinner cells (5×10^5 cells/ml) grown in SMEM *or* 20–30 150 cm^2 bottles of monolayer cells
- 10 mM Tris–HCl pH 9.0
- 1 mM Tris–HCl pH 9.0
- 36% (w/v) sucrose in 10 mM Tris–HCl pH 9.0
- 15–40% (w/v) sucrose gradients in 10 mM Tris–HCl pH 9.0
- Tight fitting Dounce homogenizer
- Beckman ultracentrifuge, SW27 swing-out rotor and tubes (or equivalent equipment)

A. *Preparation of crude virus stocks*

1. Seed eight 150 cm^2 tissue culture bottles each with 5×10^7 HeLa S3 spinner cells (ATCC) in MEM and allow to settle overnight.
2. Take half of the progeny virus from a small stock and incubate with 0.1 vol. of 2.5 mg/ml trypsin for 30 min at 37°C.
3. Dilute the virus to 16 ml with PBS containing 0.1% BSA. Add 2 ml of the diluted virus to each tissue culture bottle. This should represent a fairly low p.f.u. to cell ratio (i.e. 0.1–0.01). Incubate for 1 h at 37°C.
4. Add 40 ml MEM containing 5% FCS per bottle and incubate for a further 48 h.
5. Recover the infected cells by scraping them into the medium and pelleting at low speed in a bench-top centrifuge.
6. Resuspend the cells in 2 ml per monolayer (i.e. 16 ml total for eight bottles) of MEM and subject the pooled suspension to three cycles of freeze–thawing to release the virus.
7. Titrate the crude virus stock (see *Protocol 1*). These preparations should have titres of between 2×10^9 and 10^{10} p.f.u./ml and although much lower amounts of virus can be used for most techniques stocks are best kept at this titre as they are more stable upon freezing than low titre stocks.

B. *Preparation of purified virus stocks*

1. Grow 2–10 litres of HeLa S3 spinner cells[a] to a density of 5×10^5 cells/ml in SMEM. Alternatively 20–30 bottles (150 cm^2 area each) of monolayers can be used.
2. Incubate the appropriate amount of crude virus stock (from part A, step 7) with 0.1 vol. 2.5 mg/ml trypsin at 37°C for 30 min. The amount of stock should be sufficient to infect the spinner cells at a multiplicity of 5 p.f.u./cell.
3. Concentrate the spinner cells to 5×10^6 cells/ml by low speed centrifugation and resuspension in culture medium.
4. Add the trypsin treated crude virus stock to the concentrated spinner cells and incubate in suspension for 1 h at 37°C.

5. Dilute with culture medium to give 5×10^5 cells/ml. Incubate for a further 48 h at 37°C.
6. Harvest the infected cells by centrifugation and resuspend them in 10 mM Tris–HCl pH 9.0 at 4°C using 2 ml of this buffer per 2×10^7 infected cells from a monolayer or 2 ml per 100 ml of suspension culture. This and subsequent steps should be carried out on ice.
7. Homogenize with 15–20 strokes of a tight fitting Dounce homogenizer. Check for complete cell disruption by microscopy.
8. Pellet the nuclei by centrifugation at 750 *g* for 5 min at 4°C. Recover and keep the supernatant on ice.
9. Resuspend the pellet in 10 mM Tris–HCl pH 9.0.
10. Repeat step 7. Re-centrifuge and combine the supernatant with that from step 8.
11. Add 0.1 vol. of trypsin (2.5 mg/ml) and incubate at 37°C for 30 min with frequent vortexing.
12. Layer the treated supernatant on to an equal volume of 36% (w/v) sucrose in 10 mM Tris–HCl pH 9.0 in a 14 ml centrifuge tube for the Beckman SW27 swing-out rotor or its equivalent. Centrifuge at 13 500 r.p.m. in the SW27 swing-out rotor (25 000 *g*) for 80 min at 4°C.
13. Discard the supernatant and resuspend the pellet (virus and cell debris) in 2 ml of 1 mM Tris–HCl pH 9.0. Add 0.1 vol. 2.5 mg/ml trypsin and incubate at 37°C for 30 min.
14. Overlay 2 ml of virus suspension on to continuous sucrose gradients (15–40% in 1 mM Tris–HCl pH 9.0) in Beckman SW27 tubes or equivalent. Centrifuge at 18 750 *g* in the SW27 swing-out rotor for 40 min at 4°C.
15. Collect the banded virus. Dilute it 1:3 with 1 mM Tris–HCl pH 9.0. Pellet the virus by centrifugation at 25 000 *g* for 60 min at 4°C.
16. Resuspend the virus in 1 mM Tris–HCl pH 9.0. Store aliquots at −70°C. Virus titres will drop more quickly with time at −20°C, however for periods of up to a year −20°C is adequate.
17. Thaw an aliquot and titrate (*Protocol 1*).

[a] Recently confluent monolayer cells such as CV-1, Vero, BSC-1, or Hela TK^- have also been used. However, even at confluence the number of cells is likely to be two- to threefold less than for HeLa spinner cells and therefore tends to give lower yields of virus.

The basic requirement for preparing large stocks of virus is to have reasonable quantities of cells. For most purposes, 10–20 150 cm^2 tissue culture flasks of cells such as CV-1, RK13, or BSC-1 will yield enough virus to work with. If larger quantities are required (and spinner facilities are not available) it is possible to grow HeLa S3 cells as a semi-adherent monolayer. The HeLa S3 cells are subcultured from two or three 150 cm^2 tissue culture flasks into 30 such flasks and allowed to come to confluence. The medium is replaced with 50 ml of fresh media and the cells allowed to grow for a further three to four days. During this time, many of the cells detach from the monolayer and float in the medium (each flask will contain as many as 10^8 cells). In order to infect both the cells that have detached and those left attached to the flask, the medium (containing the detached cells) is spun at low speed in a bench-top centrifuge and the cell pellets are resuspended in diluent (5 ml/flask containing 0.25 p.f.u./ml of virus) and added back to the flasks. Recovery and purification of virus is then carried out as for *Protocol 4*, step 4 onwards.

6. Isolation of recombinant vaccinia virus

6.1 Transfection methods

Several protocols for transfecting mammalian cells have already been described in Chapter 1. Two procedures readily applied to vaccinia virus, calcium phosphate precipitation and cationic lipid transfection, are described in *Protocols 5* and *6* respectively. The latter protocol uses Lipofectin™, Lipofectace™, LipofectAMINE™ (Gibco-BRL Cat. No. 18292–011, 18301–010, 18324–012, respectively), or a similar product such as DOTAP (Boehringer Mannheim Cat. No. 1202 375). Each is a synthetic cationic lipid which interacts spontaneously with DNA to form lipid–DNA complexes. The lipid facilitates fusion of the complex to the plasma membrane of cells resulting in the uptake of DNA. Although calcium phosphate precipitation produces vaccinia recombinants very well, the Lipofectin™ protocol has several advantages:

(a) Depending upon the cell line used, Lipofectin™ is 5–> 100-fold more efficient than the calcium phosphate procedure.

(b) Unlike the calcium phosphate technique, at the end of the Lipofectin™ protocol, plaques are clearly visible on the transfected monolayer.

(c) Only one cell line need be used for transfection, plaquing, and growth of virus.

(d) We have also found that miniprep DNA will reliably generate recombinants when using Lipofectin™ or DOTAP.

The use of DOTAP instead of Lipofectin™ allows serum-containing medium to be used rather than OPTI-MEM serum-free medium. Serum is said not to affect the efficiency of transfection if DOTAP is used. Overall, we now routinely use DOTAP rather than Lipofectin™ because it generates recombinants with equal efficiency yet serum can be used in the medium. DOTAP is packaged under nitrogen in 400 μl sterile aliquots. Once opened, these have a finite life-span.

Protocol 5. Calcium phosphate transfection

Reagents

- CV-1 cells (obtainable from the American Type Culture Collection (ATCC) as CCL-70)
- MEM (Gibco) containing 5% (v/v) FCS, 100 U/ml penicillin, and 100 μg/ml streptomycin
- Wild-type (WT) vaccinia virus (for example the WR strain obtainable from the ATCC as VR-119)
- Recombinant plasmid DNA containing the foreign gene cloned downstream from a vaccinia virus promoter (see section 3.1)
- Carrier DNA (1 mg/ml sonicated salmon sperm or calf thymus DNA)
- 10 × Hepes-buffered saline (HBS): 10 × HBS stock is 1.4 M NaCl, 250 mM Hepes, 85 mM Na_2HPO_4 and is stored at 4°C in 50 ml aliquots. For transfection, the 10 × stock is used to make a 2 × HBS solution, adjusted to pH 7.12 with NaOH, and filter sterilized (the correct pH is crucial for successful transfection)
- 2 M $CaCl_2$ filter sterilized and stored at 4°C

Method

1. Grow a CV-1 cell monolayer in MEM containing 10% (v/v) FCS in a 25 cm^2 flask to approximately 80% confluence.
2. Aspirate the medium and infect the cells with WT vaccinia virus at a multiplicity of infection (m.o.i.) of 0.05 p.f.u./cell. Gently rock the flask at 37°C for 2 h.
3. Prepare a calcium phosphate precipitate of the recombinant plasmid and carrier DNA as follows. Warm the $CaCl_2$ and HBS solutions to room temperature. In one tube mix 1 μg of recombinant plasmid DNA and 19 μg of carrier calf thymus DNA with 62 μl of $CaCl_2$ and make up to 500 μl final volume. Add dropwise the contents of this tube to a second tube containing 500 μl of 2 × HBS. A precipitate of calcium phosphate–DNA forms immediatly which should appear translucent. If it is opaque then the pH of the HBS should be checked.
4. Remove the virus inoculum and add the calcium phosphate precipitated DNA. Then incubate the flask for 30 min at 37°C.
5. Add 9 ml of MEM containing 5% FCS and incubate the flask for a further 3–4 h at 37°C.
6. Replace the medium with fresh MEM containing 5% FCS and incubate at 37°C.
7. Scrape the cells into the medium at 48 h post-infection (p.i.). Pellet them by centrifugation at 2000 *g* for 5 min.
8. Resuspend the cell pellet in 1 ml of MEM containing 5% FCS. Aliquot and store the cells at −70°C.

Protocol 6. Cationic lipid transfection protocol for producing recombinants

Reagents

- Cell monolayers: HeLa TK⁻ 143B, CV-1, or RK13 cells (ATCC or ECCAC)
- Lipofectin™[a]: prepare 1 μg/ml Lipofectin solution
- MEM containing 5% (v/v) FCS, 100 U/ml penicillin, and 100 μg/ml streptomycin
- Wild-type (wt) vaccinia virus (see *Protocol 5*)
- Carrier DNA (see *Protocol 5*)
- Recombinant plasmid DNA (see *Protocol 5*): mix 10 μg DNA in 25 μl sterile water
- Serum-free medium (DMEM or MEM) or OPTI-MEM low serum content medium[b]
- MEM containing 5% (v/v) FCS
- PBS containing 1% FCS and 0.5% BSA
- Polystyrene tubes (15 ml conical-bottomed Falcon tubes Code No. 2087)

Method

1. Grow monolayers of HeLa TK⁻ 143B, CV-1, or RK13 cells (ATCC or ECCAC) in MEM containing 10% (v/v) FCS until 90–95% confluent. Remove the medium and add the appropriate volume[b] of titred wild-type (wt) virus inoculum. Incubate for 1 h at 37°C.
2. Gently mix 10 μg DNA (in 25 μl water) and 25 μl Lipofectin (1 μg/ml)[c] in a polystyrene tube (the DNA complex can adhere to polypropylene tubes). Allow to stand at room temperature for 15 min.
3. Remove the virus inoculum from the cells and wash the monolayers twice with serum-free medium or OPTI-MEM low serum content medium.[d,e]
4. Add 50 μl of Lipofectin™–DNA complex to 3 ml of OPTI-MEM. Mix gently and then add the mixture to the monolayers. Incubate overnight or for 24 h at 37°C in a humidified atmosphere containing 5–10% CO_2.
5. In the morning, replace the medium with 5 ml of MEM containing 5% FCS and incubate at 37°C for a further 24 h.
6. Remove the growth medium and harvest the cells by scraping into 1 ml of PBS containing 1% FCS and 0.5% BSA. Store at −70°C. Release the virus from the cells by three cycles of freeze–thawing when required.

[a] Other cationic lipids can also be used; see section 6.1.
[b] Although low p.f.u./cell ratios (0.01–0.1) are desirable when using BUdR and selection for a TK⁻ phenotype, positive selection for *gpt* (see section 6.2.4, *Protocol 11*) allows much higher p.f.u./cell ratios to be used.
[c] BRL recommend using 1–20 μg of DNA and 30–50 μg of Lipofectin™ depending on the cell type used.
[d] Serum decreases the transfection efficiency; BRL recommend OPTI-MEM for the washes.
[e] This step is not necessary when using DOTAP instead of Lipofectin.

6.2 Selection of recombinants

6.2.1 Selection of *tk*⁻ virus

The most widely used method of generating recombinants is to insert the foreign DNA into the *tk* locus and select *tk*⁻ recombinants, as described in *Protocol 7* (1, 34). However, despite its popularity, this method has some limitations:

- it requires a *tk*⁺ parental virus
- it allows insertion only into the *tk* gene of vaccinia virus
- recombinant plaques must be selected on *tk*⁻ cells in the presence of the mutagenic compound BUdR: the BUdR is required to select against wild-type virus which possess a functional thymidine kinase capable of incorporating the BUdR into replicating DNA and causing lethal mutations
- some *tk*⁻ plaques are spontaneous mutants which do not contain the foreign gene

Protocol 7. Isolation of *tk*⁻ virus

Equipment and reagents

- Cells transfected with vaccinia virus and recombinant plasmid DNA (from *Protocol* 5 or 6)
- 5 mg/ml BUdR stock solution (filter sterilize, aliquot, and store at −20°C)
- 2% (w/v) low gelling temperature agarose stock, autoclave
- Human *tk*⁻ 143B cells grown in 50 mm Petri dishes
- MEM (Gibco) containing 10% (v/v) FCS
- MEM containing 0.01% neutral red and 1% low gelling temperature agarose[a]
- Sterile Pasteur pipettes
- Water-bath sonicator (e.g. Sonicor SC220; Orme Scientific)
- 20 mg/ml X-Gal in dimethylformamide (required only if the recombinants include the *E. coli lacZ* gene; see footnote[a])

Method

1. Thaw the transfected cells and freeze–thaw twice more. Sonicate briefly in a water-bath sonicator to complete cell lysis and disaggregate the virus.
2. Prepare tenfold dilutions of the cell lysate to a final dilution of 10^{-4} in MEM containing 10% FCS.
3. Remove the medium from confluent monolayers of *tk*⁻ 143 cells in 50 mm Petri dishes and add 0.5 ml of diluted virus to each dish. Plate out the 10^{-2}, 10^{-3}, and 10^{-4} dilutions in duplicate (10^{-1} dilution often gives confluent *tk*⁻ plaques). Gently rock the dishes for 2 h at 37°C.
4. Remove the virus and overlay the cells with 4 ml of MEM containing 1% low gelling temperature agarose, 2.5% FCS, and 25 μg/ml BUdR. Leave the dishes at room temperature in the dark until the agarose has set and then incubate them for two days at 37°C.

Protocol 7. *Continued*

5. Stain the virus plaques by overlaying the agarose in each dish with 4 ml of MEM containing 0.01% neutral red and 1% low gelling temperature agarose.[a] Incubate for 2–4 h at 37 °C.
6. Use Pasteur pipettes to pick plaques from the highest dilution at which they are visible (see *Protocol 2* for procedure).
7. Store the plaques at −70 °C in a small volume of MEM containing 10% FCS.

[a] If the recombinant is derived from vector pSC11 or other plasmids which insert the *E. coli lacZ* gene into *tk*, include 200 μg/ml X-Gal. This compound is converted by β-galactosidase encoded by the *lacZ* gene into a product which has a deep blue colour, allowing recombinants to be distinguished from spontaneous tk^- mutants.

Spontaneous tk^- mutants may be distinguished from recombinants by a variety of means, including DNA hybridization (*Protocol 8*), immunological screening for expression of the foreign antigen, or PCR (*Protocol 9*). The latter approach is fast and uses oligonucleotide primers that bind to either side of the insertion site within *tk* (42).

Protocol 8. Screening tk^- plaques by DNA hybridization

Equipment and reagents

- HeLa tk^- 143B cell monolayers in 24-well plates
- Phosphate-buffered saline (PBS): 170 mM NaCl, 3.4 mM KCl, 10 mM Na_2HPO_4, 1.8 mM KH_2PO_4 pH 7.2
- Nitrocellulose membrane (e.g. Hybond-C, Amersham)
- 20 × SSC stock: 3 M NaCl, 0.3 M trisodium citrate
- 50 × Denhardt's stock solution: 1% (w/v) BSA (fraction V), 1% (w/v) Ficoll, 1% (w/v) polyvinylpyrrolidone
- tk^- plaques isolated in *Protocol 7*
- 10% (w/v) SDS stock
- 1 M Tris–HCl pH 7.5
- 6 × SSC containing 5 × Denhardt's solution
- 2 × SSC, 0.1% SDS
- ^{32}P-labelled DNA probe specific for the foreign gene
- MEM (Gibco) containing 2.5% FCS and 25 μg/ml BUdR
- Water-bath sonicator (e.g. Sonicor SC220; Orme Scientific)

Method

1. Use half of the yield from tk^- plaques (isolated as in *Protocol 7*) to infect HeLa tk^- 143B monolayers in 24-well plates. Incubate the plates for 2 h at 37 °C.
2. Add 1 ml of MEM containing 2.5% FCS and 25 μg/ml BUdR. Incubate the plates for two days at 37 °C.
3. Check the cells by microscopy for cytopathic effect. If more than 50% of the cells show cytopathic effect, scrape the cells into the medium using the plunger from a 1 ml syringe.

4. Transfer the cells to a 1.5 ml microcentrifuge tube and centrifuge at 12 000 *g* for 1 min.
5. Discard the supernatant and resuspend the cells in 200 μl of PBS. Freeze–thaw three times and sonicate briefly.
6. Dot 50 μl of cell extract on to a nitrocellulose membrane. Place the membrane for 3 min on a filter paper soaked with 0.5 M NaOH. Then transfer for 3 min to a second filter paper soaked with 1 M Tris–HCl pH 7.5, and then for 3 min to a third filter paper soaked with 2 × SSC. Air dry the membrane and bake it for 2 h at 80°C in a vacuum oven.
7. Hybridize the membrane with the radioactive probe as follows:
   (a) Incubate the filter in 6 × SSC containing 5 × Denhardt's solution for 4 h at 65°C.
   (b) Denature the probe by boiling for 2 min, add it to the solution and incubate for 12–18 h at 65°C.
   (c) Wash the membrane twice for 15 min each in 2 × SSC, 0.1% SDS at 65°C, and twice for 15 min each in 0.2 × SSC at 65°C.
   (d) Air dry the membrane and expose it to X-ray film.
8. After a suitable exposure time, develop the film according to the manufacturer's instructions. Select positive isolates by reference to the developed film.
9. Plaque purify the recombinants further (see *Protocol 3*). A total of three cycles of plaque purification usually gives homogeneous isolates.
10. Prepare and titrate virus stocks (*Protocols 1–4*). Store them at −70°C.

Protocol 9. Screening *tk*⁻ plaques by PCR

Equipment and reagents

- *tk*⁻ plaques isolated in *Protocol 7*
- 20 μM solutions of oligonucleotide primers 1 and 2 (flanking either side of the insertion site within *tk*)
- 10 × PCR buffer: 100 mM Tris–HCl pH 8.3, 0.5 M KCl, 15 mM $MgCl_2$, 0.01% (w/v) gelatin
- 5 U/ml *Taq* polymerase
- dNTP solution: 2 mM each of dATP, dCTP, dGTP, and dTTP
- Water-bath sonicator (e.g. Sonicor SC220; Orme Scientific)
- Thermocycler for PCR (e.g. Perkin Elmer DNA thermal cycler 480)

Method

1. Carry out *Protocol 8*, steps 1–4. Resuspend the transfected cells in 400 μl of water. Freeze–thaw the cells twice and sonicate them.
2. Heat the disrupted cells at 95°C for 5 min and then place them on ice.

Protocol 9. *Continued*

3. Set-up the PCR in a final volume of 50 μl by mixing the following:
   - transfected cell extract 3 μl
   - oligonucleotide primer 1 2.5 μl
   - oligonucleotide primer 2 2.5 μl
   - dNTP solution 5 μl
   - 10 × PCR buffer 5 μl
   - *Taq* polymerase 0.25 μl
   - water 32 μl
4. Incubate the mixture at 92°C for 1 min, 60°C for 1 min, and 72°C for 1.5 min. Repeat this cycle 24 times. The lowest hybridization temperature depends upon the sequences of the primers used.
5. Analyse 10–20 μl of the PCR product by electrophoresis on a 1% (w/v) agarose gel including DNA size markers. The size of the PCR product indicates whether the plaque is a spontaneous tk^- mutant or a recombinant with exogenous DNA inserted into *tk*.
6. Carry out *Protocol 8*, steps 9 and 10.

6.2.2 Expression of β-galactosidase

To facilitate the distinction between tk^- recombinants and spontaneous tk^- mutants selected in the presence of BUdR, some plasmid insertion vectors contain *lacZ* linked to a vaccinia virus promoter which may be co-inserted into *tk* (17, 42). In the presence of the chromogenic substrate X-Gal, plaques formed by recombinants can be distinguished visually from spontaneous mutants by their blue colour (see *Protocol 7*). Indeed, expression of β-galactosidase can be used alone to detect recombinants without the need for other selection methods (42). Moreover, subsequent replacement of *lacZ* by another foreign gene enables the selection of recombinants on the basis that they are the plaques which no longer give a blue colour with X-Gal (42).

6.2.3 Selection of tk^+ virus

An alternative method for selecting recombinants is to insert the herpes simplex virus type 1 (HSV-1) *tk* gene linked to a vaccinia virus promoter into any non-essential site of a tk^- vaccinia virus genome (34). Recombinants are tk^+ and will form plaques in tk^- cells in the presence of aminopterin (methotrexate), as described in *Protocol 10*. A second foreign gene may be inserted simultaneously with HSV-1 *tk* and selected in the same way (43). Unlike the selection of tk^- virus using BUdR, this system produces only true recombinants and does not require the use of mutagenic compounds. However, it is dependent upon the use of a tk^- cell line.

Protocol 10. Isolation of tk^+ virus

Reagents

- 40 × TAGG stock: 0.6 mM thymidine, 2 mM adenosine, 2 mM guanosine, 0.4 mM glycine (filter sterilize, aliquot, and store at −20 °C)
- MEM containing 1% low gelling temperature agarose, 2.5% FCS, 0.1 mM non-essential amino acids, 1 × TAGG, 1 μM aminopterin
- 1 mM aminopterin (methotrexate) stock (filter sterilize, aliquot, and store at − 20 °C)
- A tk^- vaccinia virus, an appropriate insertion plasmid (e.g. pVPHTK2), plus all other reagents listed in *Protocol 5* or *6*
- Reagents required for *Protocol 2*, steps 3–5

Method

1. Complete *Protocol 5* or *6* using a tk^- vaccinia virus and an appropriate plasmid (such as pVPHTK2: see ref. 34).
2. Remove the virus inoculum and overlay the cells with MEM containing 1% low gelling temperature agarose, 2.5% FCS, 0.1 mM non-essential amino acids, 1 × TAGG, 1 μM aminopterin. Leave the plates at room temperature in the dark until the agarose has set and then incubate them for 2 days at 37 °C.
3. Carry out *Protocol 2*, steps 3–5.
4. Re-plaque the virus in a selective medium to ensure clonality of virus (i.e. repeat steps 1–3 above).

6.2.4 Expression of the *E. coli* guanine phosphoribosyl transferase gene (*Ecogpt*)

The *E. coli gpt* gene (*Ecogpt*) is a selectable marker gene which is now widely used for recombinants and for experimental manipulation of the vaccinia virus genome. The basis of the selection is as follows. Mycophenolic acid (MPA) inhibits inosine monophosphate dehydrogenase and results in blockage of purine synthesis and inhibition of vaccinia virus replication in most cell lines. The blockage may be overcome by expressing *Ecogpt* from a constitutive vaccinia virus promoter and providing the substrates xanthine and hypoxanthine (19, 35), as described in *Protocol 11*. Some workers have found it necessary to include aminopterin, to block *de novo* synthesis of purines, and thymidine (35). The system is versatile and overcomes the deficiencies of selecting tk^- viruses with BUdR (see section 6.2.1). It is not restricted to insertion into a particular site in the virus genome, does not require the use of specific cell lines or mutagenic compounds, and allows plaque formation by recombinants but not by spontaneous mutants.

An additional advantage of *Ecogpt* is that it is possible to select against its expression, as described in *Protocol 12*, so that $Ecogpt^-$ recombinants may be derived from an $Ecogpt^+$ parent (44). Reverse selection along these lines requires the use of a hypoxanthine phosphoribosyl transferase negative (*hprt*$^-$) cell line, since the nucleoside analogue 6-thioguanine is toxic for mammalian cells expressing *hprt*. Combined with positive selection for *Ecogpt*, reverse *Ecogpt* selection enables a mutation to be introduced into a specific site without leaving the selectable marker in the vaccinia virus genome. Consequently, multiple mutations may be built sequentially into the same recombinant.

Protocol 11. Selection of recombinants expressing *Ecogpt*[a]

Equipment and reagents

- Cell monolayers in 50 mm Petri dishes; use CV-1, HeLa *tk*$^-$ 143B, or BSC-1 cells (BSC-1 cells are available from the ATCC as CCL-26)
- 10 mg/ml MPA stock (filter sterilize, aliquot, and store at −20°C)
- 10 mg/ml xanthine stock in 0.1 M NaOH (filter sterilize, aliquot, and store at − 20°C)
- MEM containing 1% low gelling temperature agarose, 2.5% FCS, 25 μg/ml MPA, 250 μg/ml xanthine, 15 μg/ml hypoxanthine
- 10 mg/ml hypoxanthine stock (filter sterilize, aliquot, and store at −20°C)
- Reagents needed for *Protocol 7*, step 5 and *Protocol 8*, steps 9–10
- MEM containing 2.5% FCS, 5 μg/ml MPA, 250 μg/ml xanthine, 15 μg/ml hypoxanthine
- Plasmid containing *Ecogpt* linked to a vaccinia virus promoter (derived, for instance, from pgpt ATA 18–2: see *Table 1*) inserted into the vaccinia virus gene of interest

Method

1. Carry out *Protocol* 5 or *6* to obtain recombinants containing *Ecogpt*.
2. Prepare monolayers of CV-1, BSC-1, RK13, or HeLa *tk*$^-$ 143B cells in 50 mm Petri dishes. Replace the growth medium with MEM containing 2.5% FCS, 5 μg/ml MPA, 250 μg/ml xanthine, and 15 μg/ml hypoxanthine. Incubate the dishes for 12–24 h at 37°C.
3. Remove the medium and infect the cells with appropriate dilutions of the virus (10^{-2}, 10^{-3}, and 10^{-4} dilutions if the virus is obtained directly from *Protocol 5* or *6*; a dilution of 10^{-1} will give too many plaques). Gently rock the dishes for 2 h at 37°C.
4. Remove the virus inocula and overlay the cells with MEM containing 1% low gelling temperature agarose, 2.5% FCS, 25 μg/ml MPA, 250 μg/ml xanthine, and 15 μg/ml hypoxanthine. Leave the dishes at room temperature in the dark until the agarose has set and then incubate them for two to three days at 37°C.
5. Carry out *Protocol 7*, step 5.
6. Carry out *Protocol 8*, steps 9 and 10.

[a] Other investigators have reported that the selection system does not work on CV-1 cells in the absence of aminopterin (0.2 μg/ml) and thymidine (4 μg/ml).

Protocol 12. Selection against *Ecogpt* expression

Reagents

- 1 mg/ml 6-thioguanine stock (filter sterilize, aliquot, and store at −20°C)
- Reagents needed for *Protocol 7*, step 5 and *Protocol 8*, steps 9–10
- MEM containing 1% low gelling temperature agarose, 2.5% FCS, 1 μg/ml 6-thioguanine
- An *hprt*$^-$ cell line such as HeLa D98R or murine STO (available from the Sir William Dunn School of Pathology, Oxford, UK, or the ATCC (CRL 1503), respectively)
- Plasmid containing a sequence to replace *Ecogpt* present in a recombinant obtained in *Protocol 11*

Method

1. Carry out *Protocol 5* to obtain recombinants lacking *Ecogpt*.
2. Grow monolayers of *hprt*$^-$ cells in 50 mm Petri dishes. Remove the growth medium and infect the cells with the virus yield from step 1, using a range of virus dilutions to obtain an appropriate number of *Ecogpt*$^-$ plaques. Gently rock the dishes for 2 h at 37°C.
3. Remove the virus inocula and overlay the cells with MEM containing 1% low gelling temperature agarose, 2.5% FCS, and 1 μg/ml 6-thioguanine. Allow the agarose to set at room temperature and incubate them for three days at 37°C. (Plaques formed on HeLa D98R cells are smaller than on those formed on BSC-1 cells, and it is therefore necessary to incubate infected D98R cells for a minimum of three days.)
4. Carry out *Protocol 7*, step 5.
5. Carry out *Protocol 8*, steps 9 and 10.

6.2.5 Transient dominant selection

On occasions it is desirable to generate recombinants that express more than one gene. If the *E. coli gpt* gene has been used to confer resistance to MPA then it will not be possible to use this same selection system subsequently to generate a recombinant expressing further genes. As described above, *Protocol 12* describes how one might eliminate the *gpt* gene which would allow MPA to be used in subsequent steps but this is an unnecessary complication. More simply, transient dominant selection allows the *gpt* gene to be used to select recombinants, and then, on passage of the virus in the absence of MPA the *gpt* gene is lost (*Figure 3*). This method stems from the discovery that recombinants can be formed by a single homologous recombination event so that an entire plasmid can be inserted into the vaccinia genome (45, 46). Recombinant genomes formed in this way are unstable due to direct DNA repeats and can be preserved only if selected by a dominant selection. On passage in the absence of selection, one of two alternative cross-over events can occur. The first possible event removes the construct and yields wild-type

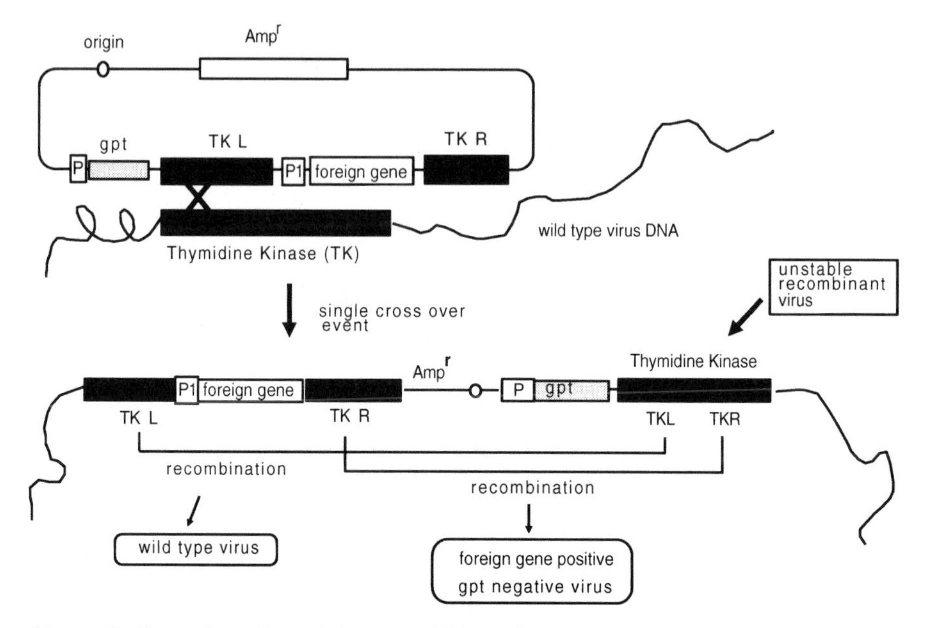

Figure 3. Formation of vaccinia recombinants by transient dominant selection. The plasmid contains a vaccinia virus gene, in this example thymidine kinase (TK), interrupted by the foreign gene under control of a vaccinia virus promoter (P1) and a dominant selectable marker such as *E. coli gpt* under the control of a second promoter (P). (If it is desired to generate a deletion in vaccinia virus then a vaccinia gene with an internal region deleted can be used.) Recombination with the homologous sequence at the left-hand side of TK (TK_L) generates an intermediate recombinant with direct repeats of the left- (TK_L) and right-hand (TK_R) sides of TK. If selection is removed, the intermediate virus resolves by recombination between the direct repeats producing one of two possible viruses. If the recombination is across TK_L a wild-type vaccinia virus will be produced, if the recombination event occurs between TK_R then a recombinant expressing the foreign gene of interest (but lacking *gpt*) will be generated.

virus. The second event removes the *gpt* gene but leaves the gene of interest (see *Figure 3* for details).

The plasmid required for this approach, containing the gene of interest driven by one promoter flanked by vaccinia DNA and the *gpt* gene driven by a separate promoter outside the regions of homology to vaccinia (*Figure 3*), are used in an identical manner to that described in the generation of MPA-resistant virus (*Protocol 11*). Selection of recombinants is achieved by at least two rounds of plaque purification of the virus in the absence of any selective agent.

During resolution of the intermediate virus, plasmids are formed in addition to the progeny virus genomes. These represent the original plasmids or the wild-type gene rescued from the virus. As all plasmids replicate in vaccinia virus cells irrespective of whether they contain vaccinia virus DNA, these plasmids formed during resolution may be amplified and then recombine with the virus causing virus heterogeneity. A screen for plasmid DNA sequences should therefore be undertaken to determine if any of the plasmid sequences persist.

7. Detection of foreign gene products

A variety of methods have been used to analyse the foreign gene product made by vaccinia virus recombinants. The three most common techniques (immunoprecipitation, immunofluorescence, and immunoblotting) are described in the following sections. However, it should be noted that a wide range of other functional and analytical procedures have also been used to study the biological properties of recombinant proteins including immune electron microscopy for subcellular localization, fusion of cells (47, 48), voltage measurements across membranes of infected cells (49), enzyme assays (50), analysis of the *trans*-activating properties of protein products (51), and a variety of immune responses (52–54).

All these methods work well with foreign genes expressed from the 7.5K promoter. If genes are expressed from weaker promoters, some difficulties may be encountered. If the 11K promoter is used it is possible to decrease the multiplicities of infection to as little as 0.1 p.f.u./cell and still detect reactivity using an alkaline phosphatase-conjugated secondary antibody.

7.1 Immunoprecipitation

A reliable procedure for detecting (and quantifying) foreign gene expression by immunoprecipitation is described in *Protocol 13*.

Protocol 13. Detection of foreign gene products by immunoprecipitation

Equipment and reagents

- Cell monolayers (CV-1 cells in 25 cm^2 bottles)
- MEM containing 0.01 mM methionine (prepare this by mixing nine parts methionine-free medium: one part normal medium)
- 50–100 μCi [^{35}S]methionine (100 Ci/mmol)
- Phosphate-buffered saline (PBS): 170 mM NaCl, 3.4 mM KCl, 10 mM Na_2HPO_4, 1.8 mM KH_2PO_4 pH 7.2
- 0.1 M NaCl, 0.1 M Tris–HCl pH 8.0, 0.1% aprotinin
- Antiserum to the foreign protein (either purified IgG at 1 mg/ml or hyperimmune sera)
- Rabbit pre-immune serum
- Protein A Sepharose
- 0.5 M Tris–HCl pH 7.5, 0.15 M NaCl, 0.1% SDS, 1% Triton X-100, 1% sodium deoxycholate
- LUT buffer: 0.4 M LiCl, 2 M urea, 10 mM Tris–HCl pH 8.0
- 0.06 M Tris–HCl pH 6.8, 1% SDS, 5% β-mercaptoethanol, 10% glycerol, 0.002% bromphenol blue
- 10% SDS–polyacrylamide gel (prepared using the Laemmli buffer system; refs 54, 66) plus appropriate electrophoresis buffer (54, 66) and vertical electrophoresis apparatus

Method

1. Infect monolayers of CV-1 cells growing in 25 cm^2 bottles with vaccinia virus (wild-type or recombinant) at 30 p.f.u./cell in Eagle's medium containing 0.01 mM methionine.

Protocol 13. *Continued*

2. After 2 h, add 50–100 μCi of [^{35}S]methionine (100 Ci/mmol).[a]
3. Harvest the cells at 12 h post-infection by washing the cells three times with PBS, scraping them into PBS and pelleting at low speed in a bench-top centrifuge.
4. Resuspend the cells in 0.2 ml of 0.1 M NaCl, 0.1 M Tris–HCl pH 8.0, 0.1% aprotinin. Keep at 4°C for 10 min.
5. Pellet the nuclei at 800 *g* in a bench-top centrifuge. Recover the supernatant into a 1.5 ml microcentrifuge tube.
6. Incubate the supernatant with 25 μl of rabbit pre-immune serum at 4°C, preferably overnight, with agitation. Add 50 μl of a 20% slurry of Protein A Sepharose and incubate for 30 min at 4°C with agitation.
7. Centrifuge for 2 min in a microcentrifuge. Collect the supernatant and incubate for 4 h at 0°C with 25 μl of the appropriate dilution of antiserum (determined empirically for purified IgG 1:100–1:1000 dilution of a 1 mg/ml stock) against the foreign protein.
8. Add 50 μl of a 20% solution of Protein A Sepharose. Incubate for 30 min at 4°C with agitation.
9. Spin in a microcentrifuge for 2 min.
10. Wash the pellet twice with 0.05 M Tris–HCl pH 7.5, 0.15 M NaCl, 0.1% SDS, 1% Triton X-100, 1% sodium deoxycholate.
11. Wash the pellet once with LUT buffer.
12. Transfer the pellet to a fresh microcentrifuge tube and wash again with LUT buffer.
13. Dissolve the precipitated proteins in 100 μl of 0.06 M Tris–HCl pH 6.8, 1% SDS, 5% β-mercaptoethanol, 10% glycerol, and 0.002% bromophenol blue. Leave for 15 min at 25°C and then clarify by brief centrifugation.
14. Heat the supernatant at 100°C for 2 min, load on to a 10% SDS–polyacrylamide gel, and electrophorese. Analyse the final gel by fluorography to detect radiolabelled immunoprecipitated polypeptides (55).

[a] For more specific labelling of products expressed at late times of infection, the addition of radioisotope can be delayed until 12 h post-infection (p.i.) and cells then harvested at 16 h p.i.

Figure 4 shows the results of an immunoprecipitation assay carried out essentially as described in *Protocol 13* but delaying the time of labelling until 12 hours post-infection as described in *Protocol 13*, footnote [a]. The foreign gene expressed by vaccinia virus is a deletion derivative of the Epstein–Barr virus gp110 gene. The immunoprecipitated products were treated with the various glycosidases indicated; the data indicated that the product was trapped in the endoplasmic reticulum since it was sensitive to digestion.

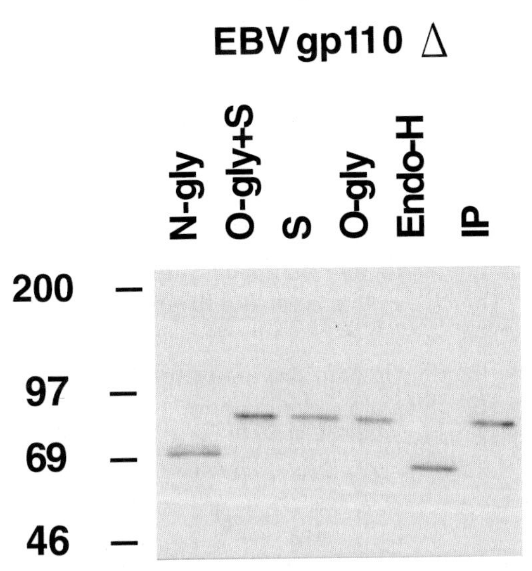

Figure 4. Immunoprecipitation of a deletion derivative of the Epstein–Barr virus glycoprotein gp110 (EBV gp110Δ) expressed by a vaccinia virus recombinant. 4×10^6 RK13 cells were infected with a vaccinia virus recombinant expressing a truncated version of the Epstein–Barr virus glycoprotein gp110, labelled at 12 h post-infection with 50 μCi of [^{35}S]methionine for 4 h. The infected cells were harvested and immunoprecipitated with a polyclonal antiserum raised in rabbits to a fragment of the gp110 gene expressed in bacteria. Precipitates were then treated overnight with the enzyme indicated. S= sialidase, O-gly = *O*-glycanase, N-gly = *N*-glycanase, Endo H = endoglycosidase H. Samples were then run on an 8% SDS–PAGE gel, the gel treated with a fluor, dried down, and exposed to X-ray film. An autoradiograph of the results is shown. Molecular weight markers were run on the same gel and their position is indicated at the left-hand side of the figure; the numbers refer to the M_r of each marker polypeptide in kilodaltons.

7.2 Immunofluorescence

Protocol 14 used in conjunction with the modification detailed at the end of the protocol allows determination of the subcellular location of the foreign gene product expressed by a vaccinia virus recombinant.

Protocol 14. Immunofluorescence assay for cell surface antigens

Equipment and reagents

- Infected cell monolayer
- 2% freshly dissolved paraformaldehyde[a]
- PBS–BSA buffer: PBS containing 0.5–1.0% BSA[b]
- Primary antibody directed against the foreign protein
- Phosphate-buffered saline (PBS): 170 mM NaCl, 3.4 mM KCl, 10 mM Na_2HPO_4, 1.8 mM KH_2PO_4 pH 7.2
- Fluorescent-conjugated second antibody
- Fluorescence microscope (e.g. Leitz Ortholux)

Protocol 14. *Continued*

Method

1. Fix the infected cells in 2% freshly dissolved paraformaldehyde[a] in PBS for 15–30 min at room temperature.
2. Wash away the formaldehyde and quench the fixation reaction by adding cell culture medium.
3. It is important to avoid the non-specific adsorption of antibodies to the monolayer. To achieve this, soak the fixed cells in PBS–BSA buffer for 30 min.
4. Incubate the fixed cells with the greatest dilution of the primary antibody which still gives a good signal, for 1 h at room temperature.
5. Wash away the primary antibody using five washes of 10 min each with PBS.
6. Soak the cells in the PBS–BSA buffer for 5 min.
7. Add the fluorescent-conjugated second antibody and incubate for 30 min at room temperature.
8. Wash the cells five times in PBS.
9. Visualize the fluorescence using a fluorescence microscope set at the appropriate wavelength of light to fluoresce the second antibody conjugate.

[a] 1 : 10 dilution of a stock 37% formaldehyde solution also works.
[b] Gelatin can be used instead of BSA.

Antiserum dilutions used for *Protocol 14* must be determined empirically. Thorough washing of the cells is necessary to avoid non-specific background and antisera should always be spun in a microcentrifuge for 5–10 min before use to remove insoluble material.

For detection of intracellular foreign protein by immunofluorescence, cells are fixed in paraformaldehyde and quenched in 50 mM ammonium chloride or culture medium. They are then permeabilized with 0.1% Triton X-100 in PBS containing 0.25% BSA for 10–30 min at room temperature, and then reacted with primary antibody and fluorescent-conjugated second antibody as for surface fluorescence (*Protocol 14*, steps 4–9). A particularly useful solution for washing cells which have been permeabilized is 100 mM NaCl, 50 mM Tris–HCl pH 7.5, 0.5% Nonidet P-40, 0.25% gelatin. This Triton X-100 method gives good preservation of cell structure and is superior to methods that use ethanol, methanol or acetone, or fixative. *Figure 5* shows typical data obtained using this procedure.

Vaccinia can be grown in a large number of cell types and hence it is possible to study transport of a variety of proteins in well differentiated cell lines using vaccinia recombinants and immunofluorescence. One lingering

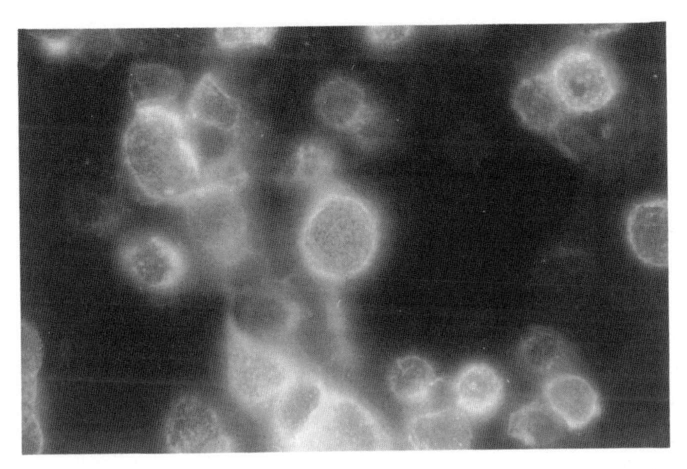

Figure 5. Surface fluorescence of influenza HA in a vaccinia virus recombinant expressing the HA. Madin Darby canine kidney cells were infected at a multiplicity of approximately 3 p.f.u. per cell with a vaccinia–HA recombinant and fixed 4 h post-infection. The subcellular localization of the HA was then detected by fluorescence using a monospecific affinity purified rabbit anti-HA IgG antiserum followed by an FITC-conjugated goat anti-rabbit antiserum.

problem, however, is that by four hours post-infection there is a significant cytopathic effect from the virus infection and this can obscure the clarity of the localization.

7.3 Immunoblotting

Detection of foreign proteins expressed by vaccinia infected cells using immunoblotting is described in *Protocol 15*. Binding of the primary antibody to the protein can be detected using either an enzyme-conjugated second antibody or ^{125}I-labelled *S. aureus* Protein A. Both procedures are described. A typical Western blot is shown in *Figure 6*.

Protocol 15. Western blotting

Equipment and reagents

- Cell monolayers (25 cm^2)
- Recombinant virus
- Polyacrylamide gel loading buffer: 60 mM Tris–HCl pH 6.8, 3% SDS, 5% 2-mercapto-ethanol, 10% glycerol, 0.002% bromophenol blue
- SDS–polyacrylamide gel, electrophoresis buffers, and vertical electrophoresis apparatus (ref. 66)
- Nitrocellulose (e.g. Hybond-C, 0.45 μm, Amersham) or Nylon membranes (e.g. Hybond-N, 0.45 μm, Amersham)
- Electroblotting apparatus (e.g. Bio-Rad mini-protean II transblot electrophoretic transfer cell; Cat. No. 170–3930)
- 4% BSA, 0.02% sodium azide in PBS
- Primary antibody against the foreign protein diluted in 4% BSA, 0.02% sodium azide in PBS[a,b]
- Phosphate-buffered saline (PBS): 170 mM NaCl, 3.4 mM KCl, 10 mM Na_2HPO_4, 1.8 mM KH_2PO_4 pH 7.2
- Electroblotting transfer buffer: 25 mM Tris–HCl pH 8.0, 0.2 mM glycine, 20% methanol

Protocol 15. *Continued*

- Enzyme-conjugated second antibody (e.g. horse-radish peroxidase-conjugated secondary antibody or alkaline phosphatase-conjugated secondary antibody) *or* [^{125}I] Protein A (0.5 μCi/ml in 4% BSA, 0.02% sodium azide in PBS)
- Diaminobenzidine/nickel chloride reagent (for detection if peroxidase-conjugated secondary antibody is used). To prepare this reagent, dissolve 6 mg of diaminobenzidine tetrachloride (Sigma) in 9 ml of 50 mM Tris–HCl pH 7.6, add 1 ml of 0.3% (w/v) nickel chloride, and then 0.1 ml of 3% hydrogen peroxide just before use.
- X-ray film, autoradiographic cassettes, and developer/fixative (for detection if [^{125}I] Protein A is used)
- BCIP/NBT (bromochloroindolylphosphate/nitroblue tetrazolium). (For detection if alkaline phosphatase secondary antibody is used.) Stock solutions of NBT and BCIP can be kept in the fridge for several weeks. Stock is prepared by dissolving 0.5 g NBT in 10 ml of 70% dimethylformamide. BCIP stock is prepared by dissolving 0.33 g of the disodium salt of BCIP in 10 ml of DMF in a glass container. Prepare the BCIP/NBT working solution 1 h before use by adding 66 μl of NBT stock to 10 ml of alkaline phosphatase buffer (100 mM NaCl, 5 mM $MgCl_2$ 100 mM Tris–HCl pH 9.5), mixing thoroughly, and then adding 33 μl of BCIP stock
- TBS (Tris-buffered saline): 10 mM Tris–HCl pH 7.2, 170 mM NaCl, 3.4 mM KCl.

A. *Protein blotting*

1. Infect 25 cm^2 cell monolayers with 10–20 p.f.u./cell of recombinant virus.
2. Harvest the infected cells in 0.5 ml of polyacrylamide gel loading buffer. Heat at 100°C for 2 min, centrifuge briefly, and electrophorese 30 μl of the extract on the appropriate percentage SDS–polyacrylamide gel.
3. After elecrophoresis, transfer the separated protein bands to a nitrocellulose filter.[b]
4. Block non-specific binding sites on the filter by incubation overnight at 4°C with 4% BSA, 0.02% sodium azide in PBS.[c]
5. Incubate with the appropriate antibody dilution in 4% BSA, 0.02% sodium azide, in PBS[c] for 2 h at 37°C.
6. Wash the filter three times for 5 min each time with excess PBS.
7. Detect bound primary antibody using *either* enzyme-conjugated secondary antibody *or* [^{125}I] Protein A as described below.[d] The latter method is the more sensitive procedure.

B. *Detection of enzyme-conjugated secondary antibody*

1. Incubate the filter with an enzyme-conjugated secondary antibody at a dilution of 1:400 in 4% BSA, 0.02% sodium azide in TBS for 1 h at 37°C.
2. Wash thoroughly for 30 min with many changes of TBS.
3. Develop the blot using a chromogenic substrate appropriate for the secondary antibody:
   (a) For peroxidase-conjugated secondary antibody, incubate the filter in diaminobenzidine/nickel chloride reagent until black bands appear representing the sites of antibody binding.

(b) For alkaline phosphatase-conjugated secondary antibody, incubate the filter in freshly made BCIP/NBT until a blue colour appears at sites of antibody binding.

4. In either development reaction, stop the reaction by washing the filter with large volumes of distilled water.

C. *Detection using [^{125}I] Protein A*[d]

1. Incubate the filter with 0.5 μCi/ml [^{125}I] Protein A in 4% BSA, 0.02% sodium azide in PBS for 1 h at 37 °C.

2. Wash the filter thoroughly for 30 min with many changes of PBS.

3. Keep the filter slightly wet and detect the bound secondary antibody by autoradiography using X-ray film.[e]

[a] Antibody concentrations need to be determined empirically. With supernatant from a monoclonal antibody, we often use a 1:1–1:5 dilution whereas, if ascitic fluid is used, dilutions of 1:100–1:400 (or even greater) are more appropriate.
[b] See ref. 56.
[c] 5% low-fat powdered milk (e.g. Marvel), 0.02% sodium azide in PBS can be used instead.
[d] If a mouse monoclonal antibody is used in the first step and its isotype is IgG1 (which fails to bind Protein A), then it is necessary *either* to use [I^{125}] Protein G as the second step detection system *or* to incorporate a further incubation with an anti-mouse antiserum followed by [I^{125}] Protein A.
[e] It may be possible to use the blot again if it is kept wet or to wash it again more thoroughly if the background is high.

8. Controlled expression of foreign genes

8.1 High level expression

As mentioned previously, the kinetics and level of expression of foreign genes depend on the promoter chosen. In general, late promoters for structural genes in vaccinia virus are stronger than the early promoters. For example, when the 11K late structural gene promoter was used to drive the β-galactosidase gene, the yields of recombinant product were in the order of 1–2 mg/litre of infected cell culture (23). The 7.5K promoter, which expresses at both early and late times in infection gives a 10–20-fold lower level of β-galactosidase than the 11K promoter. (For further features of vaccinia virus transcription and promoters, see refs 3, 8.)

Significant yields, of around 0.5 mg/litre of potential vaccine antigens such as HIV-1 gp160, have been achieved using the T7 RNA polymerase system (57, 58). In this system, two viruses are used; one expresses the T7 RNA polymerase gene and the other contains the foreign gene under control of a T7 promoter and transcription terminator. Expression of the foreign gene occurs only in cells infected with both viruses under conditions where the T7 RNA polymerase gene is expressed. Thus this system is of use if toxic genes

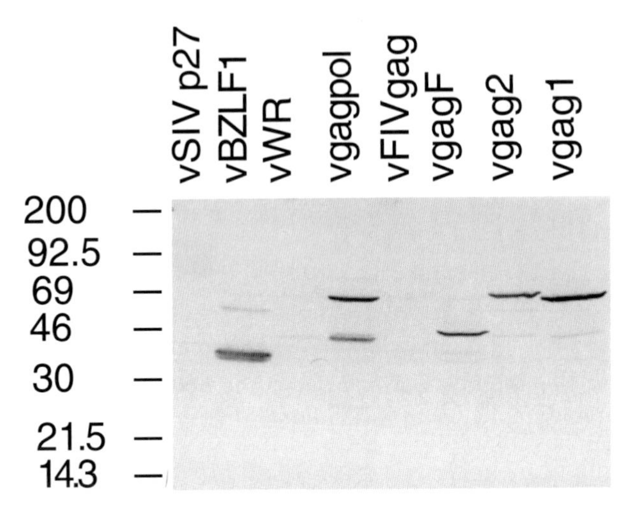

Figure 6. Western blot reactivity of anti-HIV p24 monoclonal antibody with vaccinia recombinants expressing various lentivirus gag products. Extracts of cells infected with the recombinant indicated were run on a 12% SDS–PAGE gel and transferred to nitrocellulose. Reactivity of the lentivirus gene products with a monoclonal antibody raised against HIV-1 p24 can be seen in the figure. A second monoclonal antibody raised against the Epstein–Barr virus *trans*-activator gene BZLF1 was also included as a positive control. The position of Rainbow molecular weight markers (Amersham) included on the same gel is shown at the left-hand side of the blot; numbers indicate the M_r of the marker polypeptides in kilodaltons.

are to be expressed because expression of the T7 RNA polymerase can be restricted to a late phase of cell multiplication by choosing a suitable promoter. For example, a vaccinia virus recombinant expressing the vesicular stomatitis virus (VSV) M gene from the 7.5K promoter could not be isolated presumably because toxicity of the M gene product was sufficient to prevent vaccinia replication. Linking the M gene with the T7ϕ10 promoter created a recombinant which, when infected alone into cells, did not express the gene. However, dual infection of cells with this recombinant and a T7 RNA polymerase recombinant allowed the VSV M gene to be expressed and analysed (59).

In the dual infection system, approximately 30% of the total cellular RNA at 24 hours post-infection is T7 specified. However it is mostly uncapped and only poorly translated. This problem has been overcome by the addition of the ribosome entry sequence of encephalomyocarditis virus (EMCV) at the 5′ end of the mRNA. This allows cap-independent translation of the mRNAs produced by the T7 polymerase and so considerably improves the translation of the mRNAs, transcribed by T7 RNA polymerase. In one case using this modification, the foreign gene product produced accounted for 10% of total cell protein (60). Overall, the basic T7 system gives levels of product

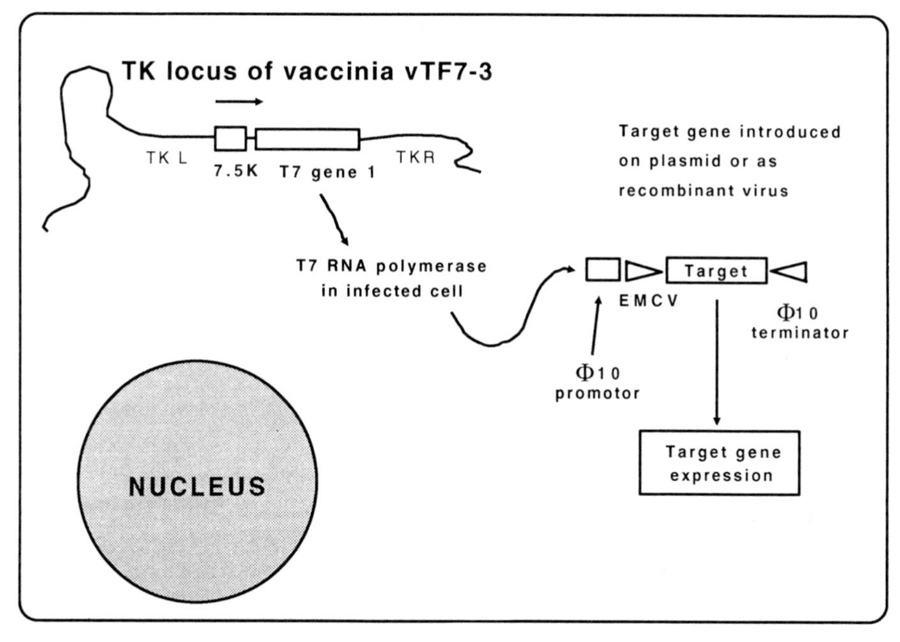

Figure 7. Transient expression of genes from vaccinia virus using bacteriophage T7 RNA polymerase. A virus is constructed, such as vTF7-3 shown here, which will express T7 RNA polymerase upon infection of a cell with expression driven by the early promoter. A plasmid is also generated which contains the target gene under control of a T7 promoter and terminator, here the gene Φ10 control elements. Upon transfection with the plasmid construct the target gene will be transcribed only in cells infected with the virus expressing T7 RNA polymerase. The levels of gene product from the transfection with plasmid are significantly less than those produced from cells infected with the dual recombinant system. EMCV is the ribosome entry sequence of encephalomyocarditis virus.

approximately equal to those derived from the 11K promoter, but inclusion of the EMCV ribosome entry site (see *Figure 7*) improves levels by a factor of at least five.

8.2 Transient expression using T7 RNA polymerase

A vaccinia virus recombinant expressing the bacteriophage T7 RNA polymerase gene has been used extensively in a transient assay system (*Protocol 16*). Cells infected with this recombinant vaccinia virus are transfected with a plasmid containing the gene of interest under control of the T7ϕ10 promoter and terminator. The T7 RNA polymerase transcribes the gene of interest from the input plasmid and terminates at the terminator sequence supplied in the plasmid. The advantage of this system over the two virus system is that it is not necessary to generate a recombinant virus—all that is required is to generate a recombinant plasmid. Sufficient protein is made from the transcripts to allow a number of biological assays to be undertaken (e.g. refs

61, 62). Vectors are available with or without the EMCV ribosome entry site (*Table 1*, ref. 28)

Protocol 16. Transient assay using vectors containing the bacteriophage T7 promoter

Equipment and reagents

- Cell monolayers of an appropriate cell line
- 5 μg plasmid construct containing the foreign gene under the control of the T7φ10 promoter and terminator
- For detection of protein: 50–100 μCi [^{35}S]-methionine (100 Ci/mmol) in medium containing 1/10 normal concentration of methionine plus 2% dialysed FCS
- Vaccinia recombinant virus vTF7–3 (57)
- Equipment and reagents listed in *Protocol 6*
- Phosphate-buffered saline (PBS): 170 mM NaCl, 3.4 mM KCl, 10 mM Na_2HPO_4, 1.8 mM KH_2PO_4 pH 7.2
- Equipment and reagents for detection of foreign protein (*Protocol 13, 14*, or *15*; see below, steps 4–5)

Method

1. Infect the appropriate cell line at a multiplicity of 10–20 p.f.u./cell for 1 h with VTF7–3 (48), a vaccinia recombinant virus that expresses the T7 RNA polymerase gene. 5×10^5 infected cells are sufficient to give significant amounts of protein.
2. Remove the virus inoculum and transfect the cells with 5 μg of plasmid construct using the Lipofectin™/DOTAP protocol (*Protocol 6*).
3. Label the proteins at 6 h post-infection by the addition of 50–100 μCi of [^{35}S]methionine (100 Ci/mmol) in medium containing 1/10 the normal amount of methionine and 2% dialysed FCS.
4. Harvest the cells at 12 h post-infection by washing them three times with PBS and then scraping them into PBS.
5. Process the cells (or supernatant if the product is secreted) as for immunoprecipitation (*Protocol 13*) or immunoblotting (*Protocol 15*).

Figure 8. Inducible expression in vaccinia virus. (a) Plasmid vectors pPR34 and pPR35 (30) contain the *E. coli lac I* gene under control of the 7.5K promoter with a 4b promoter p4b adjacent to the *lac* operator (op). pPR34 has one operator whereas pPR35 has two and allows higher levels of repression in the absence of IPTG, although upon induction yields are less than when pPR34 is used. The gene of interest (*goi*) is cloned into the plasmid vector under control of the 4b promoter–operator hybrid and a recombinant vaccinia virus is generated. The resultant recombinant expressing the *E. coli lac I* gene from the 7.5K promoter and with *goi* under the control of an inducible promoter is shown in (b). Upon infection, the repressor is synthesized, oligomerizes into an active tetramer, and binds to the operator sequence downstream of the 4b promoter (c). This blocks transcription of the gene of interest by vaccinia RNA polymerase (RNA POL). Addition of IPTG causes dissociation of the repressor from the operator and expression of the gene of interest (d). The gene of interest can be an essential vaccinia gene. Hence if the wild-type gene has been deleted, a conditionally-lethal mutation in the virus would have been generated allowing study of the function of essential vaccinia genes. B, S, K are *Bam*HI, *Sma*I, and *Kpn*I restriction sites for insertion of foreign genes respectively.

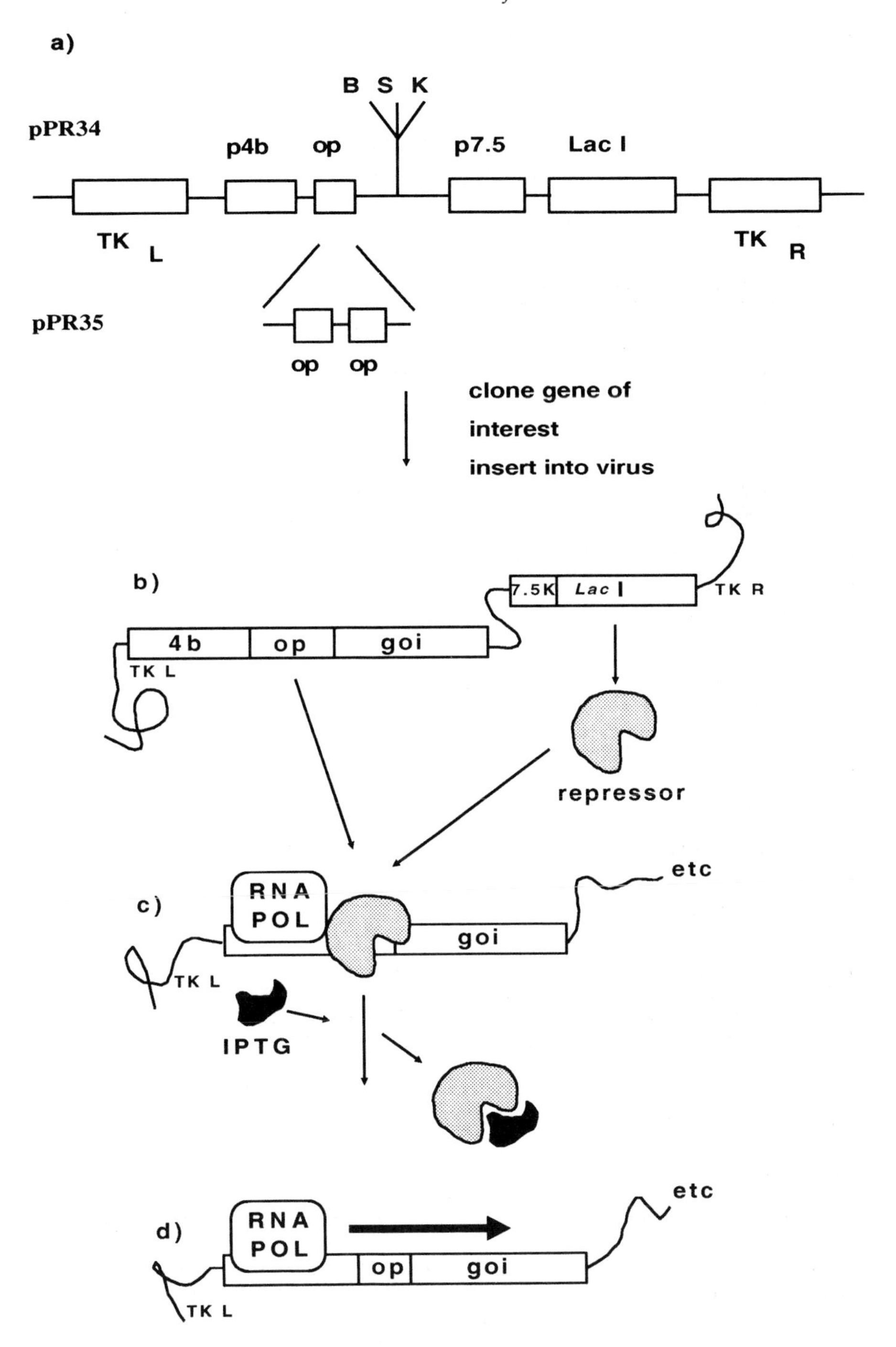
a)
B S K
pPR34
p4b
op
p7.5
Lac I
TK L
TK R
pPR35
op op
clone gene of
interest
insert into virus
b)
7.5K
Lac I
TK R
4b
op
goi
TK L
repressor
etc
c)
RNA
POL
goi
TK L
IPTG
etc
d)
RNA
POL
op
goi
TK L

Two additional points about the use of *Protocol 16* should be noted:

(a) It is not always necessary to immunoprecipitate the labelled proteins to be able to detect them. For example, the CFTR gene product (49) could be distinguished from unpurified total labelled proteins (this construct used a T7 promoter with an EMCV leader).

(b) When a T7 promoter and EMCV leader is used, an incubation in hypertonic medium (medium containing 190 mM NaCl) at the time of labelling (*Protocol 16*, step 3) can lead to preferential uptake of methionine into the target gene product (60), presumably due to the fact that translation of uncapped picornavirus-like mRNAs is favoured over cellular mRNAs.

8.3 Inducible expression

The inducible *lac* repressor/operator system from *E. coli* has also been transferred to a vaccinia virus expression vector (29, 63; see *Figure 8*). The advantage of this modification is that it allows inducible expression of the foreign gene. A recombinant virus is generated which expresses the *lacI* repressor and contains the gene of interest under the control of a hybrid *lac* operator/vaccinia promoter. Expression of the gene can be induced by the addition of IPTG to cells infected with both viruses. As yet, this system has been used only for the conditional expression of essential late genes of vaccinia virus itself but there is no reason why it should not be used to express foreign genes as well. The first step in this procedure (*Protocol 17*) is to clone the gene of interest under control of an inducible promoter such as the 4b *lac* operator hybrid (29) in plasmids pPR34 and pPR35. These vectors will insert the gene under study into the *tk* gene of vaccinia. If a late gene of vaccinia is under study, then once this is transferred to the *tk* locus the wild-type gene at its natural locus can be deleted by insertion of *E. coli gpt* into the gene. Only vaccinia late promoter–operator hybrid transcriptional control elements have so far proved to be inducible presumably because, for the early promoter to be repressed, all virus particles would have to have *lacI* bound to the *lac* operator part of the promoter (see *Figure 8*).

Protocol 17. Construction of IPTG-dependent vaccinia virus

Equipment and reagents

- Equipment and reagents listed in *Protocols 7, 9,* and *10*
- 0.1 M IPTG stock (filter sterilize the solution, divide it into aliquots, and store these at −20°C)
- PCR reagents and thermal cycler
- Plasmid vector pPR34 or pPR35 (29)
- CV-1 cell monolayers
- WT vaccinia virus
- Equipment and reagents listed in *Protocol 5* or *6* for transfection
- Plasmid containing the target gene interrupted by *Ecogpt* linked to a vaccinia virus promoter (see step 5 below)

Method

1. Select the late gene to be regulated in an IPTG-inducible manner. Produce a copy of the protein-coding region by PCR or by standard cloning techniques and clone it downstream of the IPTG-inducible promoter in plasmid vector pPR34 or pPR35 (29). These vectors allow insertion of the foreign gene into the *tk* gene and also contain *lacI* linked to the vaccinia virus 7.5K promoter.
2. Infect CV-1 cells with WT vaccinia virus and transfect with the plasmid produced in step 1, as described in *Protocol 5* or *6*.
3. Isolate tk^- plaques as described in *Protocol 7*.
4. Identify recombinants using a PCR screen (*Protocol 9*).
5. Grow a stock of recombinant virus, titrate it, and infect CV-1 cells at an m.o.i. of 0.05 p.f.u./cell. Transfect the infected cells with a plasmid containing the target gene interrupted by *Ecogpt* linked to a vaccinia virus promoter (for example for interruption of the *Sal L4R* gene use plasmid pSAD15; see ref. 64).
6. Select recombinants in the presence of MPA, xanthine, and hypoxanthine (*Protocol 10*) and 5 mM IPTG. Plaque purify the recombinants three times and grow a stock of virus in the presence of 5 mM IPTG.

Acknowledgements

Another chapter for a book in this series (65) has formed much of the basis of this chapter and I am indebted to Dr G. L. Smith for sending me a copy of the manuscript before publication. My thanks also go to Dr Simon Stacey who devised the procedure for growth of viruses in HeLa S3 cells maintained in semi-adherent culture, to Amanda Stuart for reading and commenting on the text, and to Wendy Pelham for excellent secretarial help. *Figure 4* is courtesy of Monika Madej and *Figure 6* is courtesy of Mags Gartland, both of whom were based in this laboratory. Finally, I would like to acknowledge the generous support of the Cancer Research Campaign and the patience and encouragement of my family.

References

1. Mackett, M., Smith, G. L., and Moss, B. (1985). In *DNA cloning: a practical approach,* Vol. II. (ed. D. M. Glover), pp. 191–211. IRL Press, Oxford.
2. Cadoz, M., Strady, A., Meignier, B., Taylor, J., Tartaglia, J., Paoletti, E., and Plotkin, S. (1992). *Lancet*, **339**, 1492.
3. Moss, B. (1990). In *Virology* (ed. B. N. Fields and D. M. Knipe), pp. 2079–2111. Raven Press, New York.
4. Buller, R. M. L. and Palumbo, G. J. (1991). *Microbiol. Rev.*, **55**, 80.

5. Moss, B. (1990). *Annu. Rev. Biochem.*, **59**, 661.
6. Traktman, P. (1990). *Curr. Top. Microbiol. Immunol.*, **163**, 93.
7. Smith, G. L. (1993). *J. Gen. Virol.*, **74**, 1725.
8. Binns, M. M. and Smith, G. L. (ed.) (1992). *Recombinant poxviruses*. CRC Press, Boca Raton, Florida.
9. Payne, L. G. (1980). *J. Gen. Virol.*, **50**, 89.
10. Perkus, M. E., Goebel, S. J., Davis, S. W., Johnson, G. P., Norton, E. K., and Paoletti, E. (1991). *Virology*, **180**, 406.
11. Goebel, S. J., Johnson, G. P., Perkus, M. E., Davis, S. W., Winslow, J. P., and Paoletti, E. (1990). *Virology*, **179**, 247.
12. Smith, G. L., Chan, Y. S., and Howard, S. T. (1991). *J. Gen. Virol.*, **72**, 1349.
13. Massung, R. F., Esposito, J. J., Liu, L.-I., Qi, J., Utterback, T. R., Knight, J. C., *et al.* (1993). *Nature*, **366**, 748.
14. Keck, J. G., Baldick, C. J., and Moss, B. (1990). *Cell*, **61**, 801.
15. Scheiflinger, F., Dorner, F., and Falkner, F. G. (1992). *Proc. Natl Acad. Sci. USA*, **89**, 9977.
16. Mackett, M., Smith, G. L., and Moss, B. (1984). *J. Virol.*, **49**, 857.
17. Chakrabarti, S., Brechling, K., and Moss, B. (1985). *Mol. Cell. Biol.*, **5**, 3403.
18. Davison, A. J. and Moss, B. (1990). *Nucleic Acids Res.*, **18**, 4285.
19. Falkner, F. G. and Moss, B. (1988). *J. Virol.*, **62**, 1849.
20. Perkus, M. E., Limbach, K., and Paoletti, E. (1989). *J. Virol.*, **63**, 3829.
21. Berthelot, C., Drillien, R., and Wittek, R. (1985). *Proc. Natl Acad. Sci. USA*, **82**, 2096.
22. Rosel, J. L. and Moss, B. (1985). *J. Virol.*, **56**, 830.
23. Falkner, F. G. and Moss, B. (1988). *J. Virol.*, **62**, 1849.
24. Browne, H., Churcher, M., Stanley, M., Smith, G. L., and Minson, A. C. (1988). *J. Gen. Virol.*, **69**, 1263.
25. Earl, P. L., Hugin, A. W., and Moss, B. (1990). *J. Virol.*, **64**, 2448.
26. Patel, D. D., Ray, C. A., Drucker, R. P., and Pickup, D. J. (1988). *Proc. Natl Acad. Sci. USA*, **85**, 9431.
27. Boyle, D. B., Coupar, B. E., and Both, G. W. (1985). *Gene*, **35**, 169.
28. Moss, B., Elroy-Stein, O., Mizukami, T., Alexander, A. W., and Fuerst, T. (1990). *Nature*, **348**, 91.
29. Ridriguez, J. F. and Smith, G. L. (1990). *Virology*, **177**, 239.
30. Personal communication from Dr H. Stunnenberg. EMBL Laboratories, Heidelberg. We have used this vector to express at least 25 different gene products.
31. Perkus, M. E., Limbach, K., and Paoletti, E. (1989). *J. Virol.*, **63**, 3829.
32. Coupar, B. E. H., Andrew, M. E., Both, G. W., and Boyle, D. B. (1986). *Eur. J. Immunol.*, **16**, 1479.
33. Townsend, A., Bastin, J., Gould, K., Brownlee, G., Andrew, A., Boyle, D. B., *et al.* (1988). *J. Exp. Med.*, **168**, 1211.
34. Mackett, M., Smith, G. L., and Moss, B. (1982). *Proc. Natl Acad. Sci. USA*, **79**, 7415.
35. Boyle, D. B. and Coupar, B. E. H. (1988). *Gene*, **65**, 123.
36. Franke, C. A., Rice, C. M., Strauss, J. H., and Hruby, D. E. (1985). *Mol. Cell. Biol.*, **5**, 1918.
37. Perkus, M. E., Limbach, K., and Paoletti, E. (1989). *J. Virol.*, **63**, 3829.

38. Rodriguez, J. F. and Esteban, M. (1989). *J. Virol.*, **63**, 997.
39. Redfield, R. R., Wright, D. C., James, W. D., Jones, T. S., Brown, C., and Burke, D. S. (1987). *N. Eng. J. Med.*, **316**, 673.
40. Vaccination of laboratory workers handling vaccinia and related poxviruses infectious for humans. HMSO publications. ISBN 0–11–885450–X.
41. 'A guide to genetically modified organisms (contained use) regulations 1992'. HMSO publications. ISBN 0–11–882049–4.
42. Panicali, D., Grzelecki, A., and Huang, C. (1986). *Gene*, **47**, 193.
43. Ramshaw, I. A., Andrew, M. E., Philips, S. M., Boyle, D. B., and Coupar, B. E. H. (1987). *Nature*, **329**, 545.
44. Isaacs, S. N., Kotwal, G. J., and Moss, B. (1990). *Virology*, **178**, 626.
45. Ball, L. A. (1987). *J. Virol.*, **61**, 1788.
46. Spyropolous, D. D., Roberts, B. E., Panicali, D. L., and Cohen, L. K. (1988). *J. Virol.*, **62**, 1046.
47. Stephens, E. B., Compans, R. W., Earl, P., and Moss, B. (1986). *EMBO J.*, **5**, 237.
48. Lifson, J. D., Feinberg, M. B., Reyes, G. R., Rabin, L., Banapour, B., Chakrabarti, S., *et al.* (1986). *Nature*, **323**, 725.
49. Rich, D. P., Anderson, M. P., Gregory, R. J., Cheng, S. H., Paul, S., Jefferson, D. M., *et al.* (1990). *Nature*, **347**, 358.
50. Mackett, M., Smith, G. L., and Moss, B. (1984). *J. Virol.*, **49**, 857.
51. Falkner, F. G., Fuerst, T. R., and Moss, B. (1988). *Virology*, **164**, 450.
52. Koup, R. A., Robinson, J. E., Nguyen, Q. V., Pikora, C. A., Blais, B., Roskey, A., *et al.* (1991). *AIDS*, **5**, 1309.
53. Bennink, J. R. and Yewdell, J. W. (1990). *Curr. Top. Microbiol. Immunol.*, **163**, 153.
54. Laemmli, U. K. (1970). *Nature*, **227**, 680.
55. Laskey, R. A. (1992). Review 23, Amersham International plc.
56. Towbin, H., Staehelin, T., and Gordon, J. (1979). *Proc. Natl Acad. Sci. USA*, **76**, 4350.
57. Fuerst, T., Earl, P. L., and Moss, B. (1987). *Mol. Cell Biol.*, **7**, 2538.
58. Barrett, N., Mitterer, A., Mundt, W., Eibl, M., Gallo, R. C., Moss, B., *et al.* (1989). *Aids Res. Hum. Retroviruses*, **5**, 159.
59. Li, Y., Luo, L. Z., Snyder, R. M., and Wagner, R. R. (1988). *J. Virol.*, **62**, 776.
60. Elroy-Stein, O., Fuerst, T., and Moss, B. (1989). *Proc. Natl Acad. Sci. USA*, **86**, 6126.
61. Berger, E. A., Fuerst, T. R., and Moss, B. (1988). *Proc. Natl Acad. Sci. USA*, **85**, 2357.
62. Crise, B. and Rose, J. K. (1992). *J. Virol.*, **66**, 2296.
63. Fuerst, T. R., Fernandez, M. P., and Moss, B. (1989). *Proc. Natl Acad. Sci. USA*, **86**, 2549.
64. Duncan, S. A. and Smith, G. L. (1992). *J. Virol.*, **66**, 1610.
65. Smith, G. L. (1993). In *Molecular virology: a practical approach* (ed. A. J. Davison and R. M. Elliot), pp. 257–283. IRL Press, Oxford.
66. Powell, K. L. and Courtney, R. J. (1975). *Virology*, **66**, 217.

3

Use of vectors based on gene amplification for the expression of cloned genes in mammalian cells

CHRISTOPHER BEBBINGTON

1. Introduction

1.1 Strategies for expression of cloned genes in mammalian cells

The ability of cloned genes to function when introduced into mammalian tissue culture cells has proved to be invaluable in studies of gene expression. It also provides a means of obtaining, in large quantities, proteins which are otherwise scarce or which are the completely novel products of gene manipulation. The great advantage in obtaining human or other mammalian proteins from mammalian cells is that they are generally correctly folded, and are more frequently appropriately modified and completely functional than when expressed in microbial or insect cells. Secretion of proteins containing signal peptides may also be more efficient in mammalian cells than insect cells.

For many studies of mammalian gene expression, analysis of transient expression, which occurs in a proportion of cells within a few hours of introduction of DNA, is most appropriate. In other cases, when larger amounts of product are needed, it is necessary to identify cell clones in which the vector sequences are retained during cell proliferation, either by maintenance of an episome, or as a consequence of integration of the vector into the host cell's DNA. Any foreign DNA can integrate, at a low frequency, in apparently random sites in the host chromosomes and the resultant clones can be identified by the use of a suitable selectable marker gene. Various selectable vectors and methods for introducing them into tissue culture cells have already been described in Chapter 1. The present chapter focuses on a class of selectable markers which can be used to obtain markedly increased efficiency of gene expression from integrating vectors and which provide some of the most efficient mammalian expression systems currently available.

Genes expressed in eukaryotic vectors can consist of genomic sequences containing natural promoter elements and RNA processing signals, or alternatively, cDNAs can be used provided that they are equipped with a heterologous promoter and splicing and polyadenylation signals. Some of the most efficient transcription units can be constructed using viral regulatory signals, many of which function in a variety of host cell types. However, the integration of such vectors into random chromosomal locations leads to a high degree of variation in expression levels between different transfected clones because of a profound influence of chromosomal location on transcription levels, a phenomenon known as 'position effect'. Consequently, efficient transcription units on their own rarely provide the maximum possible levels of expression; higher levels are generally obtained if the vector copy number can be increased. This is done by selection for gene amplification.

1.2 Gene amplification

When cultured cells are subjected to appropriate concentrations of certain toxic drugs, variant clones can often be selected which are more resistant to the drug than are wild-type cells. One common mechanism by which such variants arise is by over-production of an essential enzyme which the drug inhibits. The over-production of enzyme most commonly results from increased levels of its particular mRNA, which in turn is caused by an increase in the number of copies of its structural gene in each cell.

The first gene found to 'amplify' in this way was the dihydrofolate reductase (DHFR) gene, in variants of hamster and mouse cell lines which arose after selection for resistance to the specific inhibitor of DHFR, methotrexate (MTX), reviewed by Stark (1). When the DHFR gene was subsequently cloned and re-introduced into a cell line which lacked endogenous DHFR activity, the cloned gene was shown to be capable of amplification as well. Since then, many other genes have been shown to be amplifiable when selected with a specific inhibitor. Those which have been used as a selectable marker in an expression vector are listed in *Table 1*. In all cases tested so far, both genomic DNAs and cDNAs encoding such enzymes are amplifiable

Table 1. Amplifiable genes cloned and expressed in mammalian cells

Gene	Selective agent	Reference
Adenosine deaminase	Deoxycoformycin	11
Asparagine synthetase	Albizzin or β-aspartyl hydroxamate	12–14
Aspartate transcarbamylase	*N*(phosphonacetyl)-L-aspartate	15
Dihydrofolate reductase	Methotrexate	16–18
Glutamine synthetase	Methionine sulfoximine	5, 6
Metallothionein-I	Heavy metals, e.g. Cd^{2+}	19
Ornithine decarboxylase	α-Difluoromethyl ornithine	20
P-glycoprotein 170	Adriamycin (and others)	21
Ribonucleotide reductase	Hydroxyurea	22
Thymidine kinase (defective)	–	23, 24
Xanthine–guanine phosphoribosyl transferase	Mycophenolic acid	3, 25

when transferred into tissue culture cells in an expression vector. Thus, it is unlikely that any particular sequence within these genes is responsible for the mechanism of amplification. Although the mechanism is not fully understood, it appears that the selection protocol serves merely to identify random amplification events occurring with characteristic frequencies in all proliferating cell populations, probably as a result of a mutational event involving unequal sister chromatid exchange (2).

It is a general feature of gene amplification that the region of chromosomal DNA amplified is much greater than the enzyme's coding sequence (often more than 1000 kb of DNA is amplified). Hence, when transfected genes are selected for amplification, other sequences on the vector also show increased copy number; that is, they are co-amplified. The significance of this for maximizing gene expression is that it provides a means of progressively increasing transcription until the RNA level is no longer limiting for protein production. Consequently, this class of expression vectors includes the most efficient mammalian expression systems currently available. Moreover, by varying the concentration of the selective agents used, it is often possible to use amplifiable markers under highly stringent selection conditions to select for integration of vectors into favourable sites in the genome so that high level expression can be achieved even in the absence of gene amplification.

Despite the long list of amplifiable markers which are now available (*Table 1*), not all are equally suitable:

- In some cases inhibitor-resistant mutations can arise by mechanisms other than gene amplification.
- The concentrations of inhibitor required to achieve significant amplification may be excessive.
- Over-production of one amplifiable enzyme, XGPRT, has been shown to be toxic to the cell thereby limiting the degree of amplification which is practically useful (3).

The two most widely used amplifiable marker genes are described in detail in this chapter; dihydrofolate reductase (DHFR) and glutamine synthetase (GS), in CHO cells and rodent myeloma cells.

(a) *DHFR selection in CHO cells* is attractive because there exist a number of DHFR-deficient mutant CHO lines which can not grow without nucleosides in the culture medium. This facilitates initial selection of primary transfectants, in a basal medium without nucleosides, and reduces the likelihood of amplification of endogenous DHFR genes leading to MTX resistance. There is also now a wealth of experience of the use of MTX to select DHFR gene amplification in CHO cells to achieve efficient expression of a wide variety of genes. In some cases, DHFR selected cell lines retain the amplified vector sequences for extended periods even in the absence of continued MTX selection (4).

(b) *GS selection in rodent myelomas* is attractive for similar reasons since there are a number of such lines which are naturally deficient in GS and hence require glutamine. One such line, NS0, has been used to produce a variety of proteins from GS-based expression vectors after initial selection in glutamine-free medium and selection for amplification with methionine sulfoximine (MSX). Some GS selected NS0 cell lines have also been shown to retain stable productivity in the absence of continued MSX selection (5).

It is not, however, essential to use a cell line which is deficient in the enzyme activity of the amplifiable marker and GS selection in particular can be carried out very successfully in gs^+ CHO-K1 cells using appropriately designed vectors which express sufficient GS to confer resistance to low levels of MSX. Indeed, GS selection in CHO cells allows more rapid identification of high producing cell lines than amplification of comparable DHFR vectors in which the gene of interest is expressed from an identical promoter, since primary transfectant lines selected with GS tend to show higher vector copy numbers and express higher levels of product than primary DHFR selected transfectants (6). Generally, only a single round of selection in elevated levels of MSX is then required to achieve maximal expression whereas DHFR selection commonly requires three or four rounds of amplification to achieve comparable expression levels, each round taking four to six weeks. Indeed, amplification has not always been necessary to achieve high levels of expression using GS vectors in CHO cells (7).

A high vector copy number transfection procedure has also been described for DHFR vectors in $dhfr^-$ cells (8) but does not appear to have been widely used. DHFR vectors have also been adapted for use in cell lines containing endogenous DHFR, for instance by using a strong promoter to drive expression of the DHFR cDNA (9) or a sequence encoding a MTX-resistant DHFR enzyme (10), but such high levels of MTX are needed to select amplification of these vectors in the cell lines tested that they are of limited use.

1.3 Choice of expression system

The choice of cell type and expression system will depend on a number of factors, some of which are outside the scope of this chapter; for instance there may be particular post-translational modifications which are only carried out appropriately in certain cell types. NS0 cells can show advantages in scale up in fermenters since they adapt readily to suspension culture and, in a few cases, particular proteins appear to be secreted more efficiently from NS0 than CHO cells (unpublished results). However, many proteins have been successfully produced from CHO cell lines and, because of the wealth of experience of production of proteins in these cells, they may be the preferred choice in many instances. DHFR selection has been the most widely used but, because of the higher levels of expression in initial transfectants, GS selection is becoming increasingly popular.

2. Dihydrofolate reductase selection in CHO cells

2.1 Choice of cell line and medium

Typically, a *dhfr*$^-$ variant CHO cell line is used, such as DUKX-B11 (also called DXB11), isolated from the proline auxotroph CHO-K1 (26), or DG44 (27), isolated from the CHO (Toronto) cell line. In DXB11, only one of the alleles of the *dhfr* gene is defective whereas in DG44, both alleles are mutant. Consequently there is a low but appreciable reversion frequency of DXB11 to a *dhfr*$^+$ phenotype. The reversion frequency of DG44 is essentially undetectable.

The *dhfr*$^-$ cell lines require exogenous nucleosides and glycine for survival. DXB11 also has a requirement for proline because the CHO-K1 parent is incidentally *pro*$^-$. All of these nutritional requirements are provided by either of the non-selective media given in *Table 2*. Two alternative sets of media are given:

- based on MEM-Alpha medium
- based on Dulbecco's modified Eagle's medium (DMEM)

2.2 DHFR vector design

A variety of different vectors containing DHFR coding sequences can transform the cells to a *dhfr*$^+$ phenotype, thus allowing them to grow without added nucleosides and glycine (e.g. refs 16, 18, 28). It is preferable to use a vector with a weakly expressed DHFR gene, so that the levels of MTX required for amplification are minimized. One such plasmid is pSVM.*dhfr*, shown in *Figure 1*, in which the DHFR gene is expressed from the mouse mammary tumour virus (MMTV) promoter. An example of its use is shown in *Table 3*, to express a tissue plasminogen activator (tPA) gene, expressed from a SV40 early promoter. It can be seen that the level of MTX required to achieve high level expression of tPA in plasmid pSVM.*dhfr* was only 20 nM, whereas when a *dhfr* gene expressed from the stronger SV40 early promoter in the plasmid pSV2.*dhfr* (18) was used as the amplifiable marker, 500 nM MTX was necessary to achieve comparable levels of tPA expression.

A complete expression vector containing a strong promoter and other sequences designed to enhance expression of an inserted coding sequence and a weakly expressed DHFR sequence is pED (29). Alternatively a very efficient expression cassette for the gene of interest can be provided by the plasmid pEE6hCMV-B (*Figure 2*) used in conjunction with pSVM.*dhfr*. A single plasmid can be constructed containing both genes using the single *Bam*HI restriction site downstream of the amplifiable gene in pSVM.*dhfr*, which forms a suitable site for introducing a complete transcription unit consisting of the *Bgl*II–*Bam*HI fragment of pEE6hCMV-B. pEE6hCMV-B has a multilinker cloning site which is fused to the 5′ untranslated region of

Table 2 Media for DHFR selection[a]

Stock solutions

- 200 mM L-glutamine (Gibcol-BRL). Store at −20°C in 5 ml aliquots.
- Non-essential amino acids (NEAA; Gibco-BRL). For DMEM-based media only.
- Fetal calf serum (FCS). Heat inactivate at 56°C for 30 min and store at −20°C.
- Dialysed FCS (Gibco-BRL 063–6300). Heat inactivate at 56°C for 30 min and store at −20°C.
- For non-selective medium:
  Either (i) Ham's F12 nutrient mixture (Flow Laboratories Ltd. 12–432)
  or (ii) MEM-Alpha medium with nucleosides (Gibco-BRL 041–2571).
- For selective medium:
  Either (i) DMEM (Gibco-BRL 041–1885)
  or (ii) MEM Alpha medium *without* nucleosides (Gibco-BRL 041–2561).

Media based on DMEM
Prepare these by mixing the following stock solutions:

Non-selective medium	
Ham's F12	500 ml
FCS	50 ml
L-glutamine	5 ml

Selective medium	
DMEM	500 ml
Dialysed FCS	50 ml
NEAA	5 ml

Media based on MEM-Alpha medium
Prepare these by mixing the following stock solutions:

Non-selective medium	
MEM-Alpha with nucleosides	500 ml
FCS	50 ml

Selective medium	
MEM-Alpha *without* nucleosides	500 ml
Dialysed FCS	50 ml

[a] Two different sets of media are given. Either alternative works well.

the hCMV transcript. A cDNA can be inserted at this point provided it has its own translation initiation signal.

Vector DNA introduced into a host cell by calcium phosphate-mediated transfection can frequently become ligated into high molecular weight species so that multiple plasmid molecules ultimately integrate at the same chromosomal location. For this reason, it is not always necessary for the amplifiable gene and the non-selected gene to be on the same vector. Thus, although it is normally preferable to construct a single vector containing all the required

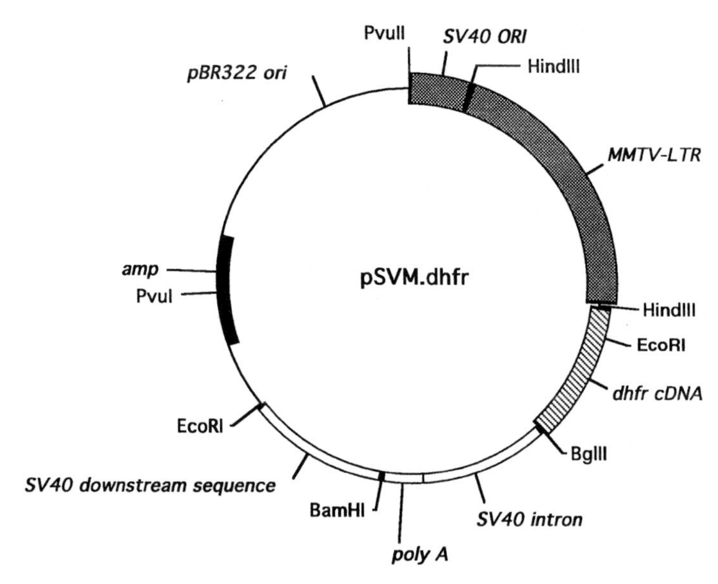

Figure 1. pSVM.*dhfr* (6.45 kb), an amplifiable vector for use in *dhfr*⁻ CHO cells (28), contains the 323 bp *Pvu*II–*Hin*dIII fragment of SV40 spanning the SV40 origin of replication (SV40 ORI); a 1.4 kb long terminal repeat of mouse mammary tumour virus (MMTV-LTR) containing the promoter; a 735 bp mouse *dhfr* cDNA (hatched box); an SV40 intron (SV40 nucleotides 4099–4710); a *Bcl*I–*Eco*RI fragment of SV40 (nucleotides 2770–1782) containing the early region polyadenylation signal (poly A) and additional SV40 downstream sequence; and the *Pvu*II–*Eco*RI fragment of pBR322 (2295 bp) containing the origin of replication and β-lactamase gene (*amp*) which confers resistance to ampicillin. Figure reproduced courtesy of Celltech Therapeutics Ltd ©.

genes, if this is not practicable, the two plasmids can be co-transfected simultaneously. A strategy which can increase the efficiency of co-transfection by selecting for integration of two vectors at the same chromosomal site has been described by Kaufman *et al.* (31). This uses a DHFR transcription unit on one vector lacking a functional transcription-enhancer element. Such a gene yields a low transfection efficiency in *dhfr*$^-$ CHO cells (3×10^{-5}) but, when co-transfected with a vector containing a SV40 enhancer, the transformation frequency is 20-fold higher. This is due to integration of the enhancer-containing vector adjacent to the DHFR vector, allowing the enhancer to act to increase the efficiency of DHFR transcription and hence the apparent frequency of transfection.

2.3 Transfection of CHO cells

Transfection can be carried out using a calcium phosphate co-precipitation procedure or by electroporation (see Chapters 1 and 2). The transfection procedure for *dhfr*$^-$ CHO cells is described in *Protocol 1*.

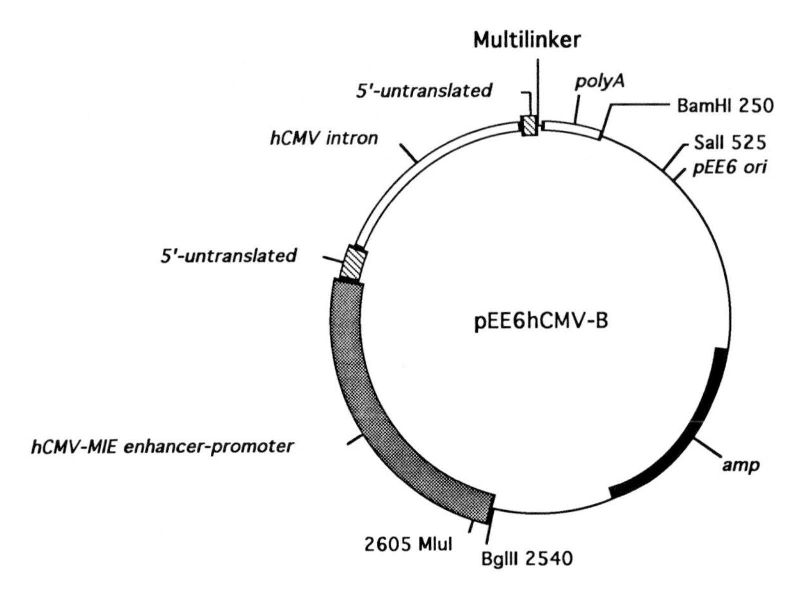

Figure 2. pEE6hCMV-B (4.7 kb) is a derivative of pEE6hCMV (30) with a single *Bgl*II site introduced to facilitate plasmid manipulation. It consists of a 2.1 kb fragment from human cytomegalovirus containing the major immediate early (hCMV-MIE) enhancer–promoter and 5′ untranslated region, including an 800 bp hCMV intron; a synthetic multilinker cloning site; the SV40 early region polyadenylation signal (poly A) (*Bcl*I–*Bam*HI); and plasmid vector pEE6 (30) sequences (thin line) containing a β-lactamase gene (*amp*) and origin of replication (pEE6 ori). The unique restriction sites in the multilinker, reading clockwise, are *Hin*dIII, *Xba*I, *Xma*I, *Sma*I, *Eco*RI, and *Bcl*I. Figure reproduced courtesy of Celltech Therapeutics Ltd©.

Protocol 1. Transfection and selection of DHFR vectors in *dhfr*⁻ CHO cells

Equipment and reagents

- Circular recombinant plasmid DNA plus plasmid without the gene to be expressed (e.g. pSVM.*dhfr*) as a control. Purify the plasmid DNA by CsCl centrifugation, then precipitate with ethanol, wash the DNA pellet with 70% ethanol, and leave the DNA pellet to dry in a tissue culture laminar flow cabinet to ensure sterility. Dissolve in sterile distilled water at 0.5 mg/ml.
- Tissue culture Petri dishes (9 cm diameter)
- Non-selective and selective media (*Table 2*)
- *dhfr*⁻ CHO cell line DXB11 or DG44 (e.g. from Dr L. Chasin, Columbia University, New York) growing exponentially in non-selective medium (*Table 2*)
- Equipment and reagents for calcium phosphate transfection procedure (Chapter 1, *Protocol 5*) or electroporation (Chapter 1, *Protocol 10*)
- Equipment and reagents for trypsinizing adherent cells (Chapter 1, *Protocol 4*)

Method

1. Plate the cells at a density of approximately 10^6 cells/9 cm Petri dish, the day before transfection, in non-selective medium.
2. Transfect with 10 μg plasmid using the $CaPO_4$ method (Chapter 1,

Protocol 5) or by electroporation (Chapter 1, *Protocol 10*) and then allow the cells to recover in non-selective medium in a tissue culture incubator overnight. Also treat some dishes of cells using the identical transfection procedure, but without adding DNA, to act as negative controls. Meanwhile, count the number of cells per dish using another dish from the same batch which is not to be transfected. Trypsinize the cells for counting as described in Chapter 1, *Protocol 4*.

3. The next day, remove the medium and replace with the selective medium.
4. Change the medium again after three or four days when substantial cell death should be apparent. (The dead cells float off the bottom of the Petri dish.)
5. Change the medium again every three or four days. After seven to eight days, many small adherent colonies of cells may appear but wait until 10–12 days after transfection before scoring the number of stably transformed colonies since many colonies survive for only a few days due to transient expression of the introduced DHFR genes.
6. Calculate the transfection frequency for the stable transformants. The transfection frequency with pSVM.*dhfr* should be about $1–2 \times 10^{-4}$ transfected cells/10 μg DNA but the exact frequency will depend on the plasmid and the particular experiment.
7. Establish cell lines from individual colonies by trypsinization and assay these for expression of the gene of interest.

2.4 Selection for amplification of DHFR vectors with methotrexate (MTX)

The concentration of MTX needed for amplification of a *dhfr* gene depends on the efficiency with which the gene is expressed, which is in part determined by the vector chosen (see section 2.2). However, there is also variability between individual transfected clones. Therefore cells from each clone to be amplified should be selected using a range of MTX concentrations, using *Protocol 2*.

Protocol 2. Amplification of DHFR vectors with methotrexate

Reagents

- 100 μM MTX: prepare this as a sterile solution in water and store at −20°C in 1 ml aliquots (NB: MTX is toxic, wear gloves when handling it and observe full safety precautions including the use of a face-mask, weigh out solid MTX in a fume-hood, dispose of waste solutions safely)
- CHO cell transfectants containing a DHFR vector grown in selective medium (see *Table 2*)

Protocol 2. *Continued*

Method

1. Add MTX to tissue culture dishes to final concentrations of 10, 20, 50, 100, 200, and 500 nM. Return the dishes to the tissue culture incubator.
2. After ten days, examine the plates, by eye, for surviving colonies.
3. Isolate colonies from the dishes with the highest concentration of MTX yielding resistant colonies and analyse these individually for levels of product expression. Alternatively, pool the MTX-resistant colonies and subject them to a second round of selection at, for instance, 500 nM and higher concentrations of MTX.

Gene amplification can be up to tenfold in the first round of amplification using *Protocol 2*, as measured by gene copy number or expression of a linked gene. In subsequent rounds of amplification, gene copy number is often roughly proportional to MTX-resistance and up to 2000 copies have been reported after three or more rounds of selection. The level of production of a co-expressed protein will vary depending on the product. It depends not only on the copy number of the gene but also is often limited ultimately by the capacity of the secretion apparatus of the cell. For human tPA, the maximum reported secretion rates from CHO cells are approximately 50 pg/cell/day (equivalent to 5000 tPA units/10^6 cells/day) (31). An example of *dhfr*-mediated co-amplification of a tPA gene is given in *Table 3*.

3. Glutamine synthetase selection in CHO cells

Glutamine is a key metabolite in a number of biosynthetic and catabolic pathways and must either be provided as a medium component or must be synthesized from glutamate and ammonia by means of GS. Some mammalian

Table 3 Examples of *dhfr*-mediated co-amplification [a]

	pSVM.*dhfr*	**pSV2.*dhfr***
Number of transfectants/plate	~100	~100
Transfectants secreting tPA	20	20
Maximum tPA secretion (units/10^6 cells/day)	150	100
Maximum MTX concentration yielding colonies (nM)	20	500
Maximum tPA secretion after one round of amplification (units/10^6 cells/day)	1300	1500

[a] A gene encoding tissue plasminogen activator (tPA), expressed from the SV40 early promoter, was cloned in pSVM.*dhfr* (*Figure 1*) or pSV2.*dhfr* (18). About 10^6 DUKX-B11 cells were transfected with 5 μg of each plasmid.

cell lines, such as CHO, express sufficient GS enzyme to survive in an appropriate glutamine-free medium. Under these conditions, GS is an essential enzyme and inhibition of GS by the specific inhibitor MSX is lethal. A transfected GS gene in a mammalian expression vector can then act as a dominant selectable marker in CHO cells if it confers resistance to concentrations of MSX which are just sufficient to kill both wild-type cells and the natural MSX-resistant variants which result from amplification of the endogenous GS genes. Amplification of vector copy number can subsequently be achieved using elevated levels of MSX.

3.1 Choice of cell type and medium

The preferred cell line is CHO-K1. For routine growth of this cell line, the preferred medium is GMEM-S with 10% dialysed FCS. In this medium, the high level of glutamine present in most tissue culture media (e.g. 2 mM) is replaced by an elevated concentration of glutamate (0.5 mM). The metabolic requirement for glutamine is reduced (though not eliminated) by the addition of nucleosides and asparagine. A high level of glucose and sodium pyruvate are added as alternative carbon sources. CHO-K1 cells grow well in this medium and MSX is toxic at concentrations above 3 μM. Dialysed FCS is required since normal FCS can contain significant amounts of glutamine. It can be purchased already dialysed. Each batch should be tested for the ability to support growth of CHO-K1 and for the absence of glutamine (batches containing less than 20 μg/ml glutamine are adequate).

3.2 Vector design

The expression vector used, pEE14 (*Figure 3*), contains a GS minigene, containing a single intron of the GS gene and about 1 kb of 3′ flanking DNA, and is transcribed from the SV40 late promoter. This transcription unit is derived from pSVLGS.1 (5) and has been chosen because it typically yields transfectant cell lines with higher vector copy numbers than some other GS expression vectors (unpublished results). A number of restriction sites have been removed from the GS gene in pEE14 so that there are several convenient sites in the multilinker which now occur only once in the plasmid and can be used for insertion of gene sequences to be expressed. The rest of the vector sequences in pEE14 are from pEE6hCMV-B (*Figure 2*).

For expression of antibodies or other heterodimeric proteins, the plasmid pEE6hCMV-B is also needed. The preferred scheme for vector construction for antibody expression is generally as follows. Insert a light chain coding sequence (which must contain a translation initiation signal) into the polylinker of pEE14. Similarly insert the heavy chain coding sequence into pEE6hCMV-B. Isolate a complete transcription unit from the heavy chain vector as a *Bgl*II–*Bam*HI fragment and insert this at the *Bam*HI site of the light chain plasmid such that both genes are in the same orientation.

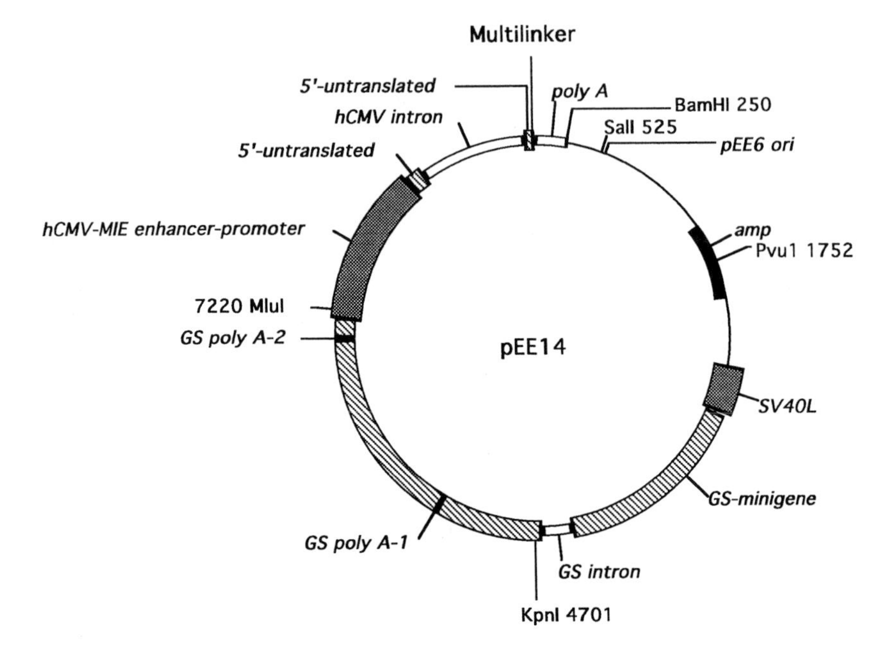

Figure 3. pEE14 (9.4 kb), an expression vector for CHO cells containing a GS-selectable marker (32). A hamster GS minigene (containing a single GS intron and polyadenylation signals from the GS gene (GS poly A-1 and poly A-2)), expressed from the SV40 late promoter (SV40L; the 344 bp *Hin*dIII–*Pvu*II fragment spanning the origin of replication) has been inserted into the *Bgl*II site of pEE6hCMV-B (*Figure 2*). The GS cassette is from pSVLGS1 (5) but a number of restriction sites have been removed to facilitate cloning into the multilinker. (For unique sites in the multilinker, see *Figure 2*.) Figure reproduced courtesy of Celltech Therapeutics Ltd©.

3.3 Cell transfection

Cells should be kept growing exponentially in GMEM-S medium (*Table 4*) for at least a few days prior to transfection and should not be allowed to reach confluence prior to transfection. The DNA used for transfection is circular plasmid DNA purified by CsCl density centrifugation and checked by agarose gel electrophoresis to ensure that it is essentially free of chromosomal DNA

Table 4 GMEM-S medium for GS selection in CHO-K1 cells[a]

Stock solutions

- Dialysed FCS (Gibco-BRL 220–6300 in US and 014–06300 in UK). Heat inactivate at 56°C for 30 min and store at −20°C. Each batch should be tested for the presence of glutamine.
- 100 mM L-MSX (Sigma). Prepare a solution of 18 mg/ml L-MSX in PBS. Filter sterilize and store at −20°C. NB: MSX is toxic. Wear gloves when handling. Weigh out the solid in a fume-hood and wear a face-mask.

Table 4 *Continued*

For JRH Biosciences formulation of GMEM-S medium:

- GMEM-S medium (JRH Cat. No. 51492)
- GS supplement (50 ×) (JRH Cat. No. 58672)

For Gibco-BRL formulation of GMEM-S medium:

- 100 × non-essential amino acids (NEAA) (Gibco-BRL 043–01140 in UK; 320–1140 in US). Store at 4°C.
- 100 × glutamate plus asparagine (G + A). Add 600 mg glutamic acid (Sigma) and 600 mg asparagine (Sigma) to 100 ml distilled water. Sterilize by passing through a sterile 2 μm filter (Nalgene) and store at 4°C.
- 100 mM sodium pyruvate (Gibco-BRL 043–01360 in UK; 320–1360 in US).
- 50 × nucleosides. Add 35 mg adenosine, 35 mg guanosine, 35 mg cytidine, 35 mg uridine, and 12 mg thymidine (each from Sigma) to 100 ml distilled water. Filter sterilize and store at −20°C in 10 ml aliquots.
- Gibco-BRL basal GMEM
  Either Gibco-BRL US formulation [GMEM (Cat. No. 320–1710 but *prepared without glutamine.*): this is not a stock item, and will need to be made to order].
  or Gibco-BRL UK formulation [10 × GMEM without glutamine (Cat. No. 042–2541), stored at 4°C.
- 7.5% sodium bicarbonate (Gibco-BRL 043–05080). Store at 4°C.

Preparation of GMEM-S medium
Add the following components in the order given using aseptic technique.

JRH formulation

GMEM-S	500 ml
GS supplement	10 ml
Dialysed FCS	50 ml

Gibco-BRL UK formulation

Autoclaved distilled water	400 ml
10 × GMEM	50 ml
7.5% sodium bicarbonate	18 ml
NEAA	5 ml
G + A	5 ml
100 mM sodium pyruvate	5 ml
50 × nucleosides	10 ml
Dialysed FCS	50 ml

Gibco-BRL US formulation

Special order GMEM (without glutamine)	500 ml
NEAA	5 ml
G + A	5 ml
100 mM sodium pyruvate	5 ml
50 × nucleosides	10 ml
Dialysed FCS	50 ml

[a] The products of two different medium manufacturers are given. The JRH Biosciences medium has recently been made available and is simpler to prepare. The alternative formulations based on Gibco-BRL reagents have been more widely used but different sets of components are available in the US from those obtainable from Gibco-BRL (UK).

and RNA. The transfection procedure is described in *Protocol 3* and selection for amplification in *Protocol 4*.

Protocol 3. Transfection and selection of GS vectors in CHO-K1 cells

Equipment and reagents

- Circular recombinant plasmid DNA, including pEE14 as a control. Purify the plasmid DNA by CsCl centrifugation, then precipitate with ethanol, wash the DNA pellet with 70% ethanol, and leave the DNA pellet to dry in a tissue culture laminar flow cabinet to ensure sterility. Dissolve in sterile distilled water at 0.5 mg/ml.
- Tissue culture Petri dishes (9 cm diameter)
- CHO-K1 cells (available from ATCC Cat. No. CCL61) growing exponentially in GMEM-S medium (*Table 4*)
- GMEM-S medium plus 10% dialysed FCS (*Table 4*)
- Equipment and reagents for calcium phosphate transfection procedure (see Chapter 1, *Protocol 5*)
- 100 mM MSX stock solution (*Table 4*)

Method

1. On the day prior to transfection, trypsinize the CHO cells and seed 9 cm Petri dishes at 10^6 cells/dish in 10 ml GMEM-S medium. Incubate in the tissue culture incubator overnight.
2. Transfect the cells with plasmid DNA using the calcium phosphate co-precipitation procedure (Chapter 1, *Protocol 5*). Also treat several plates by the calcium phosphate procedure but without DNA to act as controls. Return the dishes to the tissue culture incubator.
3. One day after transfection, add MSX to a final concentration of 25 μM in GMEM-S medium.
4. Count the number of surviving colonies two to three weeks after transfection.
5. Isolate individual colonies and analyse expression of the desired product (see section 5). When expanding cultures, maintain the MSX concentration but, after trypsinization, wait one day before adding the MSX to allow the cells to recover.

Protocol 4. Selection for GS vector amplification in CHO cells

Equipment and reagents

- Individual cell lines derived from colonies transfected with GS vector (from *Protocol 3*)
- 100 mM MSX stock solution (*Table 4*)
- GMEM-S medium plus 10% dialysed FCS (*Table 4*)
- 9 cm tissue culture dishes or 75 cm^2 tissue culture flasks

Method

1. Take several individual transfected cell lines[a] producing significant amounts of the desired product (from *Protocol 3*) and plate out each cell line on several dishes or in flasks at a density of approximately 10^6 cells per dish or flask in GMEM-S + 10% dialysed FCS. Incubate for 24 h in a tissue culture incubator.[b]
2. Replace the medium with fresh GMEM-S + 10% dialysed FCS containing various concentrations of MSX, ranging between 100 μM and 1 mM.
3. Incubate the Petri dishes for 10–14 days, changing the medium once during this time (replacing it with fresh GMEM-S plus 10% dialysed FCS containing the same concentration of MSX). After this time, considerable cell death should have occurred and colonies resistant to the higher levels of MSX should have appeared. The maximum concentration of MSX at which colonies survive will depend on the particular initial transfectant but is typically between 250 μM and 500 μM.
4. Isolate several discrete colonies which have survived at the highest MSX concentration. Either pick the colonies and assay them individually or alternatively pool all colonies from one initial cell inoculum (at step 1) and assay these together.[c]
5. Clone those amplified individual cell lines or pools which exhibit high production rates of product using limiting dilution cloning. Re-analyse the expression of product in the cloned cell lines.

[a] In our experience, independent transfectants amplify more efficiently than pools of transfectants.
[b] Whenever trypsinizing GS selected cells, leave the cells for 24 h to recover before re-applying MSX selection.
[c] The increased production rate of the desired product can be up to tenfold in this first round of amplification. It is is not normally appropriate to select for subsequent rounds of amplification because the production rate does not usually increase significantly at higher levels of MSX.

4. Glutamine synthetase selection in NS0 myeloma cells

Some mammalian cell lines, such as rodent myelomas and hybridomas, do not express sufficient GS to survive without added glutamine. In this case, a transfected GS gene can function as a selectable marker by permitting growth in a glutamine-free medium. Amplification of vector copy number can subsequently be achieved using elevated levels of MSX. The basic system (5) has been updated using modified vectors to simplify plasmid construction.

4.1 Choice of cells and medium

The preferred cell line is NS0 (33). It is highly transfectable and yields essentially no glutamine-independent variants (less than 1 in 10^8 cells). Other

Table 5 Preparation of media for GS selection in NS0 cells[a]

Stock solutions

- Dialysed FCS (see *Table 4*)
- Non-dialysed FCS from any reputable source (treat as for dialysed FCS in *Table 4*)
- 200 mM L-glutamine (e.g. Gibco-BRL 043–05030 in UK; 320–5030 in US). Store in 5 ml aliquots at −20°C.
- 100 mM L-MSX (see *Table 4*)

For JRH Biosciences formulation of DMEM base:

- Celltech DME (JRH Cat. No. 51435)
- GS supplements (JRH Cat. No. 58672)

For Gibco-BRL formulation of DMEM base:

- NEAA (*Table 4*)
- G + A (*Table 4*)
- 50 × nucleosides (*Table 4*)
- DMEM specially prepared as Cat No. 320–1965 in US (Cat No. 041–1965 in UK) but *without* glutamine, *without* ferric nitrate, and *with* sodium pyruvate at 1 mM.

For JRH Biosciences formulation of IMDM base:

- IMDM (JRH Cat. No. 51472)
- GS supplement (JRH Cat. No. 58672)

For Sigma formulation of IMDM base:

- G + A (*Table 4*)
- 50 × nucleosides (*Table 4*)
- IMDM (Sigma Cat. No. I 2762 or I 4136)

Preparation of media

Non-selective medium

DMEM *or* IMDM	500 ml
200 mM L-glutamine	5 ml
FCS	50 ml

Selective medium (G-DMEM; JRH Biosciences formulation)

Celltech DME	500 ml
GS supplement	10 ml
Dialysed FCS	50 ml

Selective medium (G-DMEM; Gibco-BRL formulation)

DMEM	500 ml
NEAA	5 ml
G + A	5 ml
50 × nucleosides	10 ml
Dialysed FCS	50 ml

Table 5 *Continued*

Selective medium (G-IMDM; JRH Biosciences formulation)	
IMDM	500 ml
GS supplement	10 ml
Dialysed FCS	50 ml
Selective medium (G-IMDM; Sigma formulation)	
IMDM	500 ml
G + A	10 ml
50 × nucleosides	10 ml
Dialysed FCS	50 ml

[a] Two different base media are described and each can be obtained from two manufacturers. DMEM is the preferred base but IMDM is adequate if DMEM is not available. The JRH Biosciences media formulations have recently been made available and are simpler to prepare. The alternative formulations based on Gibco-BRL and Sigma reagents have been more widely used but require more components to be added to the base medium.

rodent myelomas may also be used provided that they can be transfected by electroporation and do not grow or yield variants which can grow in the glutamine-free selection medium.

Two alternative selective medium formulations are described in *Table 5*, G-DMEM (based on DMEM) or G-IMDM (based on Iscove's modification of DMEM).

4.2 GS vector design

The basic expression vector used for NS0 myeloma cells is pEE12 (*Figure 4*). This is identical to pEE14 (*Figure 3*), used for CHO cells, except that the GS minigene in pEE14 has been replaced with a GS cDNA expressed from the SV40 early promoter, derived from the transcription unit from pSV2.GS (5). This GS gene gives a higher transfection efficiency in NS0 cells than the GS gene in vector pEE14. In constructing pEE12, several restriction sites have been removed from the GS gene so that sites in the multilinker remain unique.

For expression of antibodies or other heterodimeric proteins, the genes or cDNAs for the two polypeptides should be separately inserted into vectors downstream of an hCMV promoter and a single vector is then constructed containing transcription units for both polypeptide chains and the GS selectable marker. The vector pEE6hCMV-B (*Figure 2*) is useful for this purpose. The preferred scheme is generally to insert a light chain coding sequence (which should contain a translation initiation signal) at the multilinker of pEE12 and the heavy chain coding sequence with translation initiation signal at the multilinker of pEE6.hCMV-B. Then the complete hCMV–heavy chain–SV40 transcription unit is isolated as a *Bgl*II–*Bam*HI cassette and

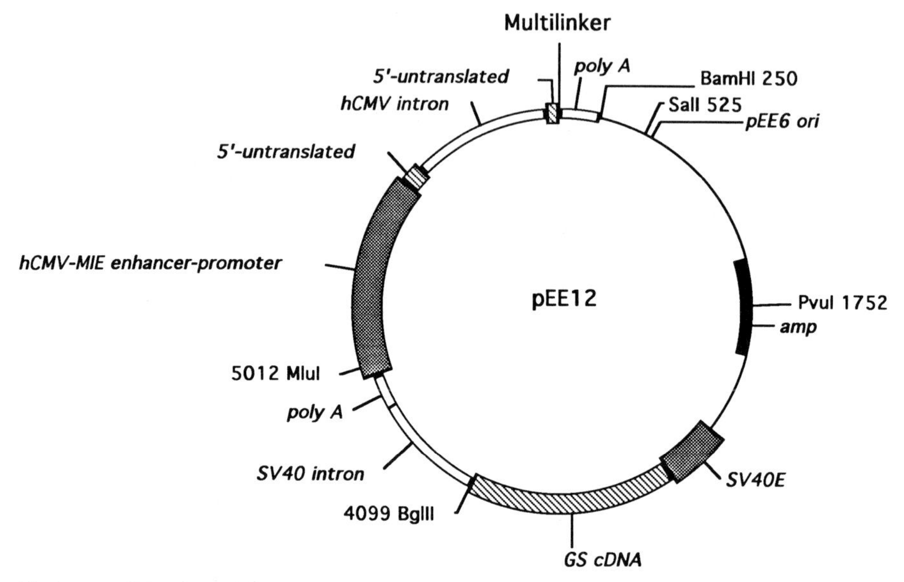

Figure 4. pEE12 (7.1 kb), a GS expression vector for myeloma cells. It consists of an SV40 early promoter (SV40E; a 344 bp *Pvu*II–*Hin*dIII fragment spanning the SV40 origin of replication); a hamster GS cDNA; and the SV40 intron and polyadenylation signal, identical to that in pSVM.*dhfr* (*Bgl*II–*Bam*HI; *Figure 1*) inserted into pEE6hCMV-B (*Figure 2*). The SV40–GS cassette is derived from pSV2.GS (5) but a number of restriction sites have been removed to facilitate cloning into the multilinker. (For unique sites in the multilinker, see *Figure 2*.) Figure reproduced courtesy of Celltech Therapeutics Ltd©.

inserted at the *Bam*HI site of the pEE12 light chain plasmid, such that the light and heavy chain genes will be transcribed in the same orientation. Light chain coding sequences can be present as cDNAs but genomic constant region segments have generally been used for the heavy chain. It is not known whether removal of the introns present in the heavy chain constant region would affect expression.

This type of plasmid has worked well for the synthesis of a number of antibodies. A single round of selection for gene amplification should suffice to provide high yielding cell lines. In some cases, the productivity of such lines has been relatively stable even in the absence of continued MSX selection.

It is not necessarily the case that a vector such as that described above will be optimal for every antibody and so, if resources permit, it may be worth testing additional vector constructs. The plasmid construction described above has the theoretical disadvantage that light chain transcription may be more efficient than heavy chain transcription due to promoter occlusion of the downstream hCMV promoter by the upstream hCMV promoter. Heavy chain expression may therefore limit productivity. Whether this is in fact the

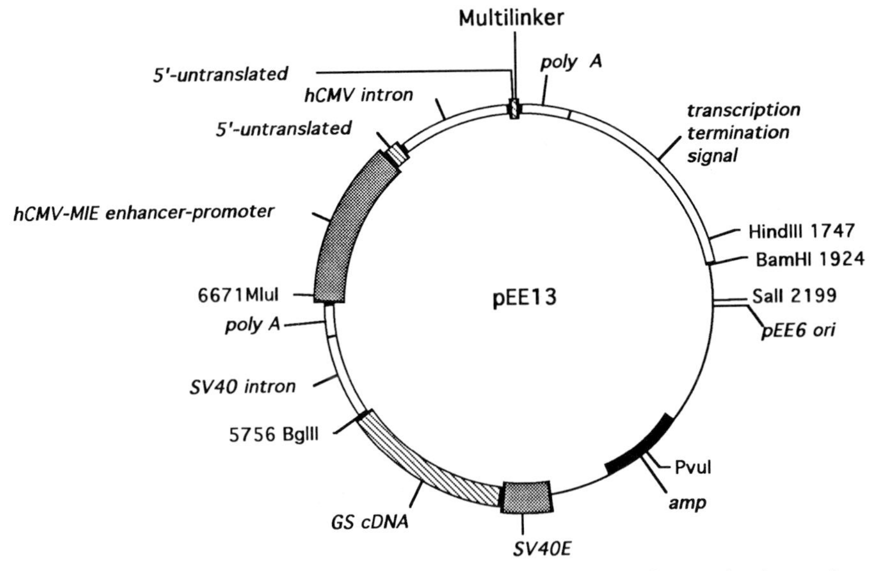

Figure 5. pEE13, a GS expression vector for expression of dimeric proteins in myeloma cells. It is identical to pEE12 but has a 1.7 kb *BclI–BamHI* sequence from the 3′ flanking region of the mouse Ig μ chain gene inserted at the *BamHI* site which acts as a transcription termination signal (34). Unique sites in the multilinker are *XmaI*, *SmaI*, *EcoRI*, and *BclI*. Figure reproduced courtesy of Celltech Therapeutics Ltd©.

case will also depend on the efficiency of translation of the mRNAs for the two chains and this may be antibody-dependent. It may therefore be advantageous to prepare and test the alternative plasmid, produced by cloning the heavy chain gene into pEE12 and the light chain gene into pEE6hCMV.B and combining the two so that the heavy chain is now upstream of the light chain. Another approach makes use of a modified pEE12 plasmid, pEE13 (*Figure 5*), which has a transcription termination signal from a mouse immunoglobulin μ chain gene downstream of the SV40 polyadenylation signal. This serves to isolate the two immunoglobulin genes in the final plasmid and minimizes the potential promoter occlusion, so leading to higher overall expression of assembled immunoglobulin in several cases (unpublished results).

4.3 Cell transfection

The NS0 cells should be kept growing exponentially in non-selective medium and have a cell viability greater than 90% prior to transfection. Cells are grown at 37°C in an atmosphere of 5% CO_2. Linearized DNA is introduced into the cells by electroporation. Transfected cells are then selected for the ability to grow in glutamine-free medium by gradually reducing the glutamine concentration (*Protocol 5*). The frequency with which glutamine-independent transfectants are isolated should be at least 1 in 10^5 cells plated, using this procedure for NS0 cells.

Protocol 5. Transfection of GS vectors into NS0 cells

Equipment and reagents

- NS0 cells available from ECACC (Cat. No. 85110503)
- Electroporation apparatus (e.g. Bio-Rad 'Gene Pulser'; note that a capacitance extender is not needed for this procedure)
- PBS (e.g. Flow Laboratories Ltd. Cat, No. 28-203-05), made up according to the manufacturer's instructions
- Suitable tissue culture equipment including humidified CO_2 incubators, set at 37°C and 5% CO_2
- pEE12 DNA as a control for transfection (prepare this using the procedure described below for the recombinant plasmid DNA)
- Non-selective medium (*Table 5*)
- Selective medium (*Table 5*)
- 96-well tissue culture trays
- 0.4 cm electroporation cuvettes (Bio-Rad Cat. No. 165-2088)
- Recombinant plasmid DNA (a GS vector based on pEE12 or pEE13). Purify the plasmid DNA by CsCl density centrifugation, ethanol precipitate the DNA, and then linearize it by digestion with a restriction enzyme which cuts once in the bacterial plasmid sequence (e.g. *Bam*HI or *Sal*I). Re-precipitate the DNA using ethanol, wash the pellet with 70% ethanol, and allow it to dry in a tissue culture laminar flow hood to ensure sterility. Resuspend the DNA in sterile distilled water at 40 μg DNA in 40 μl water.

Method

1. On the day of transfection, count the cells. 10^7 cells will be needed per plasmid transfection, plus 10^7 cells to be 'mock' transfected without DNA, and 10^7 cells to be transfected with pEE12 as a positive control.
2. Centrifuge the cells (1200 r.p.m. for 5 min in a bench centrifuge) and wash the cells once by centrifugation in 50 ml cold PBS. From this stage, maintain the cells on ice or at 4°C.
3. Resuspend the cells in PBS at 10^7 cells/ml.
4. Add 10^7 cells to each electroporation cuvette on ice. Add 40 μg plasmid DNA to one cuvette, the pEE12 DNA (positive control) to another, and 40 μl water to another cuvette containing cells to act as the mock transfected sample. Mix each sample gently with a pipette but avoid bubbles and excess liquid up the sides of the cuvette.
5. Leave the cuvettes containing the cells on ice for 5 min.
6. *Before using the electroporation apparatus, read the manufacturer's instructions and observe all safety precautions.* Wipe the outside of each cuvette dry and use the 'Gene Pulser' to deliver two consecutive pulses at 1500 V, 3 μF according to the manufacturer's instructions.
7. Return the cuvettes to ice for 2–5 min and then add each sample of cells to 30 ml of non-selective culture medium, pre-warmed to 37°C, and mix well.
8. Plate out each of the samples of transfected cells (including the pEE12

positive control transfection and the mock transfection) as follows. It is suggested that three dilutions of the original cell suspension in step 7 above are made to ensure that colonies can be picked from plates containing, on average, less than one transfected cell/well. This minimizes the risk of picking two or more colonies together.

(a) Distribute 20 ml of cell suspension from step 7 into four 96-well tissue culture plates (i.e. approx. 50 μl/well).

(b) Dilute 10 ml of cell supension with a further 30 ml of non-selective medium (i.e. 1 in 4 dilution) and distribute over another five 96-well plates (50 μl/well).

(c) Dilute 10 ml of diluted cell suspension from (b) above with 40 ml of non-selective medium (i.e. 1 in 16 dilution) and distribute over a further five 96-well plates (50 μl/well).

9. Return all the plates to a 37°C tissue culture incubator and incubate overnight.

10. The next day, add 150 μl of selective medium (*Table 5*) to each well of the transfected plates without removing the medium already there. Return the plates to the incubator and incubate until substantial cell death has occurred and discrete surviving colonies appear. This procedure allows the cell to deplete the medium of residual glutamine so that the glutamine concentration declines gradually.

11. Three weeks post-transfection, viable colonies of glutamine-independent transfectants should be visible amongst the background of dead cells. Because the selection procedure depends on the cells depleting the medium of glutamine gradually, cells in plates which have been diluted before plating out will take longer to die and may even grow to form colonies for a few days after transfection. However, by three weeks any such growth should have died off. Viable colonies can be distinguished microscopically by the occurrence of bright round cells. There should be no surviving colonies on the mock transfected plates. The frequency of viable colonies should be 2–5/10^5 cells plated for the pEE12 positive control plasmid. Certain recombinant plasmids may show a somewhat reduced transfection efficiency. Identify wells containing single healthy colonies and mark these on the lid of the plate with a pen.

12. Collect spent culture medium from the chosen wells, once the medium has begun to turn orange-yellow, and use this to assay for product secretion (see section 5).

13. Do not expand the transfectant clones too rapidly since direct inoculation of a flask from the 96-well plate may lead to reduced growth rate or significant cell death. Rather, transfer the cells from one well of a confluent 96-well plate to one well of a 24-well plate. Then, after

Protocol 5. *Continued*

several days, take the contents of this well to inoculate a small flask. At each stage, include in the transfer of cells as much as possible of the medium in which the cells were growing. As a precaution, re-feed the empty wells after transfer of cells. In this way, if the transferred cells fail to survive, residual cells in the original well can be used for a second attempt.

14. Once in flasks, maintain the cultures at between 10^5 and 10^6 cells/ml. The doubling time will vary between transfectants but should be 20–40 h. Cell lines growing more slowly than this are unlikely to be useful.

4.4 Selection for GS gene amplification in NSO cells using MSX

Amplification of the GS vector in the NS0 cells can be selected using MSX as described in *Protocol 6*.

Protocol 6. Selection for GS vector amplification in NSO cells

Equipment and reagents

- Several (e.g. at least five) *independent* transfected cell lines producing significant amounts of product from *Protocol 5*[a]
- 24-well tissue culture plates
- Selective medium (*Table 5*)
- 100 mM MSX (*Table 5*)

Method

1. Distribute the transfected cells in 24-well plates at a density of 2–5 × 10^5 cells/0.5 ml/well in selective medium.
2. Add 0.5 ml of selective medium containing MSX to each well to *final* concentrations ranging between 5 μM and 500 μM (use several wells for each concentration of MSX).
3. Incubate the plates until discrete MSX-resistant colonies appear (typically three to four weeks).
4. Isolate pools or individual colonies from each independent transfectant at the highest MSX concentration at which MSX resistance occurs. This varies widely between different transfectants.
5. Assay the MSX-resistant cell lines or pools for protein product.[b]
6. Clone the amplified pools or cell lines secreting product at the highest rate by limiting dilution cloning, maintaining MSX selection throughout. Screen the clones obtained for the highest levels of product secretion.
7. Store stocks of these frozen in liquid nitrogen.

8. Once frozen cell stocks have been secured, the stability of the production rate can be tested by growth of the cells in the presence and absence of MSX for extended periods (e.g. two months).

[a] Pools of transfectants do not amplify as efficiently as independent cell lines.
[b] The secretion rate (in μg/10^6 cells/day) should be two- to tenfold higher than the original transfectant if vector amplification has occurred. Not all primary transfectants will show significant amplification.

5. Analysis of gene amplification

5.1 Methods of analysis

The simplest and often the most effective method of analysis of amplified cell lines is to measure expression of the product since this is usually the most important parameter. Methods for measurement of levels of the enzymes encoded by the DHFR or GS selectable markers are given in *Protocols 7–9*.

Estimation of vector copy numbers can be carried out by Southern blot analysis and estimation of RNA levels by Northern blots.

Protocol 7. Preparation of cell extracts for enzyme assays

Equipment and reagents

- Transfected cells
- Bench centrifuge
- PBS (e.g. Flow Laboratories Ltd. Cat. No. 28–203–05), made up according to the manufacturer's instructions
- 0.25 M sucrose, 10 mM Tris–HCl pH 7.6, 1 mM DTT
- Reagents or a commercial kit (e.g. bicinchoninic acid kit from Pierce Inc) for protein assay

Method

1. Pellet the cells by centrifugation at 1500 r.p.m. in a bench-top centrifuge for 5–10 min.
2. Wash the cells three times by centrifugation in PBS, resuspending the cells each time in PBS using a Pasteur pipette.
3. Resuspend at a final concentration of about 5×10^7 cells/ml in 0.25 M sucrose, 10 mM Tris–HCl pH 7.6, 1 mM DTT.
4. Disrupt the cells by sonication (e.g. using three bursts of 15 sec each).
5. Centrifuge in a microcentrifuge for 10 min at 4°C.
6. Remove the supernatant (avoiding lipid material at the surface) and place on ice or store at −20°C until assayed.
7. Measure the total protein in the extract.

Protocol 8. Assay of dihydrofolate reductase

Equipment and reagents

- 100 mM potassium phosphate buffer pH 7.9
- 5 mM dihydrofolate
- 1 mM NADPH
- Cell extracts (from *Protocol 7*)
- 1 ml cuvettes (UV transparent) with 1 cm path length

Method

1. Set-up the following reaction mixture in a 1 ml cuvette:
   - 100 mM phosphate buffer 0.8 ml
   - 5 mM dihydrofolate 10 μl
   - 1 mM NADPH 100 μl
   - cell extracts up to 100 μl
   - water to 1 ml final volume
2. Set-up a control reaction as in step 1 but omitting the cell extract.
3. Incubate at 37 °C and measure the absorbance of the reaction mixture and of the control reaction at 340 nm at 1 min intervals. The enzyme activity observed can be expressed as nmol dihydrofolate reduced/mg protein/min (molar absorbance coefficient of NADP is 12 200 M^{-1} cm).

Protocol 9. Assay for glutamine synthetase

Equipment and reagents

- Dowex ion-exchange columns: prepare these as follows. Add 0.5 M NaOH to some Dowex-1-Cl (8% cross-linked, 200–400 mesh). Aspirate off the fines and wash the resin several times with distilled water. Wash with 0.5 M HCl until the pH is below pH 2.0 and then wash several times again with distilled water. Add solid imidazole to the resin–water mixture to adjust the pH to pH 7.0. Make standard size (e.g. 0.8 ml) columns in 1 ml plastic pipette tips (e.g. the blue tips suitable for a Gilson P1000 Pipetman). Cut the end of each tip and plug it with glass wool before filling with the treated Dowex resin. Rinse the columns with several millilitres of distilled water.
- 0.5 M sodium tricine pH 7.6
- 100 mM ATP pH 7.0
- 80 mM NH_4Cl
- 0.4 M $MgCl_2$
- 100 mM [^{14}C]glutamate (4 μCi/mmol)
- Cell extract (from *Protocol 7*)
- Scintillation fluid (e.g. Biofluor) and liquid scintillation counter

Method

1. Set-up the following 50 μl reaction mixtures:
   - 0.5 M sodium tricine pH 7.6 5 μl
   - 100 mM ATP pH 7.0 7.5 μl
   - 80 mM NH_4Cl 2.5 μl

• 400 mM $MgCl_2$	2.5 μl
• cell extract	various volumes up to 25 μl
• water	to 45 μl final volume
• [^{14}C]glutamate	5 μl

2. Also set-up a control reaction as in step 1 but omitting the cell extract.
3. Incubate at 37°C for various times (e.g. 5, 10, 15, 20, and 30 min).
4. Stop the reactions (including the control reaction) by adding 200 μl ice-cold distilled water to each tube and place the tubes on ice.
5. For each reaction, add the total reaction mixture to a Dowex column. Discard the flow-through.
6. Elute the column with two distilled water washes of 0.5 ml each, collecting both eluates. Glutamine (the product) is eluted and unreacted substrate (glutamate) is retained on the column.
7. Add 10 ml scintillation fluid to the pooled eluate and count in a scintillation counter. Also count the control reaction eluate and subtract these c.p.m. from the c.p.m. for each reaction.
8. From the specific activity of the [^{14}C]glutamate, calculate the nmol of glutamate converted/min for each sample.

5.2 Stability of gene amplification

In some cases, highly amplified arrays of vector sequences can be retained in the absence of the selective agent (at least for several weeks in culture) while in other cases, the amplified sequences are lost very rapidly if the selection is removed. Whether the sequences are 'stable' or 'unstable' correlates with the state of the vector sequences in the genome. Unstable amplified sequences are frequently associated with small acentric chromosome fragments called double minutes (DMs). Because DMs have no centromeres, unequal assortment of DMs at mitosis together with more rapid growth of the cells which retain fewer DMs will lead to loss of the amplified arrays from the cell population. In stable cell lines, the amplified genes can be shown by *in situ* hybridization to have remained integrated in one or more of the host chromosomes. Because the region of chromosome involved in amplification is often very large, it can be detected as an altered staining pattern by chromosome banding techniques, forming homogeneously staining regions (HSRs). However, even amplified arrays in HSRs are not always stable (4, 35, 36). For many purposes it may be acceptable to maintain cells in the presence of the selective agent but if stability in the absence of selective agent is considered important, it would be advisable to screen a number of independently derived lines to identify any which show stable productivity under such conditions.

References

1. Stark, G. (1986). *Cancer Surveys*, **5**, 1.
2. Smith, K. A., Gorman, P. A., Stark, M. B., Groves, R. P., and Stark, G. R. (1990). *Cell*, **63**, 1219.
3. Chapman, A. B., Costello, M. A., Lee, F., and Ringold, G. M. (1983). *Mol. Cell. Biol.*, **3**, 1421.
4. Weidle, U. H., Buckel, P., and Wienberg, J. (1988). *Gene*, **66**, 193.
5. Bebbington, C. R., Renner, G., Thomson, S., King, D., Abrams, D., and Yarranton, G. T. (1992). *Bio/Technology*, **10**, 169.
6. Cockett, M., Bebbington, C. R., and Yarranton, G. T. (1990). *Bio/Technology*, **8**, 662.
7. Davis, S. J., Ward, H. A., Puklavec, M. J., Willis, A. C., Williams, A. F., and Barclay, A. N. (1990). *J. Biol. Chem.*, **265**, 10410.
8. Barsoum, J. (1990). *DNA Cell. Biol.*, **9**, 293.
9. Murray, M. J., Kaufman, R. J., Latt, S. A., and Weinberg, R. A. (1983). *Mol. Cell. Biol.*, **3**, 32.
10. Simonsen, C. C. and Levinson, A. D. (1983). *Proc. Natl Acad. Sci. USA*, **80**, 2495.
11. Kaufman, R. J., Murtha, P., Ingolia, D. E., Yeung, C. Y., and Kellems, R. E. (1986). *Proc. Natl Acad. Sci. USA*, **83**, 3136.
12. Andrulis, I. L., Chen, J., and Ray, P. N. (1987). *Mol. Cell. Biol.*, **7**, 2435.
13. Cartier, M., Chang, M. W., and Stanners, C. P. (1987). *Mol. Cell. Biol.*, **7**, 1623.
14. Cartier, M. and Stanners, C. P. (1990). *Gene*, **95**, 223.
15. Ruiz, J. C. and Wahl, G. M. (1986). *Mol. Cell. Biol.*, **6**, 3050.
16. Crouse, G. F., McEwan, R. N., and Pearson, M. L. (1983). *Mol. Cell. Biol.*, **3**, 257.
17. Lee, F., Mulligan, R., Berg, P., and Ringold, G. (1981). *Nature*, **294**, 228.
18. Subramani, S., Mulligan, R., and Berg, P. (1981). *Mol. Cell. Biol.*, **1**, 854.
19. Mayo, K., Warren, R., and Palmiter, R. (1982). *Cell*, **29**, 99.
20. Chiang, T. R. and McConlogue, L. (1988). *Mol. Cell. Biol.*, **8**, 764.
21. Kane, S. E., Reinhard, D. H., Fordis, C. M., Pastan, I., and Gottesman, M. M. (1989). *Gene*, **84**, 439.
22. Thelander, M. and Thelander, L. (1989). *EMBO J.*, **8**, 2475.
23. Israel, N., Chenciner, N., and Streeck, R. E. (1987). *Gene*, **51**, 197.
24. Roberts, J. M. and Axel, R. (1982). *Cell*, **29**, 109.
25. Israel, N., Chenciner, N., Houlmann, C., and Streeck, R. E. (1989). *Gene*, **81**, 369.
26. Urlaub, G. and Chasin, L. A. (1980). *Proc. Natl Acad. Sci. USA*, **77**, 4216.
27. Urlaub, G., Kas, E., Carothers, A. M., and Chasin, L. A. (1983). *Cell*, **33**, 405.
28. Lee, F., Mulligan, R., Berg, P., and Ringold, G. R. (1981). *Nature*, **294**, 228.
29. Kaufman, R. J., Davies, M. V., Wasley, L. C., and Michnick, D. (1991). *Nucleic Acids Res.*, **19**, 4485.
30. Stephens, P. and Cockett, M. (1989). *Nucleic Acids Res.*, **17**, 7110.
31. Kaufman, R. J., Wasley, L. C., Spiliotes, A. J., Gossels, S. D., Latt, S. A., Larsen, G. R., and Kay, R. M. (1985). *Mol. Cell. Biol.*, **5**, 1730.

32. Bebbington, C. R. (1991). *Methods*, **2**, 136.
33. Galfre, G. and Milstein, C. (1981). In *Methods in enzymology* (eds Langone, J. J. and Vulnakis, H.), Vol. 73, pp. 3–46. Academic Press.
34. Law, R., Kuwabara, M. D., Briskin, M., Fasel, N., Hermanson, G., Sigman, D. S., and Wall, R. (1987). *Proc. Natl Acad. Sci. USA*, **84**, 9160.
35. Fendrock, B., Destrempes, M., Kaufman, R. J., and Latt, S. A. (1986). *Histochemistry*, **84**, 121.
36. Urlaub, G., Landzberg, M., and Chasin, L. A. (1981). *Cancer Res.*, **41**, 1594.

4

Retroviral vectors

ANTHONY M. C. BROWN and JOSEPH P. DOUGHERTY

1. Introduction

1.1 Why use retroviruses?

The exploitation of retroviruses as genetic vectors provides an important and versatile method of introducing and expressing cloned genes in eukaryotic cells, both in culture and *in vivo*. Several features of retroviruses favour their choice for this purpose, as an alternative to DNA-mediated gene transfer methods or other viral vector systems. First, the relatively small genomes of retroviruses can be easily manipulated to introduce foreign genes in such a way that any resulting defects in viral replication can be complemented in *trans*. Secondly, the viruses can be grown to high titres in cell culture. Most importantly, the efficiency of infection of susceptible cells is extremely high. Together with high titres of virus, this can permit infection of almost 100% of the target cells in a homogeneous culture. After infection, retroviral DNA becomes stably integrated in the host cell genome in a precisely defined manner such that virtually all of the infected cells can express the genes carried by the virus. Moreover, retroviruses contain powerful transcriptional enhancer elements that allow high levels of expression in a wide range of cell types.

1.2 Applications of retroviral vectors

Retroviruses can be used very effectively as general purpose expression vectors for experiments in eukaryotic cell culture. Construction of the necessary recombinant is no more complicated than for a plasmid expression vector. Although additional effort must be invested in producing a virus stock, the infection of target cells with a thawed aliquot of virus is technically quite simple and usually offers greater efficiency and reproducibility of gene expression than DNA transfection. Retroviruses are particularly useful for gene transfer into cells that are refractory to transfection, including primary cells and cells grown in suspension.

In addition to their uses in cell culture, retroviral vectors have several more specialized applications in molecular biology, genetics, developmental biology, and medicine. Some of these are listed below:—

(i) An important application of retroviruses in recent years has been as vectors for somatic cell gene therapy in animals and in humans. Typical protocols involve infection of primary cells in culture followed by their re-introduction into the organism. For example, by retroviral infection of haemopoietic stem cells in bone marrow cultures, and subsequent transplantation into irradiated mice, an animal's entire haemopoietic system can be reconstituted with cells containing a retrovirus vector (1). Among the most promising examples of retroviral gene therapy for human genetic diseases has been the introduction of the human adenosine deaminase gene (*ADA*) into T cells of children with severe combined immunodeficiency caused by an *ADA* gene defect (2).

(ii) A variant of the gene therapy concept in animals is to use retrovirus vectors to express genes ectopically in embryonic tissues in order to investigate the developmental function of specific genes. This usually requires direct microinjection of concentrated virus into embryos. Although experiments of this sort may be technically simpler in avian embryos (3), they are also applicable to mammalian systems (4).

(iii) Rather than expressing a gene that affects cellular behaviour, retroviruses can be used to introduce a biologically neutral marker gene *in vivo* which should then be stably expressed in the descendants of the initially infected cells. For this reason, replication-defective retrovirus vectors are ideally suited for cell lineage analysis in developing animals (5, 6). Studies are typically performed with low numbers of infectious particles such that individual cells are infected and the subsequent clone of marked cells can be inferred to comprise the linear descendants of a single progenitor. See ref. 6 for practical details about the use of retroviruses as lineage markers.

(iv) Retroviruses can also be used to introduce DNA into germ cells, by infecting pre-implantation embryos directly or embryonic stem cells in culture. Subsequent implantations of these provide methods of generating transgenic animals (7, 8). Poor vector expression resulting from DNA methylation in pre-implantation embryos has limited the use of germline infection as a means of expressing genes in transgenic animals (9–11), but modified vectors that overcome this block are now available (12).

(v) Retrovirus vectors can be used as insertional mutagens, either in cultured cells or in transgenic animals. Incorporation within the virus of a marker gene selectable in *E. coli* facilitates subsequent cloning of sequences from the mutated locus (13, 14). A powerful variation on this approach has been to use retrovirus vectors as 'promoter traps' in the mouse genome by infection of embryonic stem cells with a vector carrying a promoterless *lacZ* reporter gene. The reporter allows the expression pattern of the mutated gene to be visualized in mouse embryos, while the proviral insertion usually disrupts gene function so that mutant phenotypes can be characterized directly in homozygous animals (14, 15).

(vi) A more esoteric but nevertheless significant use of retroviral vectors is in the study of retrovirus biology. For example, replication-defective vectors have greatly facilitated experimental analysis of the retroviral life cycle, mechanisms of reverse transcription, and studies of retroviral mutation rates during a single round of replication (16). Such studies are of obvious importance, not only for knowledge of how to use retroviruses as gene transfer vectors, but also for our understanding of pathogenic human retroviruses such as HIV.

The purpose of this chapter is to provide a practical guide to the use of retroviruses as general purpose expression vectors, and to describe the methods of virus production, testing, and infection, which are common to most of the applications described above. Few details of more specialized applications will be presented here and for these the reader should refer to the cited literature. The retrovirus vectors in most widespread use in mammalian systems are those derived from the murine leukaemia viruses (MLV). This chapter will therefore concentrate on MLV-based vectors and techniques associated with their use. For more specialized purposes, vectors based on other viruses that infect mammalian cells are also available. These include spleen necrosis virus (SNV) (16), mouse mammary tumour virus (MMTV) (17), and human immunodeficiency virus (HIV) (18). Additional vectors have been developed specifically for avian systems (19, 20), but these will not be discussed in detail here.

The next section outlines some of the fundamental aspects of retrovirus biology which underlie the use of these viruses as vectors, and section 3 presents a guide to the principal types of retroviral vector now available. Sections 4–6 describe methods of virus production, infection of target cells, and testing of virus stocks, while section 7 considers some potential problems associated with the use of retrovirus vectors.

2. Retrovirus biology

2.1 The retrovirus life cycle

Retroviruses comprise a family of RNA viruses whose replication cycle is distinguished by the conversion of their RNA genome into a DNA form by a virus-encoded reverse transcriptase. The life cycle of the murine leukaemia virus MLV is described in detail in ref. 21 and illustrated in simplified form in *Figure 1*. The virus particles, each of which contains two copies of a single-stranded RNA genome, are composed of a central nucleoprotein core surrounded by a membrane coat which bears specific viral 'envelope' glycoproteins on its surface. Infection of target cells is initiated by the interaction

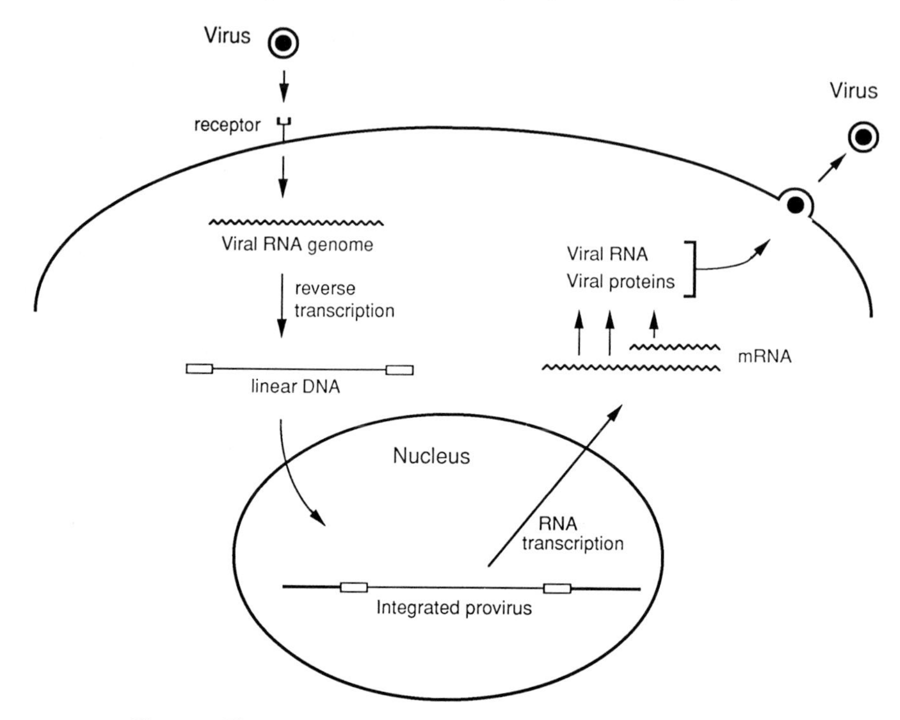

Figure 1. The retrovirus life cycle (see section 2.1 for description).

of these glycoproteins with specific cell surface receptors. The genes for several of these receptor proteins have been cloned in recent years (22, 23). After entry into the cell, the viral genome is converted into a linear double-stranded DNA form by reverse transcriptase molecules present in the infecting particles. This involves a complex series of steps that include site-specific binding of primers for reverse transcription, and serial transfers of the nascent DNA strands between templates (21). The end-product is a linear DNA molecule which, in the case of MLV, gains access to cellular chromatin after breakdown of the nuclear envelope at M-phase (24). It then becomes integrated into the host cell genome. A single infecting virus particle gives rise to only one integrated copy of viral DNA and this copy is usually stably maintained. Apart from a possible preference for transcriptionally active regions or DNase I-hypersensitive sites (25), retroviral insertion shows little or no specificity for particular sites within the target cell DNA and for most purposes may be considered to occur at random locations (26).

The integrated linear DNA form of the retroviral genome is known as a provirus and has a characteristic structure with long terminal repeats (LTRs) at either end (see section 2.2). It is this DNA structure that is manipulated in cloned form to make retroviral vectors. A wild-type MLV provirus contains three structural genes, *gag*, *pol*, and *env*, which encode the core proteins, reverse transcriptase/integrase, and envelope glycoproteins, respectively (21). The promoter in the 5′ LTR drives transcription of the integrated proviral DNA, while the 3′ LTR provides a signal for polyadenylation of the transcripts. In the case of MLV and many other retroviruses, two distinct RNAs are produced in approximately equal proportions: a full-length 'genomic' transcript, and a smaller subgenomic mRNA which has undergone splicing and serves as a template for translation of the *env* gene products. The genomic length transcripts direct translation of the *gag* and *pol* gene products, but a proportion of these also serve as RNA genomes and are packaged into viral particles. The nascent viral particles, which contain molecules of reverse transcriptase, form by budding of the plasma membrane and are released from the cell surface (21).

2.2 Structure of the retroviral genome and provirus

All retroviruses have a similar genetic organization, exemplified here by the murine leukaemia virus (MLV). The structure of an MLV provirus is shown in *Figure 2*, together with the viral RNA genome and spliced subgenomic RNA. The principal features of the provirus are as follows (21, 27):

(a) A long terminal repeat (LTR) sequence at each end of the provirus, containing sequences for transcription initiation, polyadenylation, reverse transcription of the viral RNA, and integration into the host chromosome. The LTR itself has a tripartite structure containing sequences derived from the 5′ end of viral RNA (U5), the 3′ end of viral RNA (U3), and a short repeated sequence (R) which is found at the ends of viral RNA.

(b) Short sequences (labelled PB(−) and PB(+) in *Figure 2*) required for priming of negative and positive strand synthesis of viral DNA by reverse transcription.

(c) The packaging (or 'encapsidation') sequence, ψ, which is needed for efficient packaging of viral RNA into virions. In MLV, the most essential sequences are located immediately upstream of *gag*, but for maximum efficiency of packaging and high virus titres the first 420 bp of *gag* coding sequences are also required (28, 29).

(d) The *trans*-acting viral structural genes *gag*, *pol*, and *env* (see section 2.1).

(e) The viral splice donor and acceptor sequences (S_D and S_A) which are used for the production of subgenomic *env* RNA.

2.3 Host range of virus infection

Some aspects of the retrovirus life cycle merit further consideration here because of their relevance to the host range of retroviral vectors. The host range of a wild-type retrovirus can be limited at either the extracellular or intracellular stages of its life cycle (27). Of these, the extracellular stage,

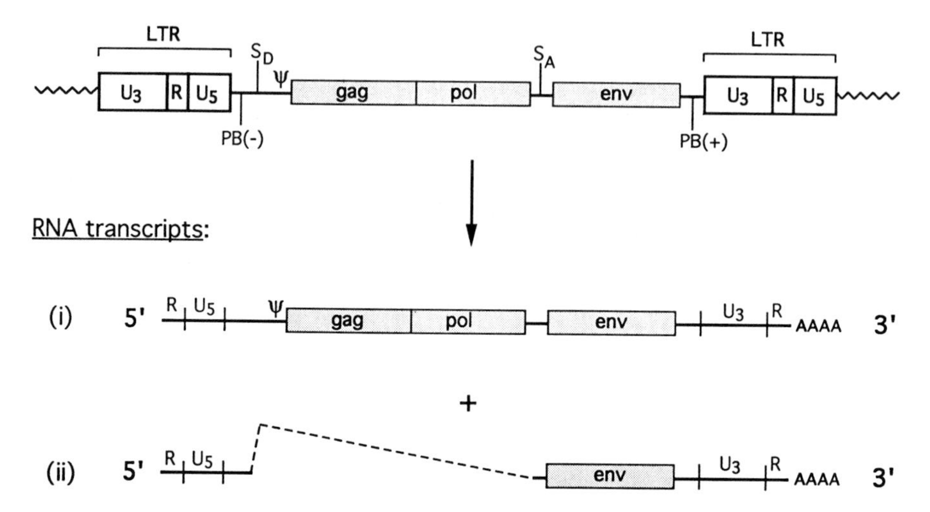

Figure 2. Structure of a retroviral provirus and its relationship to viral transcripts (not to scale). The proviral DNA is composed of two long terminal repeats (LTRs) flanking a central region that contains the three structural genes *gag*, *pol*, and *env*. The primer binding sites for reverse transcription of viral RNA, PB(−) and PB(+), are located adjacent to the LTRs. Each LTR is composed of sequences derived from the 3′ (U_3) and 5′ (U_5) ends of genomic RNA, linked by a copy of a short sequence (R) which is found at both ends of the viral RNA. The U_3 sequences contain promoter and enhancer elements. Transcription initiates within the 5′ LTR and the RNA terminates at a polyadenylation site in the 3′ LTR. (i) The full-length 'genomic' transcript serves both as the viral RNA genome and as the template for translation of the *gag* and *pol* genes. (ii) The *env* gene is translated from a subgenomic RNA formed by splicing of the primary transcript between the splice donor (S_D) and splice acceptor (S_A) sites. Note that only the genomic length RNA contains the packaging signal (ψ) required for efficient incorporation into virions.

involving interaction of viral envelope proteins with their receptors, is the most relevant here.

The envelope glycoproteins of MLV-related viruses can be divided into different classes based on the particular cell surface receptors with which they interact, and the host range or 'tropism' of the viruses is accordingly limited to animal species that express the appropriate receptors. The two most important host ranges in this context are termed 'ecotropic' and 'amphotropic'. An ecotropic murine retrovirus, such as Moloney-MLV, is limited to infecting mouse and rat cells. In contrast, an amphotropic MLV has a much broader host range since its envelope glycoproteins interact with a receptor present in a variety of species of birds, rodents, and other mammals, including humans (23, 27). Although most MLV-based vectors are derived from ecotropic Moloney-MLV, their host range can easily be extended beyond its normal limits by packaging the vector genome in virion particles bearing envelope glycoproteins from amphotropic MLV or other retroviruses (30–32).

In addition to these species-dependent tropisms, retroviral host range may also be limited by tissue-specific expression of the appropriate cell surface receptors in a given species, or by post-translational modification affecting virus receptor recognition (33). Most murine fibroblasts, lymphoid cells, and many epithelial cells are susceptible to MLV infection. However, an important proviso is that the target cells should be replicating since M-phase appears to be a prerequisite for MLV proviral integration (24).

Even if a target cell expresses the correct receptors for a particular retrovirus, it should be appreciated that those receptors may be blocked if the cell is already expressing the relevant viral envelope protein. This phenomenon is known as 'interference' (21, 27). Thus, a cell that is productively infected with a replication-competent retrovirus will be immune to superinfection by a virus with the same tropism. This resistance can be as high as 10 000-fold or greater, but is not absolute and presumably depends on the level of envelope protein expressed.

3. Choosing a vector

3.1 Principles of vector design

Certain retroviruses are naturally occurring vectors for transmission and expression of foreign genes. The acutely oncogenic retroviruses, for example, have oncogene sequences of cellular origin substituted for some of the viral sequences that encode *trans*-acting functions. With one exception (see section 3.5), such viruses are replication-defective. Viral products essential for the replication and integration of these defective viruses in nature are supplied in *trans* by a non-transforming replication-competent 'helper' virus. The most logical approach to retrovirus vector design is thus to follow the precedent set by nature and make replication-defective vectors requiring *trans*-complementation of viral functions. In principle, these functions can be supplied by a replication-competent helper virus, giving rise to a mixed virus stock of vector plus helper. Most commonly, however, the missing viral gene products are supplied by retroviral 'packaging' cells (also known as 'helper' cells), which express the viral proteins without producing replication-competent virus.

When transfected with retroviral vector DNA, packaging cells produce helper-free stocks of vector virus. Although replication-defective, such viruses are nevertheless fully infectious. By virtue of viral proteins packaged in the virions, the vector viruses will efficiently infect susceptible target cells, undergo a single round of reverse transcription, and integrate their viral DNA in the recipient cell genome to form a provirus. Transcription of the vector provirus will then commence from the viral LTR, but in the absence of helper viral gene products no virus particles will be produced from the infected cells.

The *cis*-acting viral sequences that must be retained by an MLV-based retroviral vector are the two LTRs, the priming sites for reverse transcription (adjacent to the LTRs), and the packaging sequence near the 5′ LTR (see *Figure 2*). All of the remaining sequences are dispensable, as their loss can be complemented in *trans*. The structure of the provirus is readily adapted for vector construction since the *cis*-acting sequences required for replication are clustered at the ends of the proviral DNA while the *trans*-acting sequences lie in between. This organization makes it simple to insert foreign sequences in the middle of the genome, in place of the viral structural genes.

Sections 3.2 to 3.4 describe the principal types of replication-defective vectors available. An alternative strategy, in which sequences are inserted into replication-competent vectors, is discussed briefly in section 3.4.

3.2 Single gene vectors

In the simplest retroviral vectors, *trans*-acting viral sequences are deleted and replaced with a site for insertion of new sequences. Two such vectors derived from MLV are pMX1112 (34) and MFG (35); these are represented schematically in *Figure 3A*. In pMX1112 the viral genes *gag*, *pol*, and *env* are removed entirely and replaced with a polylinker. A foreign gene inserted here is expressed from the promoter in the 5′ viral LTR. Retention of essential *cis*-acting viral sequences allows production of infectious particles from the vector in the presence of suitable helper functions. In the more recent vector, MFG, additional sequences from the viral *gag* region are retained (35). These enhance packaging efficiency and should therefore result in higher titres of vector virus.

Single gene vectors have the advantage of simplicity, and in our experience consistently produce high titre virus stocks that successfully express the inserted gene. More sophisticated vectors are more versatile but more often seem prone to problems of instability and/or poor expression. The major disadvantage of single gene vectors is lack of selection for the recombinant virus and difficulty in identifying individual infected cells. This is not a problem if the inserted gene is either a selectable marker, a gene producing a distinct phenotype in infected cells (e.g. a transforming oncogene), or one whose product can be identified by enzymatic assay or immunohistochemical staining. Examples of the latter include β-galactosidase (*lacZ*), alkaline phosphatase, luciferase, or any gene product for which there is a suitable antibody.

3.3 Double gene vectors

For transmission of genes that produce no easily scored phenotype, it is convenient to have a second gene within the vector which encodes a selectable

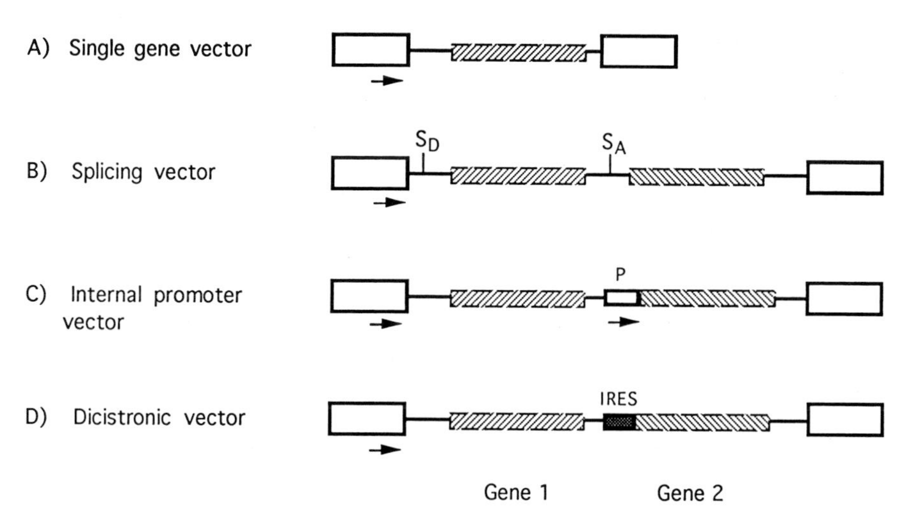

Figure 3. Schematic representation of four different designs of retrovirus vector. Vector sequences are indicated by bold lines and boxes. Genes inserted within the vectors are shown as hatched boxes with dashed lines. Promoters are indicated by arrows. S_D, splice donor site; S_A, splice acceptor. Each vector is constructed on a bacterial plasmid containing a replication origin and antibiotic resistance marker (not shown). In each construct, Gene 1 is expressed from the promoter in the 5′ LTR. Gene 2 is expressed: in (*B*) from a spliced message derived from the full-length transcript; in (*C*) from a separate transcript originating from the internal promoter (P); and in (*D*) from a dicistronic RNA with translation mediated by an internal ribosome entry site (IRES). See sections 3.2 and 3.3 for details.

or readily identifiable marker. Common examples of dominant selectable markers are neomycin phosphotransferase (*neo*), hygromycin-B-phosphotransferase (*hph*), guanine phosphoribosyl transferase (*gpt*), and histidinol dehydrogenase (*hisD*). In double gene vectors, the gene proximal to the 5′ LTR is expressed from the genomic length viral RNA. Three different strategies have been adopted to express the more distal gene:

- it can be expressed from a separate subgenomic message formed by splicing of the viral RNA
- a second promoter can be installed within the vector to initiate a second transcript
- an internal ribosome entry site (IRES) can be inserted such that both gene products are translated from a single dicistronic mRNA

Each of the these designs is discussed in more detail below.

3.3.1 Splicing vectors

The seemingly most natural approach to construction of double gene vectors is to mimic the mechanism employed by wild-type retroviruses and use the viral splice donor and acceptor sites to form a separate subgenomic RNA analogous to *env* mRNA (see *Figure 3B*). This is then the template for translation of the distal gene. The proximal gene is installed in place of the *gag* and *pol* reading frames and is translated from unspliced RNA. Two examples of double gene splicing vectors are pMX1122*neo* (34) and pZIP-Neo SV(X)1 (36). Both have cloning sites for insertion of an exogenous gene in the '*gag*' position, and contain *neo* as a selectable marker in the '*env*' position.

Although these vectors have been used successfully to express certain genes, a problem often encountered with splicing vectors is failure to transmit and express both genes simultaneously. We and others have observed that sequences installed in the upstream cloning site may affect formation of the subgenomic RNA from which the downstream gene is expressed (16, 34). The altered efficiency of viral RNA splicing can lead to poor expression of one or other of the two genes carried. Moreover, if selection is imposed for expression of the distal gene, this may result in inadvertent selection of mutant proviruses which have suffered deletion of proximal gene sequences that were inhibitory for splicing. Effects on splicing efficiency are evident with some inserts but not others, and for this reason splicing vectors can behave unpredictably. Consequently, such vectors can not now be recommended for general purposes.

3.3.2 Internal promoter vectors

A more successful strategy for the expression of two genes from a retroviral vector is to use an additional promoter inserted within the vector in order to drive expression of the more distal gene, while the proximal gene is again expressed from the viral LTR. Thus each gene has its own independent promoter (*Figure 3C*). One of the two genes is typically a utilitarian marker such as *neo* or *lacZ*, and this can be in either the proximal or distal position. The most commonly used internal promoters in MLV vectors have been the immediate early promoter of cytomegalovirus (CMV), the SV40 early promoter, and the thymidine kinase gene (TK) promoter of herpes simplex virus. Of the three, CMV is generally the strongest promoter and TK the weakest. Like the MLV-LTR, these promoters function in a broad spectrum of cell types. Promoters of cellular 'housekeeping' genes, such as phosphoglycerate kinase or β-actin, are also widely used. For certain more specialized purposes, however, it may be appropriate to choose an internal promoter that is tissue-specific.

Of the MLV-based internal promoter vectors available, at least two sets of vectors have been developed which yield high viral titres because they include the *gag* sequences needed for most efficient packaging. The 'LX' series of vectors described by Miller and colleagues (37) offers a choice of internal promoter (SV40 or CMV), a choice of selectable marker (*neo, hph*, or *hisD*), and the inserted gene can be placed in either the proximal or distal position. In the vectors described by Morgenstern and Land and designated 'pBabe'

(38), foreign genes are inserted in the proximal position such that they are expressed from the LTR. An SV40 internal promoter is in the distal position and drives expression of one of four available selectable markers (38).

Retroviral vectors using internal promoters have been used extensively and most often provide a successful strategy for expressing two genes. Naturally occurring retroviruses do not contain internal promoters, however, and a possible disadvantage of this type of vector is that potential interactions between the two promoters may result in poor expression of one of the two genes. Instances of such 'promoter interference' have been clearly documented for vectors based on other retroviruses (39), but for MLV vectors this phenomenon appears to be of minimal significance (40).

3.3.3 Dicistronic vectors

A third strategy devised for co-expression of two genes from retrovirus vectors makes use of the internal ribosome entry site (IRES) found in certain picornaviruses. This sequence permits ribosomes to initiate translation within an RNA transcript in a 5′ cap-independent manner (41). When placed between two adjacent protein coding regions, an IRES sequence can therefore allow translation of both gene products from a single dicistronic mRNA (*Figure 3D*). Several dicistronic retrovirus vectors of this sort have been described recently, including those based on MLV, and have been shown to co-express two genes with high efficiency (42–44).

The potential advantages of dicistronic vectors over internal promoter vectors are their greater simplicity, their freedom from promoter interference, and the tightly co-ordinated expression of their two genes. Such vectors are particularly suitable for ectopic gene expression studies *in vivo* in which detection of a vector-encoded histochemical marker must be taken as evidence for expression of a second gene in the same vector. For some experiments, however, the single promoter may be a limitation since a well chosen internal promoter may be more transcriptionally active than the MLV-LTR in certain target cells. For example, in many human cells the CMV promoter gives higher levels of expression than the MLV-LTR.

3.4 Self-inactivating vectors

Self-inactivating (SIN) or 'suicide' vectors are designed to eliminate effects of retroviral promoter and enhancer sequences on expression of a cloned gene under the control of an internal promoter (45). During reverse transcription and integration of these vectors, a region of the LTR containing the viral enhancer element becomes deleted (45). The proviral LTRs in infected cells are thus transcriptionally inactive. This allows unfettered expression from an internal promoter and reduces the already low probability that proviral insertion might cause detrimental effects through transcriptional activation of adjacent cellular genes. Although in principle suitable for gene therapy protocols, SIN vectors were initially reported to yield low virus titres and have not been widely used. At least one set of MLV-based SIN vectors has produced high titres, however, and specialized derivatives have been successfully used for promoter trapping in embryonic stem cells (15, 46).

3.5 Replication-competent vectors

No general purpose replication-competent mammalian retroviral vectors have yet been constructed, apparently because limitations on viral genome size and organization preclude much addition of genetic material to that required for autonomous replication. For example, the addition of a small *neo* cassette to the broad host range retrovirus SNV resulted in rapid deletion of the added sequences, presumably because of selection against the increased genome size (47). In the case of MLV, the one successful strategy has been to insert short sequences into the viral LTR, specifically the *E. coli* suppressor tRNA gene (*supF*) (48, 49). The resulting replication-competent virus has been used as a genetic and developmental marker, and for the facilitated cloning of adjacent sequences after proviral insertion. However, introduction of longer sequences within the LTRs of wild-type MLV would again have an adverse effect on genome size and has not been reported.

Although not applicable to mammalian systems, a set of general purpose replication-competent vectors is available for infecting avian cells and tissues. These vectors are based on Rous sarcoma virus (RSV), which is unique among acutely oncogenic retroviruses in being replication-competent. The organization of RSV is unusual because the oncogene *src* is expressed from its own subgenomic mRNA formed by additional splicing of the viral RNA, while the viral genes are expressed in the normal fashion (27). Hughes and colleagues have developed a series of RSV-based replication-competent vectors, designated RCAS, in which the *src* coding sequences are replaced by a site for insertion of foreign DNA (19, 20). These viruses grow to very high titres and provide an efficient means of expressing exogenous genes in chick tissues. The continuing spread of the virus may pose problems in some experiments, but this can be limited by grafting infected tissues into a recipient embryo from an RSV-resistant strain (50). Unfortunately, the restricted host range of RSV-based vectors precludes their use in mammalian systems unless cells are made susceptible to infection by ectopic expression of an RSV receptor (51).

4. Generation of recombinant virus stocks

4.1 Preparation of insert sequences

The best rule of thumb governing sequences to be inserted into a retrovirus vector is to reduce them to the minimum length required for protein expression, and so avoid unforeseen problems of expression or instability. For MLV vectors the insert must also be small enough that the total size of the viral genomic RNA to be packaged does not exceed 9–10 kb (27). The ideal arrangement is to use a fragment that encompasses the entire open reading frame, with 5′ and 3′ flanking sequences as short as conveniently possible. Particular attention should be paid to the elimination of sequences affecting transcription or processing of RNA, such as enhancers or polyadenylation signals. Additional ATG codons upstream of the intended translation start site should also be avoided as these may interfere with translational efficiency. To avoid possible complications caused by splicing, a cDNA insert is preferable to sequences containing introns. Although retrovirus vectors can accurately remove introns from genomic sequences and so be used to generate cDNA copies of cloned genes (51, 52), this process is not always reliable.

4.2 Production of helper-free stocks

To derive a virus stock from a replication-defective retroviral vector, plasmid DNA containing the proviral form of the recombinant vector is introduced into a retroviral packaging cell line. Packaging cells are typically fibroblast derivatives created by transfection with plasmids that express the viral *gag–pol* and *env* gene products from transcripts lacking the viral packaging sequence. When a vector provirus is introduced into such cells, viral proteins are thus supplied in *trans* and the vector RNA genome is packaged into virus particles. These particles bud off into the culture medium. Since the sequences encoding viral gene products should not be encapsidated, the virus stocks produced by packaging cells should be replication-defective or 'helper-free'.

4.2.1 Choice of packaging cell lines

Numerous different packaging cell lines have been described for the production of helper-free stocks of MLV-based retrovirus vectors. These are catalogued in refs 20 and 37. In choosing which packaging cell line to use, three major issues should be considered:

- the probability of producing contaminating replication-competent virus
- the desired host range of the virus vector
- the required titre of the virus vector

Although packaging cells are intended to produce helper-free virus stocks, several of the early packaging cell systems were found occasionally to produce replication-competent virus contaminating the vector stock. This apparently resulted from recombination between the vector and the viral coding sequences transfected into the cells. Several MLV-based packaging lines have now been produced in which the risk of producing helper virus is significantly lowered. To reduce homology between helper and vector sequences, the viral genes in these lines are deleted for *cis*-acting sequences at the 3′ end (31, 53, 54). In addition, the *gag–pol* and *env* genes are present on separate plasmids and are transfected sequentially rather than simultaneously. This further reduces the risk of recombination which might restore the colinear order of these genes. After transfection with retroviral vector DNA, such packaging cells would require at least three different recombination events to produce replication-competent virus. To date, packaging cells designed in this way have not been reported to produce replication-competent virus.

MLV-based packaging cells developed according to the above principles are available with a choice of three different host ranges in terms of the virus they produce.

(a) Ecotropic packaging lines, such as GP+E−86 (53), psiCRE (31), and BOSC 23 (54). As explained in section 2.3, ecotropic MLV vectors are limited to infecting murine and rat cells.
(b) Amphotropic packaging lines, such as GP-*env*Am12 (55) and psiCRIP (31), produce virus with a broader host range that includes human, simian, rodent, feline, and canine cells, as well as other species.
(c) A third host range is available from the packaging cell line PG13 (56), which produces vector virus with the host range of gibbon ape leukaemia virus (GaLV). This is similar to the broad host range of amphotropic MLV, with the notable exception of murine cells, but depends on a different cell surface receptor.

All of the packaging lines mentioned above have been reported to yield vector virus stocks with titres in excess of 1×10^6/ml after stable transfection with vector DNA. Of special note is the ecotropic packaging cell line BOSC 23 (54). This highly transfectable line has been reported to yield virus titres as high as 1.5×10^7/ml after transient transfection of vector DNA without the need to establish stably transfected lines (54). BOSC 23 cells should facilitate rapid production of high titre ecotropic virus stocks and also allow propagation of vector viruses carrying genes that are toxic to packaging cells.

4.2.2 Transfection of packaging cell lines

Since all the MLV packaging cell lines so far described have been derived from readily transfectable cells, recombinant retroviral DNA can be introduced using standard transfection procedures, such as calcium phosphate co-precipitation or DMSO/polybrene treatment. *Protocol 1* describes an example of the latter method which we have used routinely for transfection of GP+E−86, AM12, and other cell lines. With the exception of BOSC 23 cells, transient transfection procedures typically give rise to low titre virus stocks (e.g. 10^1–10^4/ml), and for higher titres (10^5–10^7) it is necessary to select stably transfected lines. This may be done using a selectable marker within the vector itself, or by co-transfection of an appropriate marker on a separate plasmid. Whichever marker gene is used, it should not be one already employed in construction of the packaging cell line itself.

When selecting transfected cells, one has the option of pooling numerous resistant cell clones or the more laborious task of isolating individual colonies.

If virus titres from pooled populations are insufficient, it is usually possible to identify cells producing higher titres by testing several different clonal lines. Individual clones of transfected producer cells will often show differences of 10–100-fold in virus titre. It is prudent to freeze two or more high titre clones at an early stage since in some cases virus titre may fall after repeated passage.

Protocol 1. DNA transfection using DMSO/polybrene

Reagents

- Packaging cells (e.g. GP+E-86)
- Purified retroviral vector plasmid DNA in sterile distilled water (at least 250 μg/ml)
- DMSO
- 10 mg/ml polybrene (Sigma) in distilled water, sterilized by filtration
- Cell culture medium: DMEM containing 10% fetal calf serum (FCS)

Method

1. One day before transfection, plate the packaging cells at a density of 3×10^5/6 cm tissue culture dish.
2. For each dish to be transfected, add 5 μg of plasmid DNA and 30 μg polybrene to 1 ml of fresh cell culture medium.
3. Aspirate the dishes and apply the DNA and polybrene mixture.
4. Incubate at 37°C with 5% CO_2 for 6 h.
5. Prepare a 25% DMSO solution by mixing 1 vol. of DMSO with 3 vol. of culture medium. Pre-warm this solution at 37°C.
6. Remove the DNA/polybrene mixture. Replace it with 2 ml of the 25% DMSO solution per dish and incubate for a further 45 sec.
7. Rinse the cells two or three times with medium before feeding with fresh medium and serum.

4.2.3 Infection of packaging cell lines

A more elaborate, but sometimes more satisfactory, procedure for generating a helper-free virus stock involves the sequential use of two different packaging lines, such that virus produced by transfection of the first is used for infection of the second. This usually requires that the vector itself carries a marker suitable for selecting the infected cells. To circumvent the intrinsic 'immunity' of the packaging cells to infection with virus of the same tropism, the virus can be passed from an amphotropic packaging line to an ecotropic one, or vice versa. If for safety reasons it is undesirable to handle amphotropic stocks of a particular recombinant, it is also possible to infect ecotropic packaging cells with ecotropic virus. In practice, the immunity to infection is not absolute, and it can be alleviated by treating the cells with sublethal doses of tunicamycin (57).

The advantages of this two-step procedure over single-step transfection of a packaging line are threefold:

(a) Proviral DNA introduced into a cell as a result of retroviral infection is often expressed at a higher level than is the identical DNA introduced by transfection (58). The reasons for this are not clear, but a consequence of the more abundant viral RNA is that viral titre from an infected packaging line can sometimes be tenfold higher than that obtained by transfection.

(b) If cells derived from individual infected colonies are used as the source of virus, the resulting stock is likely to be more genetically uniform than one originating from transfected cells. This is because transfected cells often contain multiple integrated copies of the transfected DNA, some of which may be rearranged or otherwise mutated, but all of which may contribute to the virus stock. In contrast, cells from an infected colony should contain individual copies of the proviral DNA integrated in a precisely defined manner.

(c) This method offers the opportunity to verify the structure of the proviral DNA from which all the viral genomes are immediately derived, by analysing DNA from a clonal line of infected packaging cells.

4.3 Production of replication-competent virus stocks

4.3.1 Making a helper virus stock

A stock of helper virus is made by transfecting appropriate cells with cloned DNA of a replication-competent virus in its proviral form. The transfected population is then passaged several times to allow spread of the virus until all the cells in the culture are infected. A suitable proviral clone for generating Moloney-MLV helper virus is pZAP (59). For producing high titre stocks, NIH 3T3 cells work well as hosts, as will most rapidly dividing mouse fibroblast lines. Alternatively, to avoid possible recombination with endogenous mouse retroviruses, Rat-1 cells or their relatives may be used. Production of an MLV helper stock is described in *Protocol 2*.

Protocol 2. Production of MLV helper virus stocks

Reagents

- NIH 3T3 cells plated at 3×10^5 in a 6 cm tissue culture dish
- Plasmid DNA of pZAP (59) or other plasmid clone of complete MLV genome
- Reagents for transfection (see *Protocol 1*)
- Cell freezing medium: DMEM containing 10% FCS and 10% DMSO

Method

1. Transfect NIH 3T3 cells with 5 μg pZAP plasmid DNA as described in *Protocol 1* and allow them to continue growing without selection.

2. When the cells reach confluence, split them 1:10 into two new tissue culture dishes.
3. Each time the cells become confluent, split them 1:25 and continue to passage in this manner for two weeks.
4. Harvest the virus stock as described in *Protocol 3*.
5. Freeze the cells for future use as a source of helper virus, as follows. Rinse the cells in PBS and then detach them into 1 ml of freezing medium per 6 cm dish, using a cell scraper. Pipette the cells into polypropylene freezing vials (0.5 ml/vial), place in a −20°C freezer for 2 h, and then in a −70°C freezer overnight. The following day, transfer the vials to liquid nitrogen for long-term storage.

4.3.2 Making a mixed stock of helper and recombinant virus

This can be achieved by either of two methods:

(a) A helper virus stock can be used to infect a cell line that already contains the replication-defective vector provirus (introduced either by transfection or infection), so that the vector is 'rescued' and contributes to the virus stock (see section 5 for a description of infection methodology).

(b) DNA of the helper and recombinant proviruses can be co-transfected into suitable host cells in a 1:20 molar ratio, and the cells passaged for at least ten days to allow spread of the virus throughout the culture. This will often result in an overall higher titre of recombinant virus than obtained by method (a), as well as a higher ratio of recombinant to helper virus in the mixture.

4.4 Harvesting and storage of the virus

Retrovirus stocks are obtained from packaging cell lines or productively infected cells by simply harvesting the culture medium in which the cells have been growing, as described in *Protocol 3*.

Protocol 3. Harvesting virus

Equipment and reagents

- Virus-producing cells
- Preparative ultracentrifuge with swinging bucket rotor (optional)
- Sterile 0.45 μm syringe filters (batches of filters should be tested to ensure they do not lead to reduced virus titres)

Method

1. Grow the virus-producing cells until they are 80–100% confluent.
2. Replace the culture medium with half the normal volume of fresh medium (without G418 or other selective agents).

Protocol 3. *Continued*

3. Incubate the cells at 37°C and 5% CO_2 for 12–24 h and then harvest the medium. A second harvest over the next 12–24 h can also yield high titres.
4. Remove floating cells and debris from the harvested medium by centrifugation at 5000 *g* for 5 min. Alternatively, filter the collected medium through a 0.45 μm filter. The supernatant or filtrate is now the virus stock.
5. If necessary, the virus stock can be concentrated by centrifugation at 30 000 *g* in a swinging bucket rotor for 5–16 h at 4°C. After centrifugation, carefully remove 98–99% of the supernatant with a pipette and resuspend the pellet very gently in the residual volume. A 50-fold concentration by volume will result in less than a 50-fold increase in functional titre since some viability is lost during pelleting.
6. Use the virus directly for infection (see section 5), or freeze it in aliquots at −70°C for future use. Important stocks should be stored in several small aliquots. Freezing and thawing may cause a slight drop in titre so an aliquot should be assayed after thawing to determine the working titre (see *Protocol 6*).

5. Infection of target cells

5.1 Infection with virus stocks

The brevity of this section testifies to the ease with which retrovirus vectors can be used to introduce cloned DNA into target cells, once a functional stock of virus is in hand. For fibroblasts, and many other cell types that express appropriate cell surface receptors, the efficiency of infection can be very high. The procedure is described in *Protocol 4*.

Protocol 4. Infection of target cells with a virus stock

Reagents

- NIH 3T3 cells (or other target cells) plated at 3×10^5/6 cm tissue culture dish one day before infection
- Cell culture medium
- 8 mg/ml polybrene[a] (Sigma) in distilled water, sterilized by filtration

Method

1. Make serial dilutions of the virus stock in cell culture medium containing 8 μg/ml polybrene.
2. Draw off the medium from the cells and replace with 1 ml of diluted virus per 6 cm dish. Return the dishes to the incubator for 1–2 h at 37°C.

3. Draw off the polybrene-containing medium and replace with 5 ml of fresh medium. Incubate the cells for at least 24 h (or one cell division cycle, if longer) before passaging.
4. If the virus carries a selectable marker, the cells can now be split into selective medium.

[a] Polybrene is a polycation that promotes virus binding to the cells.

5.2 Infection by co-cultivation

An alternative means of achieving efficient retroviral infection is to co-cultivate the target cells with cells that actively produce the relevant virus. This method is particularly appropriate when multiple infectious events in each cell are required, or when the efficiency of infection of the target cells is low. To prevent subsequent contamination of the target cell culture with donor cells, the latter cells can be killed by prior exposure to either mitomycin C or gamma radiation and will still produce high titre virus for several days.

Infection by co-culture is often applied when infecting primary cultures of non-adherent cells, such as haematopoietic cells, especially if the infected cells are to be returned to an animal host. This can allow efficient gene transfer without the need for lengthy drug selection *in vitro*. As a result, the ability of the cells to survive *in vivo* and home to appropriate organs is more likely to be retained. As an example, *Protocol 5* describes a procedure for infection of primary murine B cells by co-cultivation. This method has been successfully used to introduce one to five copies per cell of a retroviral vector. After adoptive transfer into recipient mice, the infected B cells persisted *in vivo*, homed to appropriate lymphoid organs, maintained the vector provirus, and expressed the exogenous gene for up to six months (60).

Protocol 5. Retroviral gene transfer into primary murine B cells

Equipment and reagents

- Primary populations of murine B lymphocytes from spleen or lymph nodes
- An MLV packaging cell line producing vector virus
- RPMI medium (Gibco-BRL) supplemented with 10% FCS and 50 μg/ml Gentamycin (Gibco-BRL)
- ^{137}Cs source for γ-irradiation (e.g. Nordion International Gammacell-40)
- 10 mg/ml polybrene (Sigma) stock solution in sterile distilled water
- 1 mg/ml lipopolysaccharide (LPS; Sigma) stock solution in RPMI medium

Method

1. Culture the murine B cells in RPMI medium supplemented with 10% FCS and 50 μg/ml Gentamycin.
2. Stimulate the cells 16 h prior to infection by adding LPS at 50 μg/ml final concentration.

Protocol 5. *Continued*

3. Irradiate confluent monolayers of the virus-producing cell line with 1600–5000 rads of γ-radiation.[a]
4. Add 10 ml of B cell suspension (10^6 cells/ml) to each 10 cm dish of irradiated virus-producing monolayers. Include 50 μg/ml LPS and 6 μg/ml polybrene in the medium.
5. Co-culture for 24 h and then recover the non-adherent B cells from the dish.
6. For *in vitro* analysis of gene transfer, culture the cells for a further 48 h in fresh medium containing 10% FCS and 50 μg/ml LPS before extracting DNA and RNA. Alternatively, the infected cells may be gently pelleted, resuspended in PBS, and used immediately for adoptive transfer into recipient mice by tail vein injection (60).

[a] The optimal radiation dose that prevents further cell division but has no effect on virus titre should be determined empirically for a given cell line.

6. Monitoring the efficiency of retroviral gene transfer

6.1 Titration of the virus stock

6.1.1 Titration by marker transduction

The most informative measures of virus titre are those which directly determine the number of infectious particles by biological assay. Stocks of vector viruses carrying a selectable or histochemical marker gene can be titred easily and accurately by determining the frequency of marker transduction upon infection of susceptible target cells. Infection is performed essentially as in *Protocol 4* but using tenfold serial dilutions of the virus stock. With selectable markers, the titre is calculated from the number of resistant colonies obtained after infection, taking into account the virus dilution and the ratio by which the infected cells were split into the selective medium. The titre is then expressed in colony-forming units (c.f.u.) per millilitre of virus. With histochemical markers such as *lacZ*, it is not necessary to passage the cells after infection but sufficient dilutions of virus should be used such that clonally derived patches of X-Gal positive cells can be clearly discerned. Virus titre is then calculated from the number of these patches multiplied by the dilution factor and is expressed as *lacZ* transducing units (TU) per millilitre.

Protocol 6. Titration of virus by marker transduction

Reagents

- NIH 3T3 cells (or other target cells), cell culture medium, 8 mg/ml polybrene (see *Protocol 4*)
- G418 (Geneticin; Gibco-BRL)
- 2% paraformaldehyde in PBS pH 7.2
- PBS: 170 mM NaCl, 3.4 mM KCl, 10 mM Na_2HPO_4, 1.8 mM KH_2PO_4 pH 7.2
- X-Gal staining solution: 1 mg/ml X-Gal, 20 mM $K_3Fe(CN)_6$, 20 mM $K_4Fe(CN)_6.3H_2O$, 2 mM $MgCl_2$, in PBS pH 7.2

A. *For vector virus carrying neo*

1. Infect NIH 3T3 or other target cells as in *Protocol 4* and continue incubating for 24–48 h before assaying for transfer of the marker gene.
2. 24–48 h post-infection, split the cells at a ratio of 1:5 and 1:25 into medium containing 400 μg/ml G418 final concentration.
3. Change the medium every three or four days.
4. Count the number of G418-resistant colonies approximately ten days after infection.

B. *For vector virus carrying lacZ*

1. Infect cells as in part A, step 1.
2. 48 h post-infection, fix the cells in 2% paraformaldehyde in PBS for 15 min at room temperature.
3. Wash the dishes three times with PBS.
4. Incubate the cells in X-Gal staining solution at 37°C for 2 h to overnight until cells containing blue precipitate are evident.
5. Count the number of clonal patches of blue cells (one to four cells each).

6.1.2 Dot blot assay for viral RNA

Virus titres can be estimated by determining the quantity of viral genomic RNA in the harvested culture medium. After concentrating the virus, this is conveniently done by a standard RNA dot blot hybridization procedure as described in *Protocol 7*. Quantitative estimates are achieved by comparing hybridization intensities with those obtained from an accurately titred control virus or from known quantities of purified RNA or DNA on the filter. By using different probes this method can also be used to estimate the titre of individual components of a mixed stock composed of vector plus helper virus.

Protocol 7. Dot blot assay for viral RNA

Equipment and reagents

- Virus-containing medium
- Nitrocellulose (or nylon) filter sheet (e.g. Schleicher and Schuell)
- 20 × SSC (SSC is 0.15 M NaCl, 15 mM tri-sodium citrate)
- Dot blot apparatus (optional), e.g. Schleicher and Schuell 'minifold' apparatus
- Radiolabelled DNA probe specific for the viral RNA
- 10 μg/ml RNase A
- Whatman 3MM filter paper
- Vacuum oven
- 5 mg/ml tRNA from yeast or *E. coli*
- STE buffer: 100 mM NaCl, 10 mM Tris–HCl pH 7.5, 1 mM EDTA, containing 50 μg/ml Proteinase K and 0.2% SDS
- TE buffer: 10 mM Tris–HCl pH 7.5, 1 mM EDTA
- 3 M sodium acetate pH 5.2
- Phenol/chloroform (1:1), equilibrated with 0.1 M Tris–HCl pH 8.0, 0.2% β-mercapto-ethanol
- Chloroform

Method

1. Clarify the virus-containing medium by centrifuging at 10 000 *g* for 10 min at 4°C.
2. Pellet the virus from the supernatant by centrifuging at 200 000 *g* for 1 h at 4°C. Drain the pellet well.
3. To each tube, add 10 μl of 5 mg/ml tRNA carrier, and resuspend the pellet in 0.4 ml of STE containing 50 μg/ml Proteinase K and 0.2% SDS (pre-incubated for 15 min at 37°C).
4. Transfer each sample to a 1.5 ml microcentrifuge tube and incubate at 37°C for 30 min.
5. Extract the sample once with phenol/chloroform and then once with chloroform.
6. Add 0.1 vol. of 3 M sodium acetate pH 5.2, and 2 vol. of ice-cold ethanol. Store at −70°C for 30 min, and then centrifuge for 10 min in a microcentrifuge to pellet the RNA.
7. Take up each RNA pellet in 20 μl of TE buffer and make a series of fivefold dilutions.[a]
8. Wet a small sheet of nitrocellulose (or nylon) filter in water, soak in 20 × SSC, blot it briefly to remove excess buffer, and place it on a sheet of Whatman 3MM filter paper.
9. Heat the RNA samples at 65°C for 5 min and spot 5 μl aliquots on to the filter. (Alternatively, a Schleicher and Schuell 'minifold' vacuum dot blot apparatus can be used.)
10. Air dry the filter and then bake for 1 h at 80°C under vacuum.
11. Hybridize by standard procedures with a probe specific for the viral sequences.

[a] To provide a control for non-specific hybridization, 10 μg/ml RNase A can be added to a duplicate sample at this stage. Allow the sample to digest at 37°C for 30 min and then process as for the other samples.

6.2 Direct analysis of DNA transfer

The overall efficiency of retroviral gene transfer into cells can be monitored by Southern blotting to determine the total quantity of proviral DNA within a population of cells after infection. The average number of proviruses per cell reflects the efficiency of retroviral infection. This can be estimated by isolating genomic DNA from infected cells 48 hours after infection and analysing 10 μg samples with a probe specific for the provirus. The average number of proviruses per cell can be determined by quantitative comparison of hybridization signals with those of control samples. The latter should contain genomic DNA from uninfected cells mixed with a range of known quantities of the retroviral vector plasmid. As an example, for a 5 kb provirus, 25 pg of cloned proviral DNA added to 10 μg of genomic DNA would correspond to approximately one provirus per cell. By using restriction enzymes that cleave the proviral DNA only within the LTRs, this analysis can also be used to verify that the proviruses are of the expected size. For vectors based on Moloney-MLV the enzymes *Xba*I and *Sac*I can frequently be used for this purpose. Note that if the infected cells are of murine origin the probe for these blots should not contain MLV sequences as these will hybridize to a complex pattern of MLV-related endogenous proviruses.

7. Potential problems

7.1 Genetic instability

An important concern with all retroviral vectors is their genetic instability. Mutations in vector DNA may arise from errors in retroviral replication, from recombination with endogenous murine retroviral sequences, from illegitimate recombination during transfection, or as a result of aberrant processing of viral RNA.

In the course of the retroviral life cycle, the genetic information in a vector genome is copied by each of three polymerases: RNA polymerase II, reverse transcriptase, and cellular DNA polymerase. There is apparently no proofreading activity associated with the first two of these enzymes and the spontaneous mutation rate in retroviruses is therefore unusually high. During a single cycle of replication of an MLV-based vector, it has been shown that a 1 kb segment of DNA is mutated in at least 1–2% of the vector proviruses (61). The mutations in question can take the form of deletions, insertions, base substitutions, frame shifts, and duplications (61). Certain sequences are prone to even higher rates of mutation, particularly direct repeats or homopolymeric stretches (61, 62). In addition to these replication errors, mutations can also result from aberrant RNA processing in retroviruses: if cryptic splice sites are present in the vector, intervening sequences may be excised, resulting in partially deleted proviruses in the infected cells.

Gross rearrangements in vector provirus DNA can be readily revealed by

Southern blotting (section 6.2) but point mutations are harder to detect and thus more insidious. The experimenter should be aware of the chance of genetic variants in a virus stock and in critical cases consider verifying the structure of a provirus in infected cells by DNA sequencing.

7.2 Expression problems

A great deal has been learned over the past decade about how to design retrovirus vectors for expression of exogenous genes, but there is always a possibility that a particular construct will not express the desired gene product as efficiently as expected. As mentioned in section 4.1, potential problems can arise from unwanted promoters, polyadenylation signals, or additional translational starts in the inserted sequences. Cryptic mutations may affect protein expression, and there may be problems of inadequate co-expression in certain double gene vectors (section 3.3). As with most expression systems, a novel construct may also behave unpredictably for unknown reasons. It is therefore prudent to check the expression of exogenous genes in retroviral vectors by analysis of their protein products, or by Northern analysis of RNA from infected cells if specific antibodies are not available. Successful expression of a marker gene is not necessarily strong evidence that a second gene in the vector is expressed.

7.3 Safety considerations

As with other infectious agents, caution should be exercised when handling retrovirus vectors in the laboratory, and safety procedures should be discussed with the appropriate institutional Biosafety committee. Potential risks are minimized by using replication-defective ecotropic vectors, since these infect rodent cells only and can not spread in an infected host. Amphotropic vectors present a greater hazard since they have the capacity to infect human cells. The inclusion of oncogenes or other biologically active sequences in a vector may add considerably to the risk factor, since even a single infected cell might have a growth advantage. It is therefore best to propagate such vectors with an ecotropic packaging cell line whenever possible.

Replication-competent retroviruses present an additional hazard because of their ability to spread in an infected host and potentially to cause insertional activation of cellular proto-oncogenes. Amphotropic stocks of murine retroviruses should be handled with particular caution. Although pathological effects of such viruses have not been documented in human hosts, it is possible that they could be oncogenic. In one report, monkeys were inadvertently infected with a broad host range MLV derivative and as a result developed T cell lymphoma (63). In view of these potential hazards, amphotropic packaging cell lines producing replication-defective vector viruses should be tested in case they are contaminated with helper virus (see below).

7.4 Contamination with helper virus

Although retroviral packaging cells are designed to propagate vector viruses in a replication-defective form, there is nevertheless a possibility that replication-competent helper virus may be produced, and such occurrences have been documented (63). This is more likely to arise if early versions of retroviral packaging cells are used (e.g. ψ2, PA12). The viral protein coding sequences in these cells were present in a single contiguous piece of DNA and a single recombination event with *cis*-acting sequences from the vector could result in a provirus capable of producing helper virus. Although the more recent packaging cells have been designed to avoid these problems (see section 4.2.1), it is wise to verify that cells producing a replication-defective vector virus are unable to produce replication-competent helper. This is particularly important:

- for safety reasons when using amphotropic vectors
- when vector stocks are to be used for cell lineage analysis or other experiments that require no virus spread
- when replication-competent MLV present in the laboratory could be a further source of contamination

7.4.1 Reverse transcriptase assay for helper virus

Since retrovirus particles contain an RNA-directed DNA polymerase, assays of reverse transcriptase activity in cell culture medium have traditionally been used to detect and titre replication-competent retroviruses. These assays are not applicable to assaying virus production from packaging cells directly, since untransfected packaging cell lines spontaneously release significant titres of polymerase-containing particles (64). However, if virus is harvested from vector-producing packaging cells and used to infect a standard fibroblast cell line such as NIH 3T3, the target cells should produce no reverse trancriptase-containing particles whatsoever unless helper virus is present. The reverse transcriptase assay is a simple method for qualitative screening of samples to see if they contain helper virus (see *Protocol 8*). For quantitative titrations, a series of known standards must be used to determine the linear range of the assay.

Protocol 8. Reverse transcriptase assay for helper virus

Reagents

- NIH 3T3 cells plated at 2×10^5/6 cm dish one day beforehand
- Cell culture supernatant or virus stock to be tested
- Cell culture medium
- 8 mg/ml polybrene (Sigma) in distilled water, sterilized by filtration
- Reaction cocktail: 60 mM Tris–HCl pH 8.3, 24 mM DTT, 0.6% Nonidet P-40, 0.7 mM $MnCl_2$, 75 mM NaCl, 6 μg/ml poly(A), 0.12 mM dTTP, 10 μCi [α-^{32}P]TTP
- DEAE paper (e.g. Whatman DE81)
- 2 X SSC: 0.6 M NaCl, 0.06 M sodium citrate
- X-ray film

Protocol 8. *Continued*

Method

1. To a dish of NIH 3T3 cells, add 1 ml of virus or medium to be tested together with 4 ml of culture medium. Add polybrene to 8 μg/ml final concentration to facilitate virus infection.
2. Incubate the cells at 37 °C overnight before changing the medium.
3. Maintain the culture for two weeks to allow the spread of any replication-competent virus. Passage the cells 1:10 as needed.
4. Harvest the virus as described in *Protocol 3* and concentrate it by centrifugation if desired.
5. Add 10 μl of the virus sample to be tested to 50 μl of reaction cocktail. Incubate at 37 °C for 1 h.
6. Spot duplicate 5 μl aliquots of each reaction mixture on to DEAE paper and allow to dry.
7. Wash the DEAE paper in 2 × SSC once for 5 min and then twice in fresh 2 × SSC for 15 min each.
8. Rinse the DEAE paper twice with 95% EtOH then allow to dry.
9. Expose the paper to X-ray film and develop the film.

7.4.2 Detection of helper virus by marker rescue

A second and more sensitive approach to detecting contaminating helper virus is by marker rescue. In this approach, supernatant from the packaging cells is used to inoculate a cell line that harbours a replication-defective vector carrying an easily assayable gene such as *lacZ* or *neo*. Supernatant from these cells is then used to inoculate fresh target cells and these are assayed for presence of the marker gene (see *Protocol 9* and *Figure 4*). Any detectable transfer of the marker indicates that helper virus is present. This procedure is orders of magnitude more sensitive than reverse transcriptase assays (such as that described in *Protocol 8*) in detecting replication-competent virus.

Protocol 9. Assay for helper virus by marker rescue

Reagents

- NIH 3T3 cell line harbouring a replication-defective retroviral vector expressing *lacZ*, e.g. BAG (6)
- Uninfected NIH 3T3 cells
- Other reagents as listed in *Protocols 3, 6,* and *8*

Method

1. Infect the *lacZ* vector-containing NIH 3T3 cells with 1 ml of virus stock to be tested, and passage the cells for two weeks, as described in *Protocol 8*, steps 1–3.

2. Harvest the culture medium from these cells as described for harvesting virus in *Protocol 3*.
3. Assay the culture medium for *lacZ*-transducing particles by infection of fresh NIH 3T3 cells followed by X-Gal staining as in *Protocol 6*.

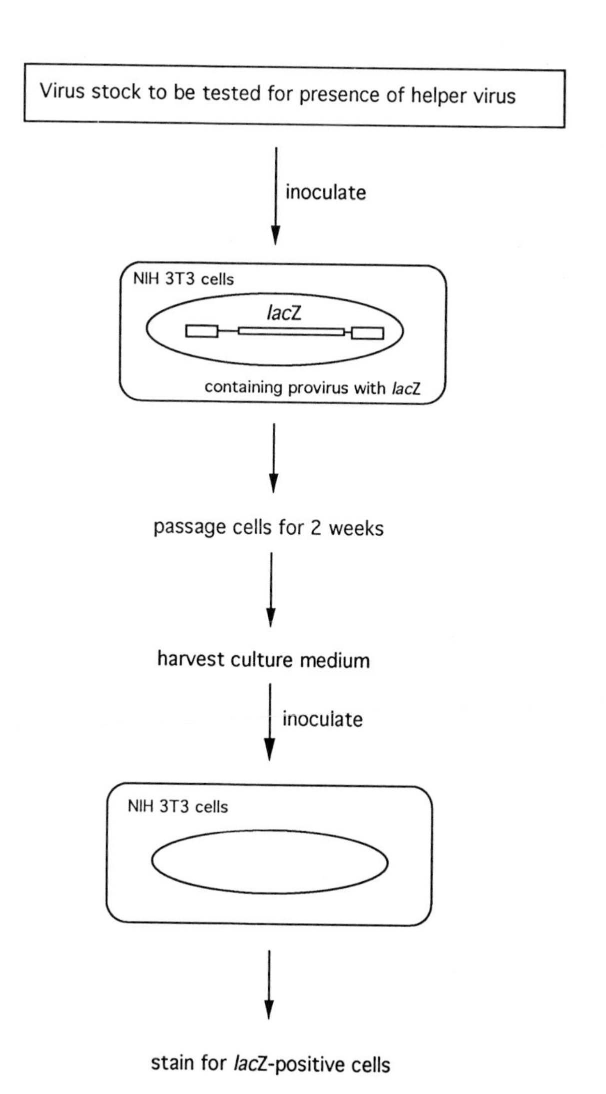

Figure 4. Marker rescue assay to determine whether a replication-defective retrovirus vector stock is contaminated with replication-competent helper virus (see section 7.4.2 and *Protocol 9* for details).

References

1. Williams, D. A., Lemischka, I. R., Nathan, D. G., and Mulligan, R. C. (1984). *Nature*, **310**, 476.
2. Anderson, W. F. (1992). *Science*, **256**, 808.
3. Morgan, B. A., Izpisua-Belmonte, J.-C., Duboule, D., and Tabin, C. J. (1992). *Nature*, **358**, 236.
4. Cepko, C. (1988). *Neuron*, **1**, 345.
5. Sanes, J. R. (1989). *TINS*, **12**, 21.
6. Cepko, C. L., Ryder, E. F., Austin, C. P., Walsh, C., and Fekete, D. M. (1993). In *Methods in enzymology* (eds P. M. Wasserman and M. L. DePamphilis), Vol. 225, pp. 933–60. Academic Press, San Diego.
7. Jahner, D., Haase, K., Mulligan, R. C., and Jaenisch, R. (1985). *Proc. Natl Acad. Sci. USA*, **82**, 6927.
8. Robertson, E., Bradley, A., Kuehn, M., and Evans, M. (1986). *Nature*, **323**, 445.
9. Jahner, D., Stuhlmann, H., Stewart, C. L., Harbers, K., Lohler, J., Simon, I., *et al.* (1982). *Nature*, **298**, 623.
10. Soriano, P., Cone, R. D., Mulligan, R. C., and Jaenisch, R. (1986). *Science*, **234**, 1409.
11. Stewart, C. L., Schuetze, S., Vanek, M., and Wagner, E. F. (1987). *EMBO J.*, **6**, 383.
12. Grez, M., Akgun, E., Hilberg, F., and Ostertag, W. (1990). *Proc. Natl Acad. Sci. USA*, **87**, 9202.
13. Gridley, T., Soriano, P., and Jaenisch, R. (1987). *Trends Genet.*, **3**, 162.
14. Friedrich, G. and Soriano, P. (1993). In *Methods in enzymology* (eds P. M. Wasserman and M. L. DePamphilis), Vol. 225, pp. 681–700. Academic Press, San Diego.
15. Friedrich, G. and Soriano, P. (1991). *Genes Dev.*, **5**, 1513.
16. Dougherty, J. P. and Temin, H. M. (1986). *Mol. Cell. Biol.*, **7**, 4387.
17. Shackleford, G. M. and Varmus, H. E. (1988). *Proc. Natl Acad. Sci. USA*, **85**, 9655.
18. Richardson, J. H., Child, L. A., and Lever, A. M. (1993). *J. Virol.*, **67**, 3997.
19. Petropoulos, C. J. and Hughes, S. H. (1991). *J. Virol.*, **65**, 3728.
20. Stoker, A. W. (1993). In *Molecular virology: a practical approach* (eds A. J. Davison and R. M. Elliott), pp. 171–97. IRL Press, Oxford.
21. Coffin, J. M. (1990). In *Virology* (ed. B. N. Fields and D. M. Knipe), pp. 1437–500. Raven Press, New York.
22. Albritton, L. M., Tseng, L., Scadden, D., and Cunningham, J. M. (1989). *Cell*, **57**, 659.
23. Miller, D. G., Edwards, R. H., and Miller, A. D. (1994). *Proc. Natl Acad. Sci. USA*, **91**, 78.
24. Roe, T., Reynolds, T. C., Yu, G., and Brown, P. O. (1993). *EMBO J.*, **12**, 2099.
25. Rohdewohld, H., Weiher, H., Reik, W., Jaenisch, R., and Breindl, M. (1987). *J. Virol.*, **61**, 336.
26. Brown, P. O. (1990). *Curr. Top. Microbiol. Immunol.*, **157**, 19.
27. Weiss, R., Teich, N., Varmus, H., and Coffin, J. (1984). *RNA tumor viruses*. Cold Spring Harbor Laboratory, Cold Spring Harbor, New York.

28. Armentano, D., Yu, S.-F., Kantoff, P. W., von Ruden, T., Anderson, W. F., and Gilboa, E. (1987). *J. Virol.*, **61**, 1647.
29. Bender, M. A., Palmer, T. D., Gelinas, R. E., and Miller, A. D. (1987). *J. Virol.*, **61**, 1639.
30. Markowitz, D., Hesdorffer, C., Ward, M., Goff, S., and Bank, A. (1990). *Ann. N.Y. Acad. Sci.*, **612**, 407.
31. Danos, O. and Mulligan, R. C. (1988). *Proc. Natl Acad. Sci. USA*, **85**, 6460.
32. Miller, A. D., Garcia, J. V., Von Suhr, N., Lynch, C. M., Wilson, C., and Eiden, M. V. (1991). *J. Virol.*, **65**, 2220.
33. Miller, D. G. and Miller, A. D. (1992). *J. Virol.*, **66**, 78.
34. Brown, A. M. C. and Scott, M. R. D. (1987). In *DNA cloning: a practical approach,* Vol. III (ed. D. M. Glover), pp. 189–212. IRL Press, Oxford.
35. Dranoff, G., Jaffee, E., Lazenby, A., Golumbek, P., Levitsky, H., Brose, K., *et al.* (1993). *Proc. Natl Acad. Sci. USA*, **90**, 3539.
36. Cepko, C. L., Roberts, B. E., and Mulligan, R. C. (1984). *Cell*, **37**, 1053.
37. Miller, A. D., Miller, D. G., Garcia, J. V., and Lynch, C. M. (1993). In *Methods in enzymology* (ed. R. Wu), Vol. 217, pp. 581–99. Academic Press, San Diego.
38. Morgenstern, J. P. and Land, H. (1990). *Nucleic Acids Res.*, **18**, 3587.
39. Emerman, M. and Temin, H. M. (1984). *Cell*, **39**, 459.
40. Emerman, M. and Temin, H. M. (1986). *Nucleic Acids Res.*, **14**, 9381.
41. Jang, S. K., Davies, M. V., Kaufman, R. J., and Wimmer, E. (1989). *J. Virol.*, **63**, 1651.
42. Adam, M. A., Ramesh, N., Miller, A. D., and Osborne, W. R. A. (1991). *J. Virol.*, **65**, 4985.
43. Morgan, R. A., Couture, L., Elroy-Stein, O., Ragheb, J., Moss, B., and Anderson, W. F. (1992). *Nucleic Acids Res.*, **20**, 1293.
44. Boris-Lawrie, K. A. and Temin, H. M. (1993). *Curr. Opin. Genet. Dev.*, **3**, 102.
45. Yu, S. F., von Ruden, T., Kantoff, T. W., Garber, C., Seiberg, M., Ruther, U., *et al.* (1986). *Proc. Natl Acad. Sci. USA*, **83**, 3194.
46. Soriano, P., Friedrich, G., and Lawinger, P. (1991). *J. Virol.*, **65**, 2314.
47. Gelinas, C. and Temin, H. M. (1986). *Proc. Natl Acad. Sci. USA*, **83**, 9211.
48. Lobel, L., Patel, M., King, W., Nguyen-Huu, M., and Goff, S. (1985). *Science*, **228**, 329.
49. Reik, W., Weiher, H., and Jaenisch, R. (1985). *Proc. Natl Acad. Sci. USA*, **82**, 1141.
50. Fekete, D. M. and Cepko, C. L. (1993). *Proc. Natl Acad. Sci. USA*, **90**, 2350.
51. Bates, P., Young, J. A., and Varmus, H. E. (1993). *Cell*, **74**, 1043.
52. Brown, A. M. C., Wildin, R. A., Prendergast, T. J., and Varmus, H. E. (1986). *Cell*, **46**, 1001.
53. Markowitz, D., Goff, S., and Bank, A. (1988). *J. Virol.*, **62**, 1120.
54. Pear, W. S., Nolan, G. P., Scott, M. L., and Baltimore, D. (1993). *Proc. Natl Acad. Sci. USA*, **90**, 8392.
55. Markowitz, D., Goff, S., and Bank, A. (1988). *Virology*, **167**, 400.
56. Miller, A. D., Garcia, J. V., Von Suhr, N., Lynch, C. M., Wilson, C., and Eiden, M. V. (1991). *J. Virol.*, **65**, 2220.
57. Miller, A. D., Trauber, D. R., and Buttimore, C. (1986). *Somatic Cell Mol. Genet.*, **12**, 175.
58. Hwang, L. S. and Gilboa, E. (1984). *J. Virol.*, **50**, 417.

59. Hoffman, J. W., Steffen, D., Gusella, J., Tabin, C., Bird, S., Lowing, D., *et al.* (1982). *J. Virol.*, **44**, 144.
60. Suteowski, N., Kuo, M.-L., Varela-Echavarria, A., Dougherty, J. P., and Ron, Y. (1994). *Proc. Natl Acad. Sci. USA*, **91**, 8875.
61. Varela-Echavarria, A., Prorock, C. M., Ron, Y., and Dougherty, J. P. (1993). *J. Virol.*, **67**, 6357.
62. Burns, D. P. W. and Temin, H. M. (1994). *J. Virol.*, **68**, 4196.
63. Donahue, R. E., Kessler, S. W., Bodine, D., McDonagh, K., Dunbar, C., Goodman, S., *et al.* (1994). *J. Exp. Med.*, **176**, 1125.
64. Mann, R., Mulligan, R. C., and Baltimore, D. (1983). *Cell*, **33**, 153.

5

Genetic manipulation of embryonic stem cells

AMY MOHN and BEVERLY H. KOLLER

1. Introduction

Embryonic stem (ES) cells are pluripotent stem cell lines established from mouse embryos cultured *in vitro* to the early post-implantation stage of development (1, 2). ES cell lines are established by transferring blastocysts to tissue culture plates seeded with a feeder layer consisting of either primary embryonic fibroblasts or cells of an established embryonic fibroblast line. Under these conditions, the embryo continues to grow, hatching from the zona pellucida and attaching to the feeder layer. As the trophectoderm cells spread out, the cells of the inner cell mass, which would normally give rise to the embryo proper, are exposed to the feeder layer. In many cases, the cells of the inner cell mass begin to proliferate upon contact with the feeder layer, forming a small clump of cells, which can be picked away from the trophectoderm cells and expanded. After several weeks of growth in culture, these cells often reach a point where they can be propagated using established tissue culture techniques (3). With proper tissue culture conditions, an ES cell line established by this technique can be maintained in a pluripotent, undifferentiated, and euploid state. This maintenance of pluripotency is critical to the applications of ES cells in biological research.

Events of early embryonic development that would otherwise be inaccessible can be visualized by allowing ES cells to differentiate *in vitro*. In the absence of embryonic fibroblasts, ES cells spontaneously differentiate into embryoid bodies at a stage of development equivalent to embryonic day six to eight, and can differentiate into specialized cell types, including heart tissues, smooth and skeletal muscle, cartilage, melanocytes, and neurones (4). By exploiting this property of ES cells, researchers can identify the genetic elements involved in aspects of differentiation and development.

Although ES cells provide a useful *in vitro* system for studying differentiation and development, their major role in biological research is dependent on the fact that, when introduced into mouse embryos at an early stage of development, they can potentially contribute to all of the tissues of the

Table 1 Applications of ES cells for *in vitro* and *in vivo* genetic analysis

Application	Reference
1. Gene targeting	
• Creation of a null mutation by gene disruption	13, 20, 21, 52, 53
• Creation of a null mutation by deletion	22
• Introduction of a subtle mutation (see section 3.3)	27, 28
• Recombinase-mediated gene targeting	30, 31
2. Introduction of yeast artificial chromosomes (YACs) into the ES cell genome	49, 50
3. Introduction of transgenes for analysis of differentiation *in vitro*	62
4. Identification of developmentally-regulated genes by gene trapping	60, 61

resulting mouse (5). Thus, any mutations that have been introduced into the ES cell genome *in vitro* can be incorporated into the genome of whole mice (6, 7). The advantage for this route of transgenesis is that rare genetic events can be identified in culture and isolated before re-introduction into embryos. For this reason ES cells are used for experiments such as gene targeting, or gene trapping, where screening procedures are required. *Table 1* provides a reference for published applications of ES cells.

This chapter will focus on the techniques used for the introduction of DNA into ES cells, a necessary step for most ES cell applications. The transfection method will vary according to the desired genetic outcome. We provide protocols for three common methods effective for ES cell transfection: electroporation, calcium phosphate co-precipitation, and lipofection. Following these are discussions and protocols for the selection, identification, and analysis of recombinant ES cell lines. Finally we describe the preparation of ES cells for microinjection into mouse embryos for the creation of chimeras.

2. Maintenance of ES cells in culture

It is the failure to maintain ES cell lines in a pluripotent state, rather than an inability to successfully implement various transfection protocols, that most often represents a stumbling block to groups beginning work in this field. Therefore, before covering the various techniques by which ES cells may be transfected, the techniques that have proven successful in maintaining ES cell in a pluripotent state will be discussed in some detail.

A number of factors influence the extent to which a given ES cell clone will contribute to germline and somatic tissues. These include:

- the mouse strains from which ES cells and host blastocysts are derived
- the method by which ES cells are introduced into the host embryo
- the sex of the ES cells and host blastocysts

- the conditions under which the ES cells are maintained in culture
- the chromosomal complement of the ES cells
- the accumulation of as yet undefined alterations with time in tissue culture

The latter three factors contribute to what is generally referred to as the 'quality' of the ES cells. While especially critical in cases where one needs germline transmission of the ES cell genome, the quality of the ES cells is also important if one wishes to generate significant numbers of chimeric mice or to study ES cell differentiation *in vivo*.

Maintenance of ES cell lines in an undifferentiated, pluripotent state is likely to depend on a number of factors, only a few of which have been identified to date. Differentiation of ES cells can be prevented by growth of cells in the presence of medium conditioned by certain cell lines, such as Buffalo rat liver (BRL) cells (8). At least one factor capable of inhibiting differentiation has been purified from BRL cell conditioned medium; originally termed differentiation-inhibition factor (DIF), this protein was found to be identical to a previously purified factor called leukaemia-inhibitory factor (LIF) (9–11). Although it has been subsequently shown that this factor alone is sufficient to prevent differentiation of ES cells (12), many laboratories, including our own, find that culture on feeder layers of primary embryonic fibroblasts yields high quality ES cells with greater consistency.

Use of established embryonic fibroblast lines such as STO cells is also common. However, STO cells from different sources may differ in their ability to maintain ES cells, and it is important to obtain these cells from a reliable source. While STO cells are easier to maintain than primary embryonic fibroblasts because they are a transformed cell line, it is likely that their ability to prevent ES cell differentiation changes with time in culture. Soriano and co-workers described the isolation of STO cells transfected with the gene encoding DIF (13). The success of this STO cell line in maintaining ES cells is probably a combination of the expression of this gene and the clonal selection of cells ideal for growth of ES cells.

Nagy and co-workers have tried to define the culture conditions that maximize the ability of ES cells to contribute to the murine germline. They compared the ability of ES cells grown on embryonic feeders, in DIF, and on a STO cell line to give rise to viable offspring when introduced into tetraploid embryos. Using this assay, no differences could be attributed to these different culture conditions (14).

Another important factor contributing to the pluripotency of ES cells is the maintenance of a normal number of chromosomes during culture. Although the normal euploid chromosome constitution is very stable in ES cells, and cells can often be maintained in culture for months with very little change, it is prudent to monitor chromosome number to ensure euploidy before returning cells to the embryo. In most cases, altered chromosome number will decrease both the ability of an ES cell line to generate chimeric mice and the efficiency with which the ES cell genome is transmitted to the progeny of chimeras. Often an abnormal karyotype yields sterile chimeric males.

Even in ES cell lines that maintain a normal number of chromosomes, there is a tendency toward loss of pluripotency as cells continue to grow and divide *in vitro*. This loss of pluripotency results in decreases in the number of chimeras born, decreases in the level of chimerism of animals that are born, and a reduced ability of chimeric mice to transmit the ES cell genome to their offspring (14). Groups working with low passage number ES cells frequently report greater success in obtaining germline chimeras than groups using later passage cultures from the same lines. This problem can often be averted by occasional subcloning and identification of cells with the same characteristics of earlier passage cells.

The best direct evidence for alterations, other than abnormal chromosome number, that affect pluripotency also comes from the work of Nagy *et al.* (14). Using tetraploid embryos, it was found that the majority of established ES cell lines gave rise to embryos that died either during development or shortly after birth. Additional experiments using a series of newly established ES cell lines indicated that the fate of embryos derived from any given line was largely dependent upon the length of time that the particular line had been maintained in tissue culture. ES cells up to approximately passage 14 were able to give rise to live offspring, although the efficiency with which this occurred varied greatly from line to line. Moreover, the decrease in pluripotency for a given ES cell line with time in culture was not an all or nothing phenomenon, since subclones which resembled low passage number ES cells in terms of pluripotency could be isolated from high passage number cultures. This supports the idea that ES cell cultures are not uniform, but instead are composed of a mixture of cells which vary in their pluripotency.

Loss of pluripotency in ES cells can be an especially frustrating problem because of the large time lag that is often involved between the occurrence of the phenomenon and its discovery. In many cases, loss of pluripotency caused by improper handling of an ES cell line does not become apparent until several months of work have been invested in failed attempts to produce germline chimeras. Many of the problems associated with loss of pluripotency can be avoided by paying careful attention to the considerations listed in *Table 2*.

3. Targeting strategies

DNA constructs used to transfect ES cells are often designed to introduce a specific type of mutation into a particular gene, and a number of strategies for achieving this type of gene targeting have been devised. In each of these strategies, the targeting construct integrates by homologous recombination to replace native sequences, the precise nature of the resulting mutation being determined by the particular design of the construct. Although the types

Table 2 Getting started: some considerations when beginning ES cell culture

Selection of feeder cells
It is necessary to decide whether ES cells will be grown on STO cells, embryonic feeder layers, or LIF. If feeder cells are used, it is important to remember that they must be able to survive in the media in which transformed ES cells will be selected. STO cells and mice carrying a neomycin-resistance transgene are available. Selection for transfectants in HAT medium requires only normal feeder cells. Mice deficient in *hprt* are available for the generation of feeders for selection of 6-thioguanine-resistant cells. The choice of feeder cells may also be influenced by the tissue culture conditions being used by the laboratory from which the ES cells are obtained and by whether germline transmission is critical to the experiments being planned. Some researchers choose to maintain ES cells on feeder layers except during periods of selection, when pluripotency is maintained by growth in medium containing LIF.

Identification of a suitable lot of fetal bovine serum
Protocols for testing serum lots for the ability to support growth of ES cells have been described (3). One should note that ES cells are required for these tests. It is therefore advisable to request at least two vials of cells on initiating these experiments. One vial, perhaps of high passage number, can be used to establish conditions before thawing and expanding the cells that will become the laboratory stock. Commercial sources of serum that has been pre-tested on ES cells will simplify this procedure.

Source of ES cells
Numerous ES cell lines have been described. We generally have used lines which proved to be successful not only in the laboratory in which they were first isolated but also in the hands of a number of investigators that obtained the lines from this source.

Testing of ES cells
It is advisable to test cells for alterations in karyotype upon receiving them and after culturing for a period of time in conditions established for your laboratory (3). The karyotyping carried out in our laboratory is designed only to establish chromosome number and will not detect small translocations or deletions. We also check the karyotype of each transfectant that is being used to generate chimeras. Mycoplasma testing is also advisable, as it will affect many aspects of work with ES cells, from isolation of transfectants to the ability to obtain chimeric animals. Many protocols, including commercial PCR kits, are available for testing for mycoplasma.

of modifications produced by gene targeting were originally limited to the creation of null mutations by gene disruption, more sophisticated manipulations of the genome have been achieved in the past. In addition to the deletion of one or more genes, current strategies allow more subtle modifications, including point mutations and tissue-specific targeting events, to be incorporated into the murine genome. This section will cover basic design of targeting constructs, as well as strategies for using different types of targeting constructs to introduce specific types of mutations into the mouse genome.

3.1 Design of targeting constructs

Because the design of an appropriate targeting construct is the most critical step in successful gene targeting, this section will deal with factors that must be taken into consideration when designing such a construct. These factors include those that determine the type of mutation produced when the construct undergoes recombination with homologous ES cell sequences, as well as those that influence targeting frequency and ability to detect targeting events.

The type of mutation that is produced by a targeting event will be determined by whether the targeting construct used is a replacement-type vector or an insertional-type vector. In both types of vectors, homologous recombination is driven by a region or regions in the incoming DNA which can be aligned directly with homologous sequences in the endogenous target gene. The particular linear arrangement of the homologous sequences in the incoming DNA will determine the outcome of the targeting experiment. The replacement-type construct is linearized outside of the region of homology prior to transfection, resulting in cross-over events in which the endogenous DNA is replaced by the incoming DNA. The insertional vector, on the other hand, is linearized in the region of homology with the endogenous locus, resulting in the insertion of all plasmid sequences into the locus. *Figure 1* demonstrates the recombination events for insertional and replacement vectors. The way in which these two types of constructs can be used to produce specific types of mutations will be discussed in more detail in subsequent sections.

While the type of targeting construct chosen will determine the nature of the resulting mutation, the actual efficiency with which a given construct can be used to obtain targeted cell lines will depend primarily on two properties of the targeting construct. The first of these is the absolute targeting frequency that can be achieved with the construct, and the second is the ease with which targeted cells can be distinguished from both non-transformed cells and non-targeted cells that have been transformed by integration of the targeting construct at random sites in the genome. Each of these properties will be considered in turn.

3.1.1 Targeting frequency

The targeting frequency has been shown to depend on a number of factors, including:

- the length of the homologous sequences in the targeting construct
- the degree of homology between the sequences in the targeting construct and the endogenous locus
- the locus or particular region of the locus being targeted
- possibly, whether the targeting vector is of the insertional or replacement type

Targeting frequency has been shown to increase as the length of sequence homology between the targeting vector and the locus increases, until a plateau

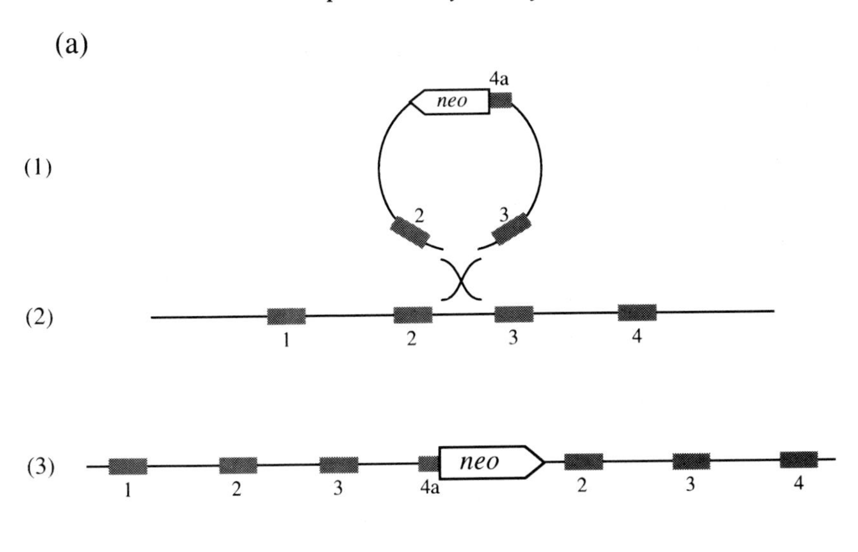

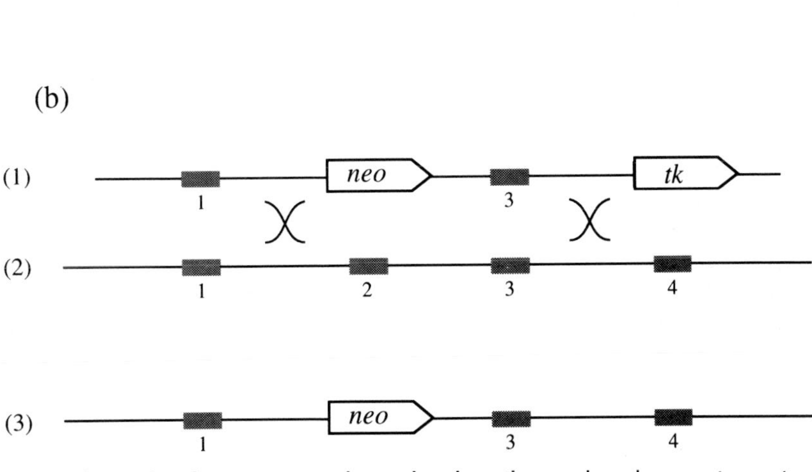

Figure 1. Strategies for gene targeting using insertion and replacement constructs. (a) Gene targeting with an insertion construct. (1) Targeting construct with two arms of homology to the locus and a neomycin resistance gene, *neo*, for positive selection of integration events. The construct is linearized within a region of homology. (2) The endogenous locus. (3) The locus following homologous recombination with the targeting construct. The gene has been inactivated by the insertion of the targeting plasmid in which exon 4 has been disrupted by *neo*. Note, however, that the duplication of exons following targeted integration can allow alternate mRNA splicing events that can yield a functional mRNA. (b) Gene targeting with a replacement construct. (1) The targeting construct with two arms of homology to the locus, *neo* for positive selection of integration events, and HSV-*tk* for selection against random integrations. (2) The endogenous locus. (3) The locus following homologous recombination with the targeting construct. The gene is inactivated when the endogenous exon 2 is replaced with *neo* from the targeting construct. The HSV-*tk* gene is not integrated during homologous recombination, keeping targeted integrants resistant to gancyclovir.

in targeting frequency is reached between 10–14 kb (15). This plateau may reflect a limit on the size of intact DNA fragments that can be introduced into the cells, rather than a limit in the effect of length of homology on targeting frequency.

Targeting frequency also depends on the degree of homology between the sequences in the targeting construct and the endogenous locus, as demonstrated by the fact that even small differences in sequence between the targeting vector and the target locus can reduce the frequency with which homologous recombination takes place (16, 17). This fact becomes important when sequences incorporated into a targeting construct are derived from a different mouse strain than that used to generate the ES cell line. Sequence variation, especially in non-coding regions, exists among different strains of mice and will affect the degree of homology between the targeting construct and its genomic counterpart. Thus, variations in the reported targeting frequencies for many loci may result from the fact that, until recently, consistent attempts were usually not made to ensure that sequences used in targeting constructs were isogenic with the ES cells into which they were transfected. Although the effect of using non-isogenic DNA may vary for different loci, it is now considered expedient to prepare targeting constructs from genomic fragments isolated from the same strain in which one hopes to achieve targeting.

Targeting frequencies for insertional and replacement vectors have also been compared, with different groups coming to different conclusions (15, 18). The most comprehensive of these studies analysed targeting frequencies in 13 replacement and 9 insertion vectors designed to correct a defective HPRT gene (15). This study found no consistent difference in frequency between the two types of vector, although it is still possible that some difference will appear between the two targeting types of constructs when other loci are analysed.

The fact that targeting frequencies can be influenced significantly by factors other than those discussed above is demonstrated by experiments in which vectors constructed with isogenic DNA of similar length and configuration were used to target different loci. The differences in targeting frequency determined in these experiments for different loci indicate that other factors affecting targeting frequency have yet to be defined.

3.1.2 Identification of targeting events

The second property of a targeting construct that will influence the efficiency with which it can be used to produce a targeted ES cell line is the set of features that it incorporates to allow identification of targeting events among the more common outcomes of transfection protocols. Except perhaps in cases where transfection is achieved by microinjection or infection with viruses, the most common outcome of transfection protocols is a failure to introduce DNA. Therefore it is essential in almost all cases that a selectable marker be incorporated into the targeting construct or co-transformed with the targeting construct to select for cells that have taken up DNA. In the case of co-transformation, a subset of the drug-resistant clones will have a targeted integration of the construct as well as a tandem or unlinked insertion of the selectable marker.

In all cases, selection for the marker should yield a mixed population of surviving clones that have incorporated the marker either by homologous recombination or by random integration, and a number of strategies have been devised to distinguish targeted cells from those that have taken up the marker by random integration. These include the use of promotorless (19) or poly A minus (20) selectable marker genes, and the use of positive-negative selection schemes (21). The advantage of the former strategy is that the marker gene is expressed only when it integrates into an expressed gene. However, use of the latter strategy is much more widespread, and it has the advantage of being applicable to all loci in which targeting is attempted. In the positive-negative selection scheme, a second, negatively selectable marker gene is incorporated at one or both ends of the transfected construct. The negative marker will be eliminated if the construct is incorporated by homologous random integration but retained in many cases where the construct is randomly integrated into the ES cell genome. This allows a percentage of the cells that have randomly integrated the targeting construct to be eliminated by negative selection, thus enriching for targeting events. Selectable marker genes and methods of selection are considered in more detail in section 4.

While the selection procedures described above will significantly reduce the number of colonies that must be screened, some scheme for the final identification of the targeted colonies must be incorporated into the design of the targeting construct. Generally, this consists of changes caused by integration of the construct into the targeted locus that can be detected either by a PCR assay or by Southern blot analysis. For most loci, the targeting frequency is such that Southern blot analysis is the most efficient means for final identification of targeted cell lines. Even in cases where PCR is used to screen for targeting events, it is still necessary to confirm targeting events by Southern analysis, and this fact should be taken into consideration when designing the targeting construct.

3.2 Creation of null mutations

Gene targeting in ES cells has been used most commonly as a means of creating mouse lines in which expression from a particular gene has been disrupted. The simplest and most common strategy for accomplishing this is to design a replacement targeting construct in which the selectable marker replaces a portion of the coding sequence of the disrupted gene. In this type of construct, the selectable marker fulfils two functions:

- it allows screening for ES cells that have integrated the targeting construct

- in cases where the targeting construct has integrated by homologous recombination, the selectable marker replaces a portion of the coding sequence of the target locus, thereby disrupting normal gene expression

The portion of the endogenous gene replaced by the selectable marker may vary from situations where none of the endogenous sequence is lost, to those in which 15 kb or more of genomic sequence is replaced (22), although it should be noted that the effects on targeting frequency of replacing extremely long stretches of the targeted locus has not yet been systematically analysed. However, it has been found that targeting frequency is not influenced significantly by the size of the marker gene (23), which may vary from 2 kb for the *neo* gene to 6 kb for the *hprt* gene.

In choosing the region of the gene to target, one should be aware that in many cases the cellular splicing machinery will splice out a disrupted exon, resulting in mRNA that lacks one exon. Thus a targeted gene may still be capable of expressing functional protein if the amino acids encoded by the disrupted exon are not essential for function. This same phenomenon makes it difficult to disrupt gene function completely using an insertional-type vector, since this type of vector creates a duplication and it is often difficult to design the vector so that the duplicated sequences do not contain one intact exon. This is especially true if exons in the targeted regions are small. While it is not impossible to create an insertional-type targeting construct that will yield a null mutation with a high degree of certainty, designing such a vector usually requires extensive knowledge of the exon/intron organization of the region to be targeted.

3.3 Introduction of subtle mutations

Often the goal of a targeting experiment is to introduce a subtle mutation into a gene, altering only a few bases instead of completely eliminating gene function. The goal is to create a mutation that in every other sense mimics the normal expression pattern for the gene. The necessity of including a selectable marker becomes problematic in these situations because the marker can not be placed within the gene, and in some cases it is not possible to design the construct to place the marker outside the gene. A number of strategies have been devised to overcome this problem for the creation of subtle mutations. They include:

(a) Introducing DNA into ES cells by microinjection rather than electroporation.
(b) Placing the selectable marker and targeting construct on separate plasmids (co-transfection).
(c) Using a two-step targeting strategy in which the selectable marker gene is removed from the targeted locus by a second homologous recombination event. Two-step targeting strategies include the 'in-out' method, in

which the marker gene is removed from the targeted locus by a spontaneous intrachromosomal recombination event, and 'double targeting', in which the marker gene is removed by homologous recombination of a second targeting construct with the targeted locus.

3.3.1 Microinjection

Unlike the other methods mentioned above, microinjection eliminates the need for a selectable marker by ensuring that virtually 100% of the cells receive DNA. However, only one instance of successful gene targeting by microinjection in ES cell has been reported, and the resulting targeted ES cells did not yield germline chimeras (24).

3.3.2 Co-transfection

Co-transfection attempts to circumvent the problem by transforming cells with a marker gene unlinked to the targeting construct. It has been demonstrated that the integration of the selectable marker can be accompanied by a targeted integration of the targeting construct for a percentage of the cells (25, 26). However, for reasons that remain unclear, cells in which gene targeting occurs are less likely than expected to randomly integrate exogenous DNA at other loci (26). Therefore, the enrichment for targeting events achieved with co-transfection is lower than that achieved by incorporation of the selectable marker into the targeting construct. It is for this reason that co-transfection has not been used extensively for introducing subtle mutations into the mouse genome.

3.3.3 Two-step targeting

The two methods which rely on two rounds of homologous recombination, 'in-out' and 'double targeting', are more widely used than co-transfection or microinjection. Both of these methods use a targeting construct that incorporates a marker gene system that allows both positive and negative selection. This system normally consists of one marker gene whose expression in ES cells can be selected for or against. It is also possible to incorporate two marker genes into the construct, one for use in positive selection and the other for use in negative selection. Cells transfected with the construct are initially subjected to positive selection to eliminate cells that have not incorporated the targeting construct. Once targeted cell lines have been isolated, the selectable marker is removed by a second round of homologous recombination. Enrichment for ES cells in which this has occurred is accomplished by negative selection to eliminate cells that contain the marker gene. The in-out method differs from the double targeting method in the design of targeting construct that is used to achieve the initial targeting event, and in the means by which the marker gene is removed during the second round of homologous recombination. *Figure 2* demonstrates the targeting strategies for subtle mutations.

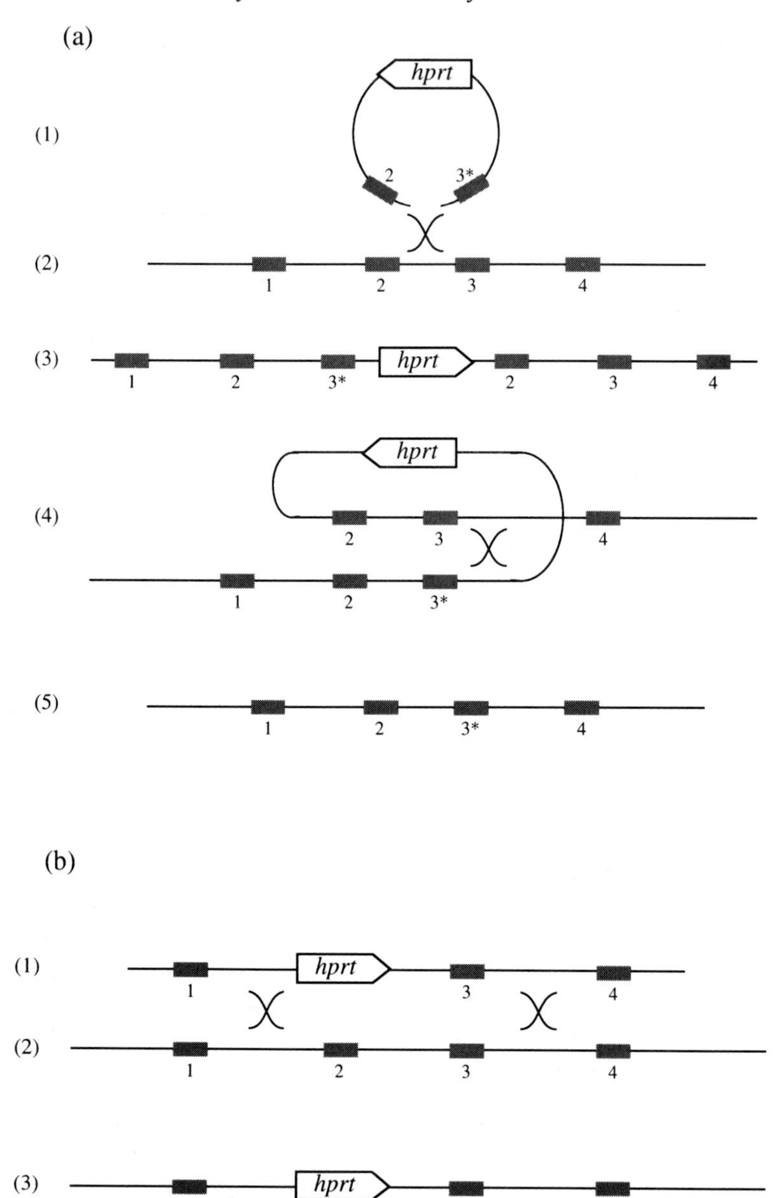

(a)
(1)
hprt
2
3*
(2)
1
2
3
4
(3)
1
2
3*
hprt
2
3
4
(4)
hprt
2
3
4
1
2
3*
(5)
1
2
3*
4
(b)
(1)
1
hprt
3
4
(2)
1
2
3
4

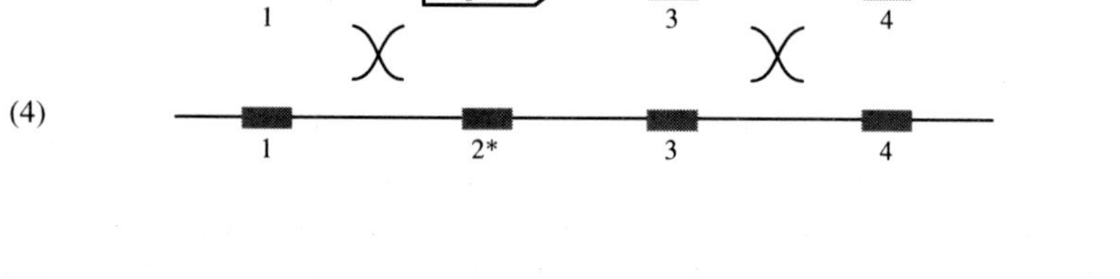

(3)
1
hprt
3
4
(4)
1
2*
3
4

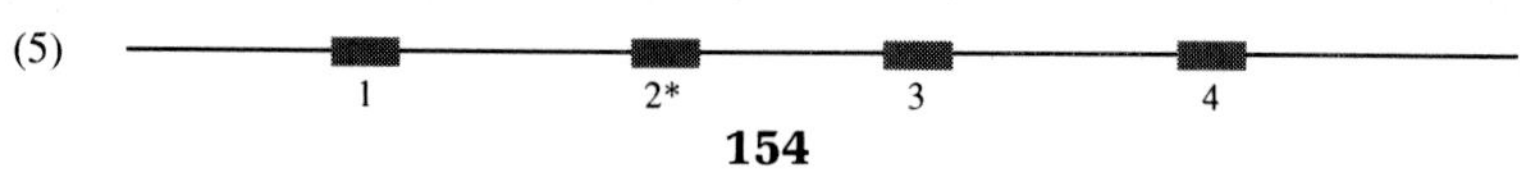

(5)
1
2*
3
4

Figure 2. Strategies for introducing subtle mutations by homologous recombination. (a) The 'in-out' strategy. (1) Insertion construct having two arms of homology to the locus and a selectable marker gene that can be used for both positive and negative selection. Although *hprt* is shown in the figure, tandem *neo* and HSV-*tk* genes can also be used for this purpose. The construct also carries a planned mutation (*) in exon 3. (2) The endogenous locus. (3) The locus following targeted integration of the insertion construct. Positive selection is used to identify recombinants carrying the selectable marker gene. (4) Intrachromosomal recombination event resulting in the removal of the selectable marker gene and duplicate sequence. Negative selection is used to identify cells that have undergone this recombination and have lost the selectable marker gene. (5) The locus following the second recombination event. Depending on the site of recombination, the locus will have either the original exon 3 or the desired mutation. (b) The 'double targeting' strategy. (1) Replacement construct having two arms of homology and a marker gene that can be used for positive and negative selection. In contrast to the 'in-out' strategy, the desired mutation does not have to be included in this construct. (2) The endogenous locus. (3) The locus following targeted integration of the construct. Positive selection is used to identify integrants that carry the selectable marker gene. The purpose of this first recombination event is to 'tag' the locus for later manipulation. (4) A second replacement construct which lacks a selectable marker gene and carries the desired mutation in exon 2 (*). (5) The locus following the second homologous recombination event. Selection against the marker gene will identify cells that have lost the marker as a result of the second recombination event. The locus now differs from the wild-type only by a subtle mutation in exon 2.

The in-out method uses an insertional-type vector for the first round of homologous recombination to introduce the desired mutation (27, 28). This leads to a duplication of the targeted sequence, as well as an insertion of the marker gene into the target locus. Targeted cell lines are isolated after enrichment by positive selection for the marker gene, in the same manner described in the previous section for lines carrying targeted null mutations. However, because of the duplication caused by the insertional vector, targeted cells will have the capability of deleting the marker gene, as well as one copy of the duplicated target sequences, by intrachromosomal recombination. Although this event will be relatively rare, enrichment for cells in which it has occurred can be accomplished by negative selection against the marker gene. In a certain percentage of cases, the deletion will leave behind the copy of the target locus which contains the desired mutation, and the frequency with which this occurs will depend on the design of the original targeting construct.

The double targeting strategy uses a replacement-type vector as described in section 3.2 to introduce a marker gene into the region where the subtle mutation will eventually be made. Once a targeted cell line has been identified, a second targeting experiment is carried out. The targeting construct is another replacement-type construct that replaces the selectable marker with the desired mutation upon targeted integration. As in the in-out method, cells in which the desired event has occurred can be enriched for by negative selection against the marker gene.

There are benefits and drawbacks to both of these systems, and the method that works best may depend on both the construct and the target locus. If the absolute targeting frequency for a locus is low, the in-out method may provide the better alternative. The rate of re-targeting in the double targeting method may be much lower than the frequency of spontaneous mutations that inactivate the marker gene, and this would generate large numbers of background ES cell clones that survive negative selection. While this background will exist with in-out constructs, the advantage is that all targeted cells can participate in the intrachromosomal event. In the double targeting method, only the small fraction of cells that are transformed with the second targeting construct are even candidates for the desired targeting event. In-out targeting is not without difficulties, however; the major drawback of the in-out method stems from difficulties inherent in designing an insertional targeting construct that will yield the desired results. The mutation must be placed close to the selectable marker so that it will be integrated during the first insertion step along with the marker, but must be far enough away that it will be left in the genome after the second recombination event removes the marker gene. In many cases, it has proven difficult to design vectors that yield both acceptable targeting frequencies and also result in the mutation being left in the target locus when the intrachromosomal recombination takes place.

3.4 Recombinase models for targeting

Mutations introduced through ES cells are transmitted in the germline and are consequently present in all cells of the ensuing progeny. While this is generally the desired effect, it does prohibit the study of the role of a mutation in one cell type independent of others. It is especially problematic when the mutation results in a lethal phenotype as a result of loss of expression in a tissue other than the one of interest. This difficulty can be overcome with tissue-specific gene targeting, a process that can be accomplished with site-specific recombinases.

Two site-specific recombinases, *S. cerevisiae* FLP recombinase and bacteriophage P1 Cre recombinase, have gained recent attention for their ability to mediate planned manipulations of the mammalian genome (29, 30). As members of the integrase family of proteins, both recombinases recognize two short stretches of DNA in a sequence-specific manner and catalyse the deletion or integration of DNA flanked by these recognition sites. The placement and orientation of these recognition sites determines the nature of the recombination event; sequence flanked by direct repeats will be deleted, while sequence flanked by inverted repeats will be inverted (31). Integration of exogenous DNA occurs precisely within a chromosomal recognition site (32). Both systems have been used successfully in ES cells; FLP recombinase and Cre recombinase have been applied in ES cells to remove the selectable marker from a targeting construct after integration (33), and the Cre/lox P system has also been used in ES cells to delete genomic sequence in a planned fashion (34).

It has also been demonstrated that Cre recombinase is capable of mediating planned recombination events in transgenic animals, since mice expressing Cre have been found to remove chromosomal sequence flanked by the appropriate recognition sites very efficiently (35, 36). The process is made tissue-specific by placing recombinase expression under the control of a tissue-specific promoter.

The predictable and efficient manipulations catalysed by these recombinases make them attractive for applications in gene targeting and transgenesis. Gu *et al.* (37) have demonstrated that these recombination systems can be applied to ES cell technology for the production of transgenic mice with a null phenotype restricted to a specific tissue type. In this way, defects that would be developmentally lethal can be examined in the limited setting of a particular organ or tissue type. Tissue-specific targeting first requires the introduction of flanking recognition sites by homologous recombination at the gene of interest in ES cells. Once the gene is tagged for deletion, a recombinase gene with a tissue-specific promoter can be introduced, either by ES cell transgenesis or by pronuclear injection. Both methods would require planned matings to generate an animal carrying two tagged genes and expressing the recombinase in restricted tissues. In these animals, the recombinase can delete both alleles of the gene in a tissue-specific way.

4. Methods of selection

Low transfection frequencies and even lower targeting frequencies in ES cells makes the use of selectable marker genes a necessity for most applications. As mentioned in section 3, targeting constructs can be fashioned with markers for positive and negative selection. Positive selection, which eliminates cells that do not carry a drug resistance gene, is generally used to identify cells that have been transfected with exogenous DNA. Marker genes that have been used for positive selection in ES cells include *neo*, *hph*, *lacZ*, and *hprt* (*Table 3*). Negative selection, which eliminates cells that carry a drug sensitivity gene, is generally used either to remove cells with an undesirable integration event or, as described previously, to select for cells that have deleted the marker gene during the second stage of in-out targeting. Marker genes that have been used for negative selection in ES cells include *hprt*, HSV-*tk*, and the diphtheria toxin A-fragment gene (*Table 3*).

Several factors must be considered in selecting a marker system for gene targeting:

(a) Marker genes expression must be driven by promoters and enhancers that have been demonstrated to function adequately in ES cells.
(b) For some marker genes, such as *hprt*, a simple cDNA clone may not

Table 3 Selectable marker genes and selection regimens for ES cells

Marker gene	**Selection**	**Drug regimen**
Neomycin phosphotransferase (*neo*)	Positive	G418 at 200 μg/ml [a]
Hygromycin-B-phosphotransferase (*hph*)	Positive	Hygromycin-B at 200 μg/ml
Puromycin *N*-acetyl transferase (*pac*)	Positive	Puromycin at 10 μg/ml
Xanthine–guanine phosphoribosyl transferase (*gpt*)	Positive	MHX medium (10 μg/ml mycophenolic acid, 15 μg/ml hypoxanthine, 25 μg/ml xanthine)
β-Galactosidase (*lacZ*)	Positive	X-Gal staining and FACS [b]
Hypoxanthine phosphoribosyl transferase (*hprt*)	Positive	HAT medium (16 μg/ml hypoxanthine, 0.175 μg/ml aminopterin, 4.8 μg/ml thymidine)
Hypoxanthine phosphoribosyl transferase (*hprt*)	Negative	6-thioguanine (6-TG) at 1.67 μg/ml
HSV thymidine kinase (HSV-*tk*)	Negative	Gancyclovir at 0.5 μg/ml (2 nM)
Diphtheria toxin A-fragment	Negative	Drug selection is not required

[a] 200 μg/ml of *active* G418.
[b] FACS; fluorescence activated cell sorting.

work, and appropriate regulatory sequences may have to be inserted into the marker construct to achieve acceptable levels of expression (38).

(c) It may be necessary to use more than one positively-selectable marker in cases where targeted ES cells are screened for the integration of a second construct. This would be the case, for instance, if it were necessary to target both alleles at a particular locus, or if it were necessary to insert a transgene into a targeted ES cell line.

(d) Feeder layers used during selection must be able to survive the selection conditions.

The most commonly used selectable marker for integration events is the neomycin phosphotransferase gene, *neo*, which confers resistance to the drug G418. The use of promoters, enhancers, and polyadenylation signals that function well in ES cells is important for effective drug resistance. Examples of promoters that have been found to work well in ES cells are those that drive expression of the RNA polymerase II (*Pol 2*) gene or phosphoglycerate kinase (*pgk-1*) gene (13).

The gene conferring hygromycin-B resistance, *hph*, has been used alone or in combination with another gene for the detection of targeting events (39). Positive selection by the application of puromycin (*pac*) and mycophenolic acid (*gpt*) are not as well documented (40, 41), but may prove to be as effective as G418 and hygromycin-B for positive selection in ES cells.

The *lacZ* gene, which codes for the enzyme β-galactosidase, can also be considered a selectable marker. While this enzyme does not confer resistance to a drug, it can be used as a marker for transfection by treating cells with the chromogen, X-Gal. Reddy and co-workers have used fluorescence activated cell sorting (FACS) to separate and hence select blue (X-Gal stained) cells from a mixed population of ES cells (42).

Since ES cells have a functional, X-linked copy of the *hprt* gene, positive selection for *hprt* must be done with an *hprt*$^-$ ES cell line. The most commonly used *hprt*$^-$ line, E14TG2a, was established by Hooper *et al.* (43); these cells have a deletion in the *hprt* locus and therefore can not revert to *hprt*$^+$. E14 cells transfected with a functional *hprt* minigene can be selected when cells are grown in HAT medium, which prevents cell growth in the absence of the enzyme hypoxanthine phosphoribosyl transferase (13).

The *hprt* gene can also be used for negative selection. The presence of *hprt* can be selected against with 6-thioguanine. The availability of *hprt* for positive and negative selection makes it an ideal marker gene for in-out targeting, as discussed in section 3.3.

Two other genes have been used for negative selection in the positive-negative selection scheme described in section 3.2. The HSV-*tk* gene has been used most commonly for this purpose. The base analogue gancyclovir is cytotoxic to cells that carry the HSV-*tk* gene and is used as the selective agent. Although gancyclovir is a potential mutagen, this does not seem to present a serious problem, since numerous laboratories have generated germline transmitting chimeras from ES cells treated with gancyclovir. A gene encoding diphtheria toxin A-fragment has also been used as the negative selection marker in a replacement construct (44). The use of this marker was shown to be effective for the enrichment of targeting events, and did not require the use of a potentially mutagenic drug selection regimen.

5. Introduction of DNA into ES cells

A number of methods have been used successfully to introduce DNA into ES cells. The best protocol for a particular experiment will depend largely on the overall goals of the experiment. If germline transmission from the transfected cells is desired, the effect of the protocol on the integrity of the ES cells must be considered. Another factor that must be considered is that protocols for introducing DNA may affect the ratio of ES cells that are transformed only transiently to those in which the introduced DNA becomes stably integrated into the genome. When stable integration occurs, the method of introduction may also influence whether the incoming DNA is incorporated by random integration or by homologous recombination. In other words, transfection methods will vary not only in the number of absolute transformants, but also in the fate of the DNA that is introduced. While the main application of ES cell transfection is the generation of homologous recombination events for the creation of transgenic mouse strains, other applications using ES cells continue to emerge. Here we discuss a number of methods for introduction of DNA into ES cells and their advantages for obtaining different genetic outcomes. However, prior to this discussion, *Protocol 1* describes procedures for preparing DNA for transfection.

Protocol 1. Preparation of DNA for transfection

Equipment and reagents

- Vector DNA. If preparing DNA for electroporation, the purity of the DNA is not critical and CsCl purification is not essential for obtaining adequate transfection efficiencies in ES cells. Even DNA prepared by alkaline lysis miniprep protocols works well. However, prepare DNA for lipofection and calcium phosphate precipitation by chromatography on a Qiagen column (Chapter 1, *Protocol 1*).
- Tissue culture laminar flow hood
- 5 M ammonium acetate
- 70% and 100% ethanol
- Sterile TE buffer: 10 mM Tris–HCl pH 8.0, 1 mM EDTA

Method

1. Linearize the DNA with an appropriate restriction endonuclease if stable integration events are desired. After digestion, either heat inactivate the restriction endonuclease or remove it by phenol extraction.
2. Ethanol precipitate the DNA as follows:
   (a) Add an equal volume of 5 M ammonium acetate to the linearized DNA and mix.
   (b) Add 2 vol. of 100% ethanol and invert several times to mix.
   (c) Microcentrifuge for 15 min at 14 000 r.p.m.
3. Remove the supernatant carefully without disturbing the pellet. Add 0.5 ml of 70% ethanol to wash salts from the DNA.
4. If the pellet lifts from the bottom of the tube, microcentrifuge for 5 min at 14 000 r.p.m.
5. Remove the supernatant with an aspirator *in the laminar flow hood.*[a]
6. Resuspend the DNA pellet in sterile TE buffer, aiming for a final concentration of 0.5–1.0 mg/ml. Due to the concentration of DNA, it may be necessary to warm the DNA to 37°C to completely resuspend it.
7. Quantify the DNA concentration by measuring its $A_{260\ nm}$; a 40 μg/ml solution of double-stranded DNA has an $A_{260} = 1$.

[a] It is important to prevent microbial contamination of DNA that will be introduced to tissue culture. For this reason, remove the 70% ethanol wash and resuspend the DNA in TE buffer while in a tissue culture hood.

5.1 Electroporation

Transfection of ES cells was first achieved by retroviral infection and calcium phosphate precipitation, but electroporation has replaced these as the method of choice for several reasons. A higher transfection efficiency can generally be achieved in mammalian cells with electroporation than with calcium phosphate precipitation, the electroporation procedure is relatively simple, and results obtained with electroporations are reproducible (45). Most important, however, is the fact that many laboratories, including our own, have found that electroporation is an effective method for the production of germline-transmitting chimeras.

The introduction of DNA into ES cells by electroporation is achieved by the application of a brief electric field that encourages the uptake of DNA, presumably by creating pores in the cell membrane (45). The 'dose' that the cells receive is a function of the voltage (V) and the time constant (τ) of the applied pulse. The time constant is a function of the resistance (R) and capacitance (C) of the system: $\tau = R \times C$. Electroporators usually allow the manipulation of the voltage and the capacitance, and sometimes also the resistance and the pulse length. Most researchers cite the voltage as V/cm to describe the cuvette size as well as the voltage, and list the capacitance in μFd, and the time constant in seconds.

The variation of voltage and capacitance in an electroporation may influence the frequency of certain recombination events. Our laboratory found that the ratio of targeted to random integrations increased when the capacitance was decreased and the voltage was increased. When using the Bio-Rad Gene Pulser we routinely use a voltage of 270 V and a capacitance of 500 μFd. When using the BTX Electro Cell Manipulator, which allows manipulation of voltage, capacitance, and resistance, we use a voltage of 270 V, a capacitance of 50 μFd, and a resistance of 360 ohms. Several companies offer electroporators for mammalian cells, and their use with ES cells can be verified by the manufacturer. The optimum voltage and capacitance for other electroporators can be determined using a *neo* gene to quantitate random integrations and an *hprt* gene that rescues the E14TG2a deletion to quantitate targeted integrations.

Table 4 describes the solutions required for electroporation of ES cells and *Protocol 2* gives the electroporation procedure. We recommend using a DNA concentration of 3 nM for electroporation. The rationale for expressing the DNA concentration by molarity is that the DNA copy number remains constant for constructs of differing sizes, and the frequencies of targeted and random integration events can be more accurately compared. It may be possible with small constructs to obtain increased transfection efficiency by increasing the concentration above 3 nM. With larger constructs an increased efficiency is possible, but large amounts of DNA can actually decrease efficiencies, probably as a result of toxicity.

Table 4 Solutions for electroporation

Supplemented DMEM-H medium for ES cells
Prepare by adding the following components to the bottle of medium:

- 500 ml of Dulbecco's modified Eagle media—high glucose (DMEM-H; Gibco-BRL Cat. No. 320–1965AJ or equivalent)
- 75 ml of ES cell tested fetal calf serum (FCS; Gibco-BRL Cat. No. 200–6140PJ)[a]
- 6 ml of 200 mM L-glutamine (Gibco-BRL Cat. No. 25030–024)
- 0.6 ml of penicillin/streptomycin diluted to 100 U/ml penicillin and 100 μg/ml streptomycin (Gibco-BRL Cat. No. 600–5145AE)
- 4 μl of 2-mercaptoethanol (Sigma Cat. No. M-7522)

Swirl to mix well and filter through a 0.2 μm bottle-top filter (e.g. Costar Cat. No. 8330) into a new sterile bottle if the sterility of any of the above components is in question. Store at 4°C and warm to 37°C before use. Shelf life is about one month.

Working trypsin solution for ES cells
Prepare a buffer solution for the trypsin by combining the following in glassware reserved for tissue culture; 2980 ml of endotoxin-free sterile water, 24.0 g NaCl, 1.2 g KCl, 3.0 g D-glucose, 1.05 g $NaHCO_3$, and 0.6 g EDTA (not disodium form). Adjust the pH to 7.0 with dropwise addition of 0.5 M NaOH; the pH will change rapidly and should be adjusted slowly. Bring the volume to 3 litres with sterile endotoxin-free water. Filter sterilize and aliquot into 500 ml sterile bottles using a 0.2 μm bottle-top filter (e.g. Costar Cat. No. 8330). Store at 4°C. For a working solution of trypsin, combine 100 ml of 0.25% trypsin (Gibco-BRL Cat. No. 15050–040) with 500 ml of buffer solution and store at 4°C. Warm the working trypsin solution to 37°C before use.

[a] Test various lots of fetal calf serum for maintenance of ES cell survival, pluripotency, and euploid karyotype (3). Heat inactivate the serum for 30 min at 56°C before use.

Protocol 2. Electroporation of ES cells

Equipment and reagents

- Mammalian cell electroporator (e.g. Bio-Rad Gene Pulser Cat. No. 165–2076 with Capacitance Extender 165–2087, or BTX Electro Cell Manipulator Cat. No. 6000)
- Electroporator cuvette (e.g. Bio-Rad Cat. No. 165–2088, or BTX Cat. No. 620)
- Centrifuge
- PBS pH 7.5: 2.7 mM KCl, 1.1 mM KH_2PO_4, 138 mM NaCl, 8.1 mM $Na_2HPO_4.7H_2O$ (e.g. Gibco-BRL Cat. No. 310–4190)
- Supplemented DMEM-H medium (see *Table 4*)
- Working trypsin solution (see *Table 4*)
- Sterile TE buffer pH 8.0 (see *Protocol 1*)
- 100 mm plate of confluent ES cells
- 100 mm plates each containing a fibroblast feeder layer of cells in supplemented DMEM-H medium
- Sterile 15 ml conical tubes
- Sterile 1.5 ml microcentrifuge tubes

Method

1. Feed the 100 mm plate of confluent ES cells with warmed supplemented DMEM-H medium 3 h before harvesting.
2. Remove the medium from the plate of ES cells by aspiration.
3. Add 2 ml of PBS pH 7.5, swirl, and remove by aspiration. Repeat.
4. Add 3 ml of working trypsin solution and return the plate to a CO_2 incubator at 37°C for 1–2 min.

5. When the cells detach from the plate, add 3 ml of supplemented DMEM-H medium to the plate and gently pipette up and down thoroughly to create a single cell suspension.
6. Add another 4 ml of supplemented DMEM-H medium, pipette mix gently, and transfer to a 15 ml conical tube.
7. Pellet the cells by centrifugation at 1000 *g* for 5 min.
8. Aspirate the medium from the cell pellet and then flick the tube sharply to loosen the pellet.
9. Rinse the cells by adding 5 ml of supplemented DMEM-H medium. Pipette up and down to create a single cell suspension.
10. Pellet again by centrifugation at 1000 *g* for 5 min. Remove the media by aspiration and then flick the pellet sharply.
11. Add 200 μl of supplemented DMEM-H medium to the cells and transfer the cell suspension to a sterile 1.5 ml microcentrifuge tube. The cell pellet occupies about 200 μl, so the total volume should be about 400 μl. If necessary, add more medium to bring the volume to 400 μl.[a]
12. Add the DNA to the cells at a concentration of 3 nM.[b]
13. Transfer the DNA and cell mixture to a sterile electroporator cuvette and place the cuvette in the electroporator, orienting the cuvette so as to complete the circuit of the electroporator. *Read the manufacturer's instructions for use of the electroporator carefully and be sure to follow all of the safety requirements for its use.*
14. Electroporate the cells with a single pulse of 270 V (675 V/cm), 50 μFd, and 360 ohms.[c]
15. Remove the electroporated cells from the cuvette with a 1 ml pipette and add to 12 ml of supplemented DMEM-H medium in a 15 ml conical tube.
16. Plate 1–2 ml of the suspended cells on 100 mm plates containing fibroblast feeder layers in supplemented DMEM-H medium.[d]
17. Allow the cells to recover overnight before adding drug selection medium.

[a] This is the volume for the BTX electroporator cuvette. If using a Bio-Rad cuvette, the volume of the cuvette is 800 μl, and so the cell volume should be doubled, and calculations for DNA molarity adjusted accordingly.

[b] To calculate the quantity of DNA required in micrograms, bear in mind that a 3 nM concentration of DNA in a 0.4 ml final volume is equivalent to 1.2 pmol of DNA. Calculate the weight in micrograms of DNA given its size; 1.2 pmol of a 1000 bp DNA molecule weighs 0.8 μg, so for every 1000 bp of DNA in the molecule, use 0.8 μg. For example, if the molecule is a 10 kb plasmid, use 8 μg of DNA.

[c] As mentioned in section 5.1, the manufacturer may have a recommendation for field strength that would be appropriate for their electroporator.

[d] It is important that the cells are plated at such a density that they will not become confluent before selective growth conditions kill off the majority of electroporated cells.

5.2 Calcium phosphate-mediated transfection

One of the first genetic manipulations of ES cells entailed the introduction of a neomycin-resistance gene by calcium phosphate precipitation (46). As demonstrated by Gossler *et al.* ES cells transfected by calcium phosphate precipitation can generate germline-transmitting chimeras (6). However, this method suffers from the following disadvantages when compared to electroporation:

- the procedure used for calcium phosphate precipitation is more laborious than that used for electroporation
- integration of multiple copies of DNA is more common with the calcium phosphate method
- lower ratios of targeted to random integrations are observed with the calcium phosphate method

Calcium phosphate precipitation is therefore not an ideal method for most applications involving gene targeting in ES cells, and it has largely been replaced by electroporation. However, calcium phosphate precipitation has been used in our laboratory for the co-transfection of a selectable marker gene with a transgene. A comparison of co-transfection efficiencies indicated that only one out of every 100 colonies carrying the selectable marker was also transfected with the transgene when electroporation was used. In contrast, four of five colonies were co-transfected when calcium phosphate precipitation was used. These results suggest that calcium phosphate precipitation may be preferable to electroporation for co-transfection experiments in ES cells. This method may also be desirable for transfection of ES cells that have been previously targeted to introduce a second transgene. When performing the co-transfection experiments, we used a commercially available kit (Stratagene Cat. No. 200285) which provides the reagents at precise pH and boasts higher transfection efficiencies. Nevertheless, for those workers who wish to use laboratory-based reagents, we describe in *Protocol 3* the traditional calcium phosphate procedure (47) as applied to ES cells.

One difficulty that must be considered is the relatively long time required for transfection and recovery before selection. During this time, ES cells can rapidly divide and cause overgrowth before selective agents can affect cell numbers, especially when G418 is the selection drug. When selection does begin, the majority of cells are killed and the heavy cell death can be cytotoxic to feeder cells and detrimental to the growth of the remaining ES cells. To prevent overgrowth, the cells must be plated at a low cell density. The use of the *hprt* gene instead of the *neo* gene as the selectable marker allows one to start with a higher cell density, since *hprt*$^-$ cells are killed within 24 hours after addition of selective media. Cells can also be trypsinized and split into several plates after the transfection procedure to prevent overgrowth, but this increases the possibility that a given transformation event is represented more than once in the colonies that are eventually analysed. However, this would only be critical in situations where a specific, rare genetic event must be identified, such as in gene targeting experiments.

Protocol 3. Calcium phosphate-mediated transfection of ES cells

Equipment and reagents

- Sterile 5 ml tube
- 2.5 M $CaCl_2$
- 2 × Hepes-buffered saline pH 7.12: 280 mM NaCl, 50 mM Hepes, 1.5 mM Na_2HPO_4
- Sterile, plugged pipette and forced air supply or Pipette-Aid (Fisher Cat. No. 13–681–15)
- Supplemented DMEM-H medium (see *Table 4*)
- PBS pH 7.5 (see *Protocol 2*)
- Supplemented DMEM-H medium containing 15% glycerol (required only if a glycerol shock is to be used at step 7)
- 5 × 10^6 ES cells seeded on a 100 mm plate with feeder cells the day before transfection
- Vector DNA for transfection (see *Protocol 1*)

Method

1. Bring 10–20 μg of DNA up to a volume of 225 μl with sterile water in a sterile 5 ml tube.
2. Add 25 μl of 2.5 M $CaCl_2$ to the DNA mix using a pipette.
3. Add 250 μl of 2 × Hepes-buffered saline to the tube by dropwise addition. While adding the Hepes, bubble air through the DNA solution using a plugged sterile pipette attached to a Pipette-Aid. A fine precipitate should form during the addition.
4. Allow the DNA precipitate to sit at room temperature for 30 min.
5. Mix the precipitate using a pipette to resuspend it immediately before adding it to the ES cells.
6. Add the precipitate to the ES cells by dropwise addition and swirl the plate to mix. The precipitate should be visible by inspection with a microscope. Return the cells to the incubator and leave the precipitate on the cells overnight.
7. If desired the cells can be treated with a glycerol shock as follows to improve transfection efficiency:
   (a) After 4 h exposure to the DNA precipitate, aspirate off the medium and add 3 ml of supplemented DMEM-H medium with 15% glycerol.
   (b) Incubate for 4 min, then aspirate off the glycerol medium.
   (c) Rinse twice with PBS and once with supplemented DMEM-H medium to remove the glycerol completely.
   (d) Add 10 ml of supplemented DMEM-H medium and incubate overnight at 37°C.
8. On the following morning, add fresh supplemented DMEM-H medium. Drug selection can be initiated 12–24 h later.

5.3 Liposome-mediated transfection

Cationic liposomes can mediate the delivery of DNA, RNA, and purified proteins into cells (48, 49). In the past, the most common use of this method was in situations where cells were not amenable to transfection by electroporation. However, this methodology has gained prominence in recent years for a number of reasons, including:

- reports of transfection frequencies that exceed those seen with calcium phosphate precipitation
- the availability of commercial reagents to carry out this procedure
- the ability to transfect using very large fragments of DNA
- the possible application of this methodology to deliver macromolecules *in vivo*

The synthetically prepared ampipathic molecules spontaneously assemble into unilaminar liposomes that carry a positive charge on their hydrophilic exterior. This positive charge allows liposomes to interact with negatively charged polynucleotides. With a proper ratio of liposome to DNA, 100% of the DNA is complexed with the liposome, as measured by sucrose gradient density. The liposome/DNA complex enters the cell by fusing with the cell membrane (50).

The early success in obtaining targeted cell lines using electroporation has made the use of lipofection rare in gene targeting experiments using ES cells. However, as methods of delivery affect the fate of the introduced DNA, we examined the possibility that lipofection would favour integration by homologous recombination over random insertion into the genome.

In our experiments, the ES cell line E14TG2a was transfected with 10 μg of a *neo* gene, or 10 μg of an *hprt* construct which rescues the E14TG2a mutation when targeted integration occurs. To identify cells with random or targeted insertions, transfected cells were treated with G418 or grown in HAT medium, respectively. Our results indicate that higher frequencies of both types of recombination are achieved with an increasing ratio of liposome to DNA. The best results were obtained with a liposome to DNA ratio of 10:1, and we have not established the upper limits for liposome toxicity.

The frequency of random integration as measured by the number of G418-resistant colonies was 3.5×10^{-4} (554 colonies/1.6×10^6 cells treated). Transfection of the same *neo* construct by electroporation resulted in a frequency of only 1.2×10^{-4} (243 G418-resistant colonies/2×10^6 cells electroporated). The frequency for random integrations is therefore almost three times greater with liposome-mediated transfection.

Surprisingly, targeted integrations were not more frequent with lipofection, but were instead less frequent. The frequency of homologous recombination events using liposome-mediated transfection was 3.8×10^{-6} (6 HAT-resistant colonies/1.6×10^6 cells treated), whereas electroporation using the same *hprt* construct resulted in a targeting frequency of 2.3×10^{-5} (136 HAT-resistant colonies/6×10^6 cells electroporated). These results indicate that electroporation can give a targeted to random integration ratio of 1:5, while lipofection using the same constructs results in a targeted to random ratio of 1:100. Based on these preliminary results, lipofection appears to be inferior to electroporation for gene targeting of ES cells. Lipofection is, however, an attractive method for random integration of DNA, for the delivery of purified proteins, and for the introduction of very large pieces of DNA, such as YACs, into ES cells.

Two laboratories have demonstrated that lipofection can introduce yeast artificial chromosome (YAC) DNA of 150 kb and 85 kb into ES cells without shearing (51, 52). These experiments also demonstrated that YAC DNA introduced by this method could stably integrate into the ES cell genome and be transmitted to subsequent generations through the germline of chimeric animals. Although it was found that many cell lines showed deletions, rearrangements, or additional copies of YAC sequence, intact YAC integration events were recovered at a frequency of 7×10^{-8} (51).

Several cationic liposomes are commercially available, such as Promega Transfectam (DOGS), Boehringer Mannheim (DOTAP), and Gibco-BRL Lipofectin (DOTMA) and LipofectAMINE (DOSPA). We have used Gibco-BRL Lipofectin and LipofectAMINE (Cat. No. 18292-011 and 18324-012) in our preliminary experiments, and found that LipofectAMINE gave higher transfection efficiencies in ES cells than Lipofectin. The procedure for transfection using LipofectAMINE with ES cells is given in *Protocol 4*.

Protocol 4. Liposome-mediated transfection of ES cells

Equipment and reagents

- Cationic liposomes (e.g. Gibco-BRL LipofectAMINE Cat. No. 18324-012)
- OPTI-MEM-I reduced serum medium (Gibco-BRL Cat. No. 320-1985AG)
- Working trypsin solution (see *Table 4*)
- PBS pH 7.5 (see *Protocol 2*)
- Supplemented DMEM-H medium (see *Table 4*)
- Supplemented DMEM-H medium without penicillin/streptomycin
- 100 mm plate of confluent ES cells
- Vector DNA for transfection (see *Protocol 1*)
- 15 ml sterile, polystyrene, conical tubes (e.g. Falcon Cat. No. 2087)
- Sterile 100 mm tissue culture plates

Method

1. Add 10 μg linearized vector DNA to 2 ml OPTI-MEM-I reduced serum medium in a polystyrene tube. To this add 100 μg of cationic liposome.
2. Invert to mix the liposomes and DNA, and allow complex formation by incubating together at room temperature for 30 min.
3. Remove the medium from the 100 mm plate of confluent ES cells (about 2×10^7 cells).

Protocol 4. *Continued*

4. Add 2 ml of PBS, swirl, and remove by aspiration. Repeat.
5. Add 3 ml of working trypsin solution. Incubate the cells in trypsin for 1–2 min in a CO_2 incubator at 37 °C.
6. When the cells have detached from the plate, add 7 ml of supplemented DMEM-H medium and gently pipette up and down to create a single cell suspension. If desired, count 10 μl of cells in a haemocytometer to give the number of starting cells.
7. Pellet the cells in a 15 ml conical tube by centrifuging at 1000 *g* for 5 min.
8. Aspirate the medium from the cell pellet and then flick the cells sharply to prevent clumping.
9. Add 10 ml of OPTI-MEM-I medium to the cells and pipette gently up and down to resuspend the cells.
10. Add the 2 ml liposome–DNA mixture (from step 3) to the tube of cells and pipette up and down to mix the liposome complex with the cells.
11. Plate the cells on a sterile 100 mm plate (without a fibroblast feeder layer).
12. Incubate the cells in a CO_2 incubator at 37 °C for 4 h. Remove the plate from the incubator every 30 min and pipette the cells up and down to prevent adherence.[a]
13. After 4 h, plate the 12 ml of cells on eight plates with fibroblast feeders in supplemented DMEM-H **without** penicillin/streptomycin.[b]
14. Allow the cells to recover overnight at 37 °C before beginning the drug selection process.

[a] During the transfection period, a number of laboratories, including our own, have found that higher transfection efficiencies are achieved by keeping the cells in suspension (51).
[b] Liposome transfected cells are more sensitive to antibiotics and should be allowed to recover overnight without penicillin/streptomycin.

5.4 Microinjection

Microinjection of ES cells has been reported only once (24). The appeal of microinjection as a method for transfecting ES cells was based on the idea that every cell would receive DNA, and therefore the inclusion of a selectable marker would not be required. This was an important consideration in early gene targeting experiments, when the expression of a selectable marker inside a silent gene or at a given chromosomal location was not assured. There are numerous technical difficulties encountered with this procedure, however, and we can not recommend it unless there are no other options available.

5.5 Retroviral infection

Retroviral infection has been used to introduce DNA stably into the ES cell genome for the production of germline-transmitting chimeras (7). The use of retroviral vectors for ES cell transfection has declined due to the low frequency of homologous recombination achieved with retroviruses (53) and the simplicity of other, more commonly used, transfection techniques. Retroviral infection is discussed in greater detail by Lovell-Badge (47).

6. Expansion and screening of ES cells

After exogenous DNA has been introduced into ES cells, individual ES cells that have integrated the DNA in the desired manner must be isolated from the random integrants. A number of protocols have been devised to accomplish this, and the expected frequency of the desired integration event will determine which method is appropriate. We will first describe the strategy that we use when the desired integration event is expected to be present in at least one out of every 300 cells that survive selection. This protocol would therefore be appropriate for screening for random integration of transgenes and most targeting events. Although quicker methods have been devised, we prefer that described here because it allows us to monitor the quality of ES cells effectively during the screening process. In our opinion, this increases the likelihood that ES cells isolated by this method will be able to colonize the germline of chimeric mice.

(a) After transfection, allow the ES cells to recover for 24 hours in supplemented DMEM-H medium on fibroblast feeder layers.

(b) After the recovery period, apply the appropriate selection drug and change the medium when it turns orange-yellow or after two days, whichever is first. If feeders with the appropriate resistance gene are not available, cells can be plated on dishes coated with gelatin in selective medium supplemented with LIF.

(c) Selection will result in heavy cell death, and during the 7–12 days of selection the medium must be changed to remove the dead cells. It may also be necessary to add additional fibroblast feeder cells to the plate when the original feeder layer becomes thin as a result of the ES cell death.

(d) After 7–12 days in selection medium, individual colonies can be visualized with an inverted microscope. These colonies will generally represent the clonal expansion of a single transfected ES cell.

(e) Once colonies become visible (*Figure 3*), pick them from their original plate using a drawn-out capillary pipette attached to latex tubing and transfer to individual wells of a 24 multiwell plate with fibroblast feeder

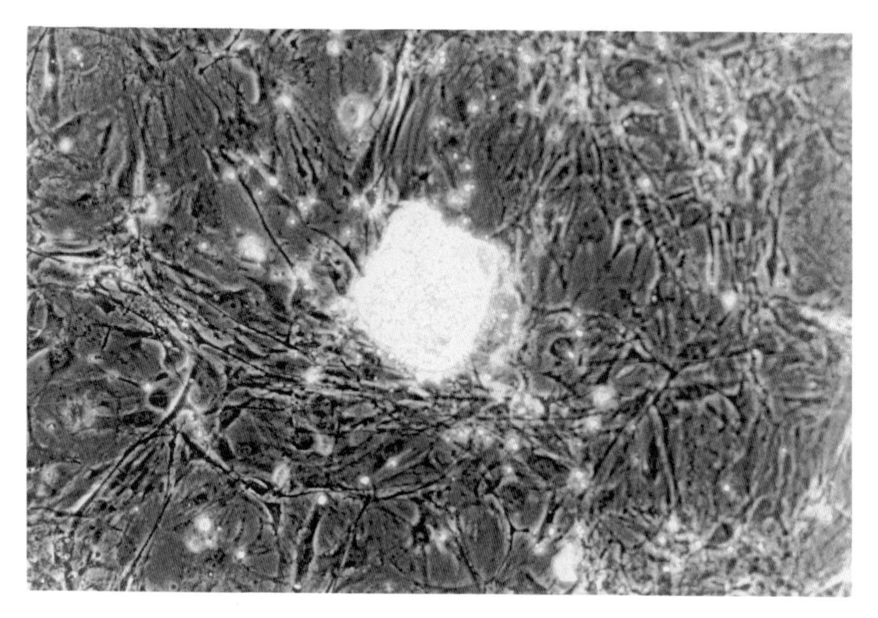

Figure 3. Phase contrast photomicrograph of an ES cell colony on an irradiated fibroblast feeder layer. After 7–12 days of growth in selective medium, ES cell colonies similar to that shown can be picked for expansion. It is preferable to pick undifferentiated colonies when germline transmitting chimeras are the desired end. An undifferentiated ES cell colony will have a distinct border rather than one that blends into the feeder layer. Individual cells within a colony should not be easily distinguishable. The colony should have depth instead of appearing flattened. This depth makes the colony more light refractile, and the colony should appear shiny when viewed with a phase contrast microscope.

layers to decrease the likelihood of infecting the colony with mycoplasma, use of a plugged mouthpiece or the placement of a filter in the line is recommended. When picking the colony, it is important to try to break up the colony by scoring or sectoring it with the pipette before transferring it to the new well. This will prevent the transferred cells from forming one large colony which can become necrotic and/or begin to differentiate. There should not be a bias towards picking colonies of a certain size; a large or small colony may have a growth advantage or disadvantage resulting from genetic manipulations, and these genetic events should be represented during later analysis. Once colonies from a plate have been picked, discard the plate since disrupted cells from a picked colony can form new colonies. If these new colonies are picked at a later date, effort will be wasted in re-screening the same ES cell clone. When picking larger numbers of colonies, we use the same capillary pipette to pick all colonies. Although this means that cells from one colony can contaminate the subsequent picks, any significant contamination can be removed by subcloning colonies at a later date if deemed necessary. If relatively few colonies are being picked, the capillary pipette can be cleaned after picking each colony by quickly rinsing in ethanol and then in media.

(f) After two or three days, trypsinize the cells in the 24-well dish and re-plate them to further disperse the cells. To accomplish this, aspirate the tissue culture medium, and rinse each well twice with PBS pH 7.5, and trypsinize by adding of 250 μl of working trypsin solution. After a 1–2 min incubation in a CO_2 incubator at 37°C, add 250 μl of media to inhibit the trypsin activity. Disperse the cells with a P1000 Gilson Pipetman, then add 1 ml of medium to each well. Then supplement the wells with additional feeder cells (about one-third of the amount originally plated).

(g) After an additional two to three days, the cells will reach confluency. When this occurs, trypsinize cells as before and transfer them to a 6-well dish with fibroblast feeders.

(h) When ES cells in the 6-well dish become confluent, trypsinize them as described above for the 24-well plate, except use 500 μl each of trypsin and medium. Resuspend the cells by pipetting. Pellet half of the cells by centrifugation and resuspend the pellet in freezing medium (50% fetal calf serum, 11% DMSO, 39% supplemented DMEM-H medium) for long-term storage at −70°C. Pellet the remaining cells for DNA preparation and screening by Southern analysis (*Protocol 6*).

The following modifications of the procedure described above can be used to increase the number of colonies screened and to implement screening by PCR:

(i) When colonies are picked initially [step (e) above], transfer half of each colony to a well in a 24-well dish and use the other half for PCR screening. If large numbers of clones must be screened, the cells may be pooled for PCR screening (54).

(ii) During the initial trypsinization of cells in the 24-well plate (step (f) above), an aliquot of cells can be removed and analysed by PCR. Maintain the remaining cells as described above. If more cells are needed to detect a PCR signal, an aliquot of cells can also be removed during the second trypsinization of cells in the 24-well plate (step (g) above). Transfer the remaining cells to a 6-well plate as described in the unmodified protocol.

(iii) After electroporation, cells can be plated directly into multiwell dishes at a density which will result in numerous colonies growing in each well. Trypsinize the wells containing colonies, remove an aliquot for PCR analysis, and re-plate the remaining cells. If a positive PCR result is obtained for a well, trypsinize the cells in the well and plate at low dilutions on 100 mm dishes with feeders. Pick individual colonies, which arise after five to seven days, and screen as in the unmodified protocol to identify the ES cell clones with the desired integration event (55). If this strategy is used, it may be necessary to carefully optimize electroporation conditions to obtain a reasonable number of colonies per well without plating ES cells at a density that will overgrow the well before selection is complete.

(iv) Sufficient DNA for Southern analysis can be obtained from a single confluent well of a 24-well plate. To accomplish this, replate ES cell colonies into 24-well plates after the first trypsinization in the unmodified procedure (step (f)). However, after the second trypsinization, use the majority of cells to prepare DNA, and re-plate the remaining cells on the 24-well plate rather than on a 6-well plate as in the unmodified procedure. Cells can be frozen in the plate while analysis is carried out (56). Because of the smaller amount of DNA obtained with this modification, it is usually necessary to collect DNA by centrifugation rather than by spooling. The DNA obtained is sufficient for one Southern blot.

7. Analysis of transfectants

7.1 Determination of the manner of DNA integration

Once ES cells have been selected for integration of exogenous DNA, they are usually analysed by PCR or Southern blot to determine whether the exogenous DNA has integrated in the desired manner. This section will concentrate on the relative merits of PCR and Southern analysis for screening transfected ES cells.

The major factor in determining whether ES colonies will be screened by PCR or Southern blot is the expected frequency of the desired recombination event. Because PCR screening can be performed rapidly on relatively small amounts of genomic DNA, it is generally preferred for detection of recombination events that are expected to be relatively rare. The sensitivity of the PCR technique also allows colonies to be pooled before screening, which considerably increases the number of clones that are initially screened. Once a positive pool is identified, the clones included in the pool can then be screened individually as described in the preceding section.

Although the advantages listed above make PCR screening the method of choice for low frequency recombination events, PCR screening suffers from several drawbacks that make it less advantageous in cases where it is estimated that relatively few ES cell colonies will be screened. One such drawback is the effort that must be invested in the process of designing an appropriate PCR assay. This process usually involves a number of steps, which may include:

- obtaining sequence data for use in designing PCR primers
- design and synthesis of primer sets

- design and construction of a positive control construct for use in optimizing reaction conditions
- testing a variety of primer sets and reaction conditions for specificity and sensitivity in detecting the desired recombination event

When screening for targeting events, PCR analysis also suffers from the drawback that the PCR reaction must amplify one of the homologous arms of the targeting construct, and shortening the arm to obtain reliable PCR will often decrease targeting frequency. A final disadvantage of PCR screening is that it does not provide a complete picture of the genomic configuration of the integrated DNA. Although PCR screening can be used to detect the juxtaposition of exogenous DNA with specific chromosomal sequences that occurs during a targeting event, as well as deletions or disruptions that interfere with amplification of endogenous sequences, all integration events should be subsequently verified by Southern analysis. Thus, since a Southern analysis protocol must be set-up even in cases where initial screening is carried out by PCR, it is often inefficient to set-up a PCR assay if relatively few colonies will be screened. In terms of offering general guidance, we have found it worthwhile to implement a PCR screening protocol for recombination events that are expected to occur at a frequency of less than about one in 200–300 transfected ES cell colonies.

The major advantage of Southern analysis is the fact that it allows determination of the precise nature of recombination events, including those that result in insertions, deletions, or rearrangements. An additional benefit of Southern analysis is that results obtained using one restriction enzyme can be confirmed by performing independent digests with other enzymes. It can then be verified that the sizes of the restriction fragments obtained with a variety of enzymes are consistent with the expected recombination event.

When Southern analysis is used to identify targeting events, two basic criteria should be considered in deciding which genomic sequences will be used as a probe:

- The DNA used for the probe must not be taken from sequences included in the targeting construct, or it will have the potential to recognize novel restriction fragments resulting from random integration events.
- The sequences recognized by the probe must lie on the same restriction fragment as targeted sequences, or it will be impossible to detect changes in the length of restriction fragments caused by the targeting event.

Protocol 5 describes procedures for the preparation of a crude cell lysate from ES cells that can be used for PCR analysis. This method can be applied at an early stage in the expansion process since it requires relatively few cells and is more efficient than preparing genomic DNA; cells can be harvested and screened by PCR in one day. Chapter 7 (*Protocol 5*) describes the procedure for PCR amplification of genomic DNA, and this can also be used for PCR of the crude ES cell lysates. The PCR can be carried out with cell lysate either from individual colonies or pools of colonies.

Southern analysis requires more DNA than PCR, and the DNA must be relatively intact. To collect the required amount of genomic DNA, cells are harvested later than for PCR screening. *Protocol 6* describes the isolation of genomic DNA from one-half of a confluent 6-well dish of ES cells, but can be scaled up for more cells as necessary.

Protocol 5. Isolation of cell lysate DNA for PCR analysis

Equipment and reagents

- Confluent ES cells in 24-well plates
- 6-well tissue culture dishes with fibroblast feeder layers
- PBS pH 7.5 (see *Protocol 2*)
- Supplemented DMEM-H medium (see *Table 4*)
- Working trypsin solution (see *Table 4*)
- P1000 Gilson Pipetman (or equivalent)
- 10 mg/ml Proteinase K (Boehringer Mannheim Cat. No. 745–723) in water
- 95°C heating block or boiling water-bath
- 55°C water-bath

Method

1. When the ES cells growing in a 24-well plate have reached confluency, remove the medium from the wells by aspiration and add 0.5 ml of PBS.
2. Remove the PBS from the wells by aspiration. Repeat the PBS rinse.
3. Remove the PBS by aspiration and add 0.25 ml of working trypsin solution. Incubate for 1–2 min in a 37°C incubator.
4. Add 0.25 ml of supplemented DMEM-H medium to each well. Using a Gilson P1000 Pipetman, pipette up and down five or six times using a fresh, sterile pipette tip for each well.[a]
5. Transfer 0.25 ml of the cell suspension from each well to a 6-well plate with fibroblast feeder layers in supplemented DMEM-H medium for further expansion.
6. Transfer the remaining 0.25 ml of the cell suspension from each well to a 1.5 ml microcentrifuge tube.[b]
7. Pellet the cells by centrifugation at 14000 r.p.m. for 10 sec.
8. Remove the medium by aspiration and rinse the cell pellet with 0.5 ml of PBS.
9. Pellet the cells by centrifugation at 14000 r.p.m. for 10 sec. Remove the PBS by aspiration.
10. Resuspend the cells in 100 μl of PBS and 200 μl of sterile water.
11. Lyse the cells by heating to 95°C for 10–15 min.
12. Allow the cells to cool briefly by leaving at room temperature for

5 min, and add 10 μl of 10 mg/ml Proteinase K to each sample. Incubate at 55°C for 1 h.

13. Inactivate the Proteinase K by heating to 95°C for 10 min.

[a] These steps should be performed in a timely manner so that the cells do not remain in undiluted trypsin for a long period of time.

[b] At this point the cell number can be determined with a haemocytometer. Generally one-half of a confluent well from a 24-well dish yields 5×10^5 cells. This protocol can be used for up to 10^6 cells without adjustment of reagent volumes.

Protocol 6. Isolation of genomic DNA from ES cells[a]

Equipment and reagents

- PBS pH 7.5 (see *Protocol 2*)
- Working trypsin solution (see *Table 4*)
- Proteinase K buffer: 100 ml of 10% SDS, 200 ml of autoclaved 0.5 M EDTA pH 8.0, 20 ml of autoclaved 5 M NaCl, 50 ml of 1 M Tris–HCl pH 8.0, autoclaved deionized water up to 1 litre final volume (mix well and store at room temperature)
- Saturated NaCl: add 175 g NaCl to 500 ml of distilled, deionized water and stir overnight with a magnetic stirrer; in the morning, allow the remaining NaCl to settle to the bottom of the flask and decant the saturated solution to a new bottle (store at room temperature)
- 10 mg/ml Proteinase K (Boehringer Mannheim Cat. No. 745–723) in water
- 100% ethanol
- 70% ethanol; prepare 1.5 ml microcentrifuge tubes each containing 1 ml of 70% ethanol for washing the DNA (see step 14 below)
- TE buffer pH 8.0 (see *Protocol 1*)
- Glass Pasteur pipettes: prepare a fused Pasteur pipette for each sample by holding the tip of the pipette directly in a Bunsen burner flame
- Freezing medium: 50% FCS, 11% DMSO, 39% supplemented DMEM-H medium

Method

1. Remove the medium from the well of confluent ES cells by aspiration.
2. Add 0.5 ml of PBS to rinse the cells, swirl, and remove the PBS by aspiration. Repeat.
3. Add 0.5 ml of working trypsin solution and incubate at 37°C for 1–2 min.
4. Add 0.5 ml of supplemented DMEM-H medium and pipette up and down gently with a Gilson P1000 Pipetman.
5. Transfer 0.5 ml of the suspended cells into a 1.5 ml microcentrifuge tube and pellet in a microcentrifuge for 30 sec at 14000 r.p.m.[b]
6. Remove the medium from the pelleted cells and rinse the pellet with 0.5 ml of PBS.
7. Pellet the cells again, remove the PBS, and then flick the pellet sharply to prevent clumping.
8. Resuspend the cells in 200 μl Proteinase K buffer by gentle inversion.

Protocol 6. *Continued*

9. Add 10 μl of 10 mg/ml Proteinase K and mix by pipetting. Incubate overnight at 55°C.
10. Add 90 μl of saturated NaCl and shake vigorously for 15 sec. Centrifuge for 10 min at 14000 r.p.m.
11. Remove the supernatant carefully, avoiding removal of the pelleted cellular debris. Transfer the supernatant (approx. 260 μl) to a new microcentrifuge tube.
12. Add 520 μl of 100% ethanol to the tube and invert two or three times to precipitate the DNA.
13. Spool the DNA on to a heat-sealed Pasteur pipette.
14. Dip the spooled DNA four or five times into a tube with 1 ml of 70% ethanol to rinse away salts. Repeat in a second tube of 70% ethanol.[c]
15. Allow the DNA to air dry for 1–2 min. Use a diamond pen to score the glass of the Pasteur pipette above the spooled DNA. Then break off the end of the pipette bearing the spooled DNA into a microcentrifuge tube containing 60 μl of TE buffer. Alternately, submerge the tip of the intact glass pipette in the TE for 5 min to resuspend the DNA and then lift out and discard the pipette.
16. Use the DNA for Southern analysis or PCR after it has resuspended completely. The DNA concentration should be approximately 0.2–0.7 μg/μl; 15–25 μl of the DNA is sufficient for digestion with restriction endonucleases for Southern analysis.

[a] Adapted from Miller *et al.* (57).
[b] The remaining cells can be frozen by pelleting, resuspending in 1 ml of freezing medium, and then storing at −70°C in cryotubes.
[c] If the DNA is to be analysed by PCR, it is important to use separate sets of 70% ethanol tubes for each cell line to prevent contamination.

7.2 Expression analysis

Applications that utilize ES cells *in vitro* to answer developmental questions may require expression analysis of recombinant cell lines as well as genomic analysis. Although gene expression may be analysed at the level of RNA or protein, this chapter will cover only RNA analysis. Once RNA has been isolated, it can be used for Northern analysis, RNase protection assays, or reverse transcriptase PCR (RT–PCR) to quantify particular mRNAs.

7.2.1 Isolation of RNA

The methods for isolation of RNA (and protein) from ES cells are no different than for other tissue culture cells, and protocols published elsewhere can be used (58, 59). For the recovery of total RNA, we use a modified guanidinium isothiocyanate procedure (*Protocol 7*) based on the method of Chomczynski and Sacchi (60) and developed by Tel-Test. This procedure uses RNAzol B (which contains guanidinium isothiocyanate and phenol at the correct pH) to separate RNA from DNA and protein with minimal effort, and yields high quality total RNA. The protocol given was developed by Tel-Test for mammalian tissue culture cells, and required no adaptations for use with ES cells.

Protocol 7. Isolation of total RNA from ES cells with RNAzol B

Equipment and reagents

- PBS (see *Protocol 2*)
- Working trypsin solution (see *Table 4*)
- Supplemented DMEM-H medium (see *Table 4*)
- 100 mm tissue culture plate of confluent ES cells
- RNAzol B (Tel-Test Cat. No. CS-105)
- Chloroform
- Isopropanol
- DEPC-treated water
- 75% ethanol made with DEPC-treated water
- RNase-free 10 ml polypropylene tubes, microcentrifuge tubes, and pipette tips
- 15 ml polypropylene, conical tubes (e.g. Falcon Cat. No. 2096)

Method

1. Remove the medium from a 100 mm plate of confluent ES cells by aspiration.
2. Add 2 ml of PBS to the cells, swirl, and remove the PBS by aspiration.
3. Add 2 ml of working trypsin solution and incubate at 37°C for 1–2 min.
4. Add 8 ml of supplemented DMEM-H medium and pipette up and down to resuspend the cells.
5. Transfer the cells to a 15 ml polypropylene conical tube and pellet them by centrifuging at 1000 *g* for 5 min.
6. Remove the medium by aspiration and then flick the pellet sharply.
7. If feeder cell RNA is not desired, feeder cells can be removed as follows:
   (a) Add 3 ml of supplemented DMEM-H medium to the cell pellet and resuspend thoroughly by gentle pipetting.
   (b) Add an additional 10 ml of supplemented DMEM-H medium and plate the resuspended cells on a tissue culture plate without a feeder layer.
   (c) Incubate at 37°C for 20 min. During this time most of the feeder cells will adhere to the bottom of the plate, while the ES cells remain in the supernatant. A longer incubation time will remove a greater percentage of the feeder cells, but will also result in decreased ES cell recovery.

Protocol 7. *Continued*

(d) Recover the ES cells by pipetting them into a 15 ml conical tube and pelleting them by centrifugation at 1000 *g* for 5 min.

(e) Remove the supernatant and then flick the pellet sharply.

8. Add 3 ml of RNAzol B to the cells and pipette the mixture up and down to lyse the cells.
9. Add 0.3 ml of chloroform and shake well to mix the phases. Keep on ice for 5 min.
10. Centrifuge for 15 min at 12 000 *g*.
11. Remove the clear upper phase (which contains the RNA) and transfer this to an RNase-free 10 ml tube. The volume of the upper phase should be about 1.5 ml. Once the RNA has been extracted, special precautions must be taken to prevent RNase contamination.[a]
12. Add an equal volume of isopropanol (about 1.5 ml) and invert to mix. Keep on ice for 5 min, then centrifuge at 12 000 *g* for 15 min to pellet the RNA.
13. Pour off the supernatant and rinse the RNA pellet with 3 ml of the RNase-free 75% ethanol.
14. Centrifuge for 10 min at 12 000 *g* and pour off the ethanol.
15. The RNA can be stored in ethanol as a precipitate, or can be suspended in DEPC-treated water. If suspending in water, be sure to air dry the RNA pellet free of ethanol before adding the water. In either instance, store the RNA at −70°C.

[a] For this procedure, it is important to wear disposable gloves and change them frequently to prevent RNase contamination. It is also important that microcentrifuge tubes and pipette tips are kept RNase-free by handling with disposable gloves before autoclaving as well as after sterilization. Plasticware that is prepared for RNA isolation should be kept separate from the rest of the laboratory stock to further prevent contamination. If glass pipettes are used, they must be baked at 180°C for 8 h or longer. (If individually wrapped plastic pipettes are used, the investigator should be aware that polystyrene is susceptible to the solvents phenol and chloroform.)

7.2.2 Northern blotting

Northern blot analysis entails the separation of RNA by size on a denaturing agarose gel (61), usually with formaldehyde or glyoxal as the denaturing agent (58). After separation, the RNA is transferred to a membrane and probed for the mRNA of interest. Total RNA can be used for Northern analysis when the mRNA is present in sufficient amounts; purified mRNA is used when detecting rare mRNAs. The methods and applications for Northern analysis are described in detail in Chapter 7 of this book.

7.2.3 Reverse transcriptase PCR (RT–PCR)

RNA isolated from ES cells can be used for qualitative expression analysis by RT–PCR (*Protocol 8*). In this procedure, reverse transcriptase is used to generate single-stranded cDNA from total RNA or mRNA, and the cDNA is then used as a template for PCR. RT–PCR is usually applied to determine the presence or absence of a specific mRNA, although it can also be used to quantify mRNAs. It may be necessary to remove contaminating DNA from the RNA preparation by DNase I treatment before use in reverse transcription. Although RNA prepared as in *Protocol 7* is largely free of DNA, even traces of DNA contamination can become problematic during the PCR steps. Thus if PCR primers hybridize within the same exon, contaminating genomic DNA may be amplified, resulting in a false positive signal.

The availability of commercial kits which supply the necessary reagents at the proper concentrations makes RT–PCR very simple. We have successfully used a commercial kit, Invitrogen cDNA Cycle Kit (Cat. No. L1310–02).

Protocol 8. RT–PCR analysis using total RNA from ES cells

Equipment and reagents

- RNase-free microcentrifuge tubes and pipette tips
- DEPC-treated autoclaved water
- Water-baths at 37°C, 60°C, and 95°C
- AMV reverse transcriptase (e.g. Boehringer Mannheim Cat. No. 1495 062)
- 5 × AMV reverse transcriptase buffer (included with enzyme): 250 mM Tris–HCl pH 8.5, 40 mM $MgCl_2$, 150 mM KCl, 5 mM DTT
- Oligo dT primer (e.g. Boehringer Mannheim Cat. No. 814261; 0.2 μg/μl in DEPC-treated sterile water) *or* random hexamer primer (e.g. Boehringer Mannheim Cat. No. 1034731; 1 μg/μl)
- Deoxynucleotide mixture: 25 mM each of dATP, dCTP, dTTP, dGTP; Boehringer Mannheim Cat. No. 1277049)
- 10 U/μl RNase inhibitor (e.g. Boehringer Mannheim Cat. No. 799017)
- *Taq* DNA polymerase (e.g. Perkin Elmer Cetus AmpliTaq Cat. No. N801–0060)
- 10 × PCR buffer: 0.67 M Tris pH 8.8, 100 mM 2-mercaptoethanol, 166 mM ammonium sulfate, 67 μM EDTA, 0.5 μg/ml BSA (store at −20°C)
- 0.1 M $MgCl_2$ (store at 4°C)
- Forward and reverse primers for PCR (each at 0.4 μg/μl)
- PCR machine (e.g. Perkin Elmer Cetus Cat. No. N801–0001)
- PCR tubes compatible with machine (e.g. Perkin Elmer Cetus Cat. No. N801–0533 and N801–0534)
- Mineral oil (optional)

A. *Reverse transcription of RNA*

Just before starting the experiment, thaw the primers, nucleotides, and 5 × reaction buffer on ice, but keep the RNase inhibitor and reverse transcriptase at −20°C.

1. To an RNase-free microcentrifuge tube on ice, add 1 μg of total RNA (or 0.1 μg of mRNA)[a] and sterile, DEPC-treated water to a final volume of 12 μl.
2. Add 1 μl of *either* 0.2 μg/μl oligo dT primer *or* 1 μg/μl random hexamer primer mix using the micropipettor.
3. Heat to 60°C for 15 min to denature the RNA.

Protocol 8. *Continued*

4. Add the following reagents in sequential order:
   - RNase inhibitor 1 μl
   - 5 × reverse transcriptase buffer 4 μl
   - deoxynucleotide mixture 1 μl
   - AMV reverse transcriptase 1 μl

   Incubate at 37°C for 2 h.
5. Inactivate the reverse transcriptase by heating to 95°C for 15 min. Use the cDNA immediately for PCR amplification (see below), or store it at −20°C for later use.[b] The reverse transcription reaction yields approximately 100 ng of first strand cDNA in 3 μl of the reaction.

B. *PCR amplification of cDNA*

1. Add the following reagents to a PCR tube and mix:
   - approx. 100 ng of cDNA (from part A, step 5) 3.0 μl
   - 0.4 μg/μl forward primer 2.0 μl
   - 0.4 μg/μl reverse primer 2.0 μl
   - deoxynucleotide mixture 1.0 μl
   - 0.1 M $MgCl_2$ 1.5 μl
   - 10 × PCR buffer 5.0 μl
   - *Taq* polymerase 0.8 μl
   - distilled, autoclaved water 34.7 μl
2. If the PCR machine does not have a heating plate for the top of the tubes as well as the bottom, the high temperatures will cause the reactions to evaporate and condense at the top of the tube. This will change the concentration of the PCR reagents and prevent amplification. To prevent evaporation when using such a PCR machine, overlay the reaction with 25 μl of mineral oil.
3. Amplify the cDNA for 30 to 40 cycles; the following cycling conditions are provided as an example[c]:
   - denature for 50 sec at 94°C
   - anneal primers for 2 min at 60°C
   - extend for 3 min at 72°C

[a] If the mRNA of interest is rare, increase the amount of total RNA used to 2–3 μg.
[b] A phenol/chloroform extraction followed by ethanol precipitation may be performed before storage at −20°C, but we have not found this to be necessary.
[c] The annealing temperature will depend upon the size and base composition of the primers used and on the desired stringency of hybridization. The exact conditions that prove optimal for primer pairs need to be determined empirically. The specificity of primer binding can be improved by increasing the annealing temperature and by decreasing the extension time. If amplified products of PCR are not detected, it may be necessary to reduce the stringency of hybridization by decreasing the annealing temperature. Increasing the extension time should also be tried.

8. Preparation of ES cells for microinjection

The preparation of ES cells for microinjection follows essentially the same methodologies as described previously in the chapter for the passage and maintenance of ES cells. Although very few cells are actually required for blastocyst injection, a 60 mm or 100 mm plate of cells is maintained. This allows for evaluation of the ES cell culture to predict in part the cell line's ability to produce chimeric animals. Cultures that are 'past their peak' should be avoided; these cultures will have exhausted their media or have ES cells sluffing off from the colonies. Cultures with extensive differentiation should also be avoided; these cultures have ES cell colonies in which large, flattened individual cells can be seen. Generally targeted ES cells can be injected up to approximately passage eight to ten without appreciable changes in the chimerism of the animals, with passage one being the number assigned to the first freeze of the targeted cell line. Preparation of an ES cell line for microinjection is described in *Protocol 9*. If the colonies of ES cells are small and the fibroblast feeder layer is relatively 'heavy', removal of the fibroblasts is advised to facilitate injection. This is done by resuspending the cells in supplemented DMEM-H medium following trypsinization and returning the cells to the incubator for 20 minutes. During this time, the fibroblast cells preferentially adhere to the plate while the ES cells remain in suspension. After 20 minutes the plate is gently swirled and the non-adherent cells are collected and prepared as described in *Protocol 9*. However, removal of fibroblasts may not be necessary, since fibroblasts can be distinguished from ES cells under the microscope during microinjection due to their 'rough' membranes. This is in distinct contrast to the smooth membrane of the ES cell.

We generally inject about 15 ES cells into the blastocoel of each embryo. However, we have encountered situations where this proves to be excessive and results in early death of highly chimeric pups. In these cases success can often be achieved by reducing the number of cells injected.

Protocol 9. Preparation of cells for microinjection

Equipment and reagents

- 100 mm plate of ES cells for microinjection
- Supplemented DMEM-H medium (see *Table 4*)
- PBS pH 7.5 (see *Protocol 2*)
- 15 ml conical tube (e.g. Falcon Cat. No. 2087)
- Working trypsin solution (see *Table 4*)
- CO_2-independent medium (Gibco-BRL Cat. No. 18045–021)
- Dimethylpolysiloxane (Sigma Cat. No. DMPS-2X)
- 100 mm tissue culture dish

Method

1. Passage a 100 mm plate of confluent ES cells at a 1:5 or 1:10 dilution two to three days prior to injection.

Protocol 9. *Continued*

2. Feed the cells with warmed supplemented DMEM-H medium 3–4 h before injection.
3. Rinse the cells with 2 ml of PBS and remove from the cells by aspiration. Repeat.
4. Add 2 ml of working trypsin solution to the cells and incubate at 37 °C for 1–2 min.
5. Add an additional 2 ml of working trypsin solution to the cells and resuspend the cells with gentle pipetting.
6. Add 10 ml of supplemented DMEM-H medium to the cells and transfer the cell suspension to a 15 ml conical tube.
7. Pellet the cells by centrifugation at 1000 *g* for 5 min.
8. Wash the pellet with 10 ml of supplemented DMEM-H medium and then pellet again by centrifugation.
9. Resuspend the pellet in 1 ml of CO_2-independent medium.
10. Transfer 100 μl of the cell suspension to a 1.5 ml microcentrifuge tube and dilute the cells tenfold by adding 900 μl of CO_2-independent medium.
11. Prepare a microinjection droplet by adding 100 μl of CO_2-independent medium to the bottom of a 100 mm tissue culture dish. Submerge the droplet by pouring a layer of dimethylpolysiloxane over the droplet.
12. Add 20 μl of the diluted cells and the three and a half day embryos to the microinjection droplet.

Acknowledgements

We would like to acknowledge the contributions of people in both our laboratory and Oliver Smithies laboratory for the development of the protocols described. We give special thanks to Tom Doetschman and Anne Latour for tissue culture protocols. We would also like to thank John Snouwaert for critical review of the manuscript and Andrew Mohn for assistance with illustrations. This work was supported by Research Grant 1R01 DK46003–01 with equivalent cofunding by the NHLBI and by NIH Grant P01–DK38103 (B.H.K.).

References

1. Evans, M. J. and Kaufman, M. H. (1981). *Nature*, **292**, 154.
2. Martin, G. R. (1981). *Proc. Natl Acad. Sci. USA*, **78**, 7634.
3. Robertson, E. J. (1987). In *Teratocarcinomas and embryonic stem cells: a practical approach* (ed. E. J. Robertson), pp. 71–112. IRL Press, Oxford.

4. Doetschman, T. C., Eitstetter, H., Katz, M., Schmidt, W., and Kemler, R. (1985). *J. Embryol. Exp. Morph.*, **87**, 27.
5. Bradley, A., Evans, M. J., Kaufman, M. H., and Robertson, E. J. (1984). *Nature*, **309**, 255.
6. Gossler, A., Doetschman, T., Korn, R., Serfling, E., and Kemler, R. (1986). *Proc. Natl Acad. Sci. USA*, **83**, 9065.
7. Robertson, E. J., Bradley, A., Evans, M. J., and Kuehn, M. R. (1986). *Nature*, **323**, 445.
8. Smith, A. G. and Hooper, M. L. (1987). *Dev. Biol.*, **121**, 1.
9. Williams, R. L., Hilton, D. J., Pease, S., Willson, T. A., Stewart, C. L., Gearing, D. P., *et al.* (1988). *Nature*, **336**, 684.
10. Smith, A. G., Heath, J. K., Donaldson, D. D., Wong, G. G., Moreau, J., Stahl, M., *et al.* (1988). *Nature*, **336**, 688.
11. Moreau, J. F., Donaldson, D. D., Bennett, F., Witek-Gianotti, J., Clark, S. C., and Wong, G. G. (1988). *Nature*, **336**, 690.
12. Nichols, J., Evans, E. P., and Smith, A. G. (1990). *Development*, **110**, 1341.
13. Soriano, P., Montgomery, C., Geske, R., and Bradley, A. (1991). *Cell*, **64**, 693.
14. Nagy, A., Rossant, J., Nagy, R., Abramow-Newerly, W., and Roder, J. (1993). *Proc. Natl Acad. Sci. USA*, **90**, 8424.
15. Deng, C. and Capecchi, M. R. (1992). *Mol. Cell. Biol.*, **12**, 3365.
16. te Riele, H., Robanus Manandag, E., and Berns, A. (1992). *Proc. Natl Acad. Sci. USA*, **89**, 5128.
17. van Deursen, J. and Wieringa, B. (1992). *Nucleic Acids Res.*, **20**, 3815.
18. Hasty, P., Rivera-Perez, J., and Bradley, A. (1991). *Mol. Cell. Biol.*, **11**, 4509.
19. Charron, J., Malynn, B. A., Robertson, E. J., Goff, S. P., and Alt, F. W. (1990). *Mol. Cell. Biol.*, **10**, 1799.
20. Joyner, A. L., Skarnes, W. C., and Rossant, J. (1989). *Nature*, **338**, 153.
21. Mansour, S. L., Thomas, K. R., and Capecchi, M. R. (1988). *Nature*, **336**, 348.
22. Mombaerts, P., Clarke, A. R., Hooper, M. L., and Tonegawa, S. (1991). *Proc. Natl Acad. Sci. USA*, **88**, 3084.
23. Mansour, S. L., Thomas, K. R., Deng, C., and Capecchi, M. R. (1990). *Proc. Natl Acad. Sci. USA*, **87**, 7688.
24. Zimmer, A. and Gruss, P. (1989). *Nature*, **338**, 150.
25. Davis, A. C., Wims, M., and Bradley, A. (1992). *Mol. Cell. Biol.*, **12**, 2769.
26. Reid, L. H., Shesely, E. G., Kim, H. S., and Smithies, O. (1991). *Mol. Cell. Biol.*, **11**, 2769.
27. Valancius, V. and Smithies, O. (1990). *Mol. Cell. Biol.*, **11**, 1402.
28. Hasty, P., Ramirez-Solis, R., Krumlauf, R., and Bradley, A. (1991). *Nature*, **350**, 243.
29. Barinaga, M. (1994). *Science*, **265**, 26.
30. Sauer, B. (1993). In *Methods in enzymology* (eds Wasserman, P. M. and DePamphilis, M. L.), Vol. 225, pp. 890–900.
31. Sauer, B. and Henderson, N. (1989). *Nucleic Acids Res.*, **17**, 147.
32. Sauer, B. and Henderson, N. (1990). *New Biologist*, **2**, 441.
33. Jung, S., Rajewsky, K., and Radbruch, A. (1993). *Science*, **259**, 984.
34. Gu, H., Zou, Y. R., and Rajewsky, K. (1993). *Cell*, **73**, 1155.

35. Orban, P. C., Chui, D., and Marth, J. D. (1992). *Proc. Natl Acad. Sci. USA*, **89**, 6861.
36. Lasko, M., Sauer, B., Mosinger, B., Lee, E. J., Manning, R. W., Yu, S. H., *et al.* (1992). *Proc. Natl Acad. Sci. USA*, **89**, 6232.
37. Gu, H., Marth, J. D., Orban, P. C., Mossmann, H., and Rajewski, K. (1994). *Science*, **265**, 103.
38. Reid, L. H., Gregg, R. G., Smithies, O., and Koller, B. H. (1990). *Proc. Natl Acad. Sci. USA*, **87**, 4299.
39. te Riele, H., Robanus Maandag, E., Clarke, A., Hooper, M., and Berns, A. (1990). *Nature*, **348**, 649.
40. de la Luna, S. and Ortin, J. (1992). In *Methods in enzymology* (ed. Wu, R.), Vol. 216, pp. 376–385.
41. Bautista, D. and Shulman, M. J. (1993). *J. Immunol.*, **151**, 1950.
42. Reddy, S., Rayburn, H., von Melchner, H., and Ruley, H. E. (1992). *Proc. Natl Acad. Sci. USA*, **89**, 6721.
43. Hooper, M., Hardy, K., Hyside, A., Hunter, S., and Monk, M. (1987). *Nature*, **326**, 292.
44. Yagi, T., Ikawa, Y., Yoshida, K., Shigetani, Y., Takeda, N., Mabuchi, I., *et al.* (1990). *Proc. Natl Acad. Sci. USA*, **87**, 9918.
45. Chu, G., Hayakawa, H., and Berg, P. (1987). *Nucleic Acids Res.*, **15**, 1311.
46. Lovell-Badge, R. H., Bygrave, A. E., Bradley, A., and Robertson, E. J. (1985). *Cold Spring Harbor Symp. Quant. Biol.*, **50**, 707.
47. Lovell-Badge, R. H. (1987). In *Teratocarcinomas and embryonic stem cells: a practical approach* (ed. E. J. Robertson), pp. 153–182. IRL Press, Oxford.
48. Felgner, P. L. and Ringold, G. M. (1989). *Nature*, **337**, 388.
49. Baubonis, W. and Sauer, B. (1993). *Nucleic Acids Res.*, **21**, 2025.
50. Felgner, P. L., Gadek, T. R., Holm, M., Roman, R., Chan, H. W., Wenz, M., *et al.* (1987). *Proc. Natl Acad. Sci. USA*, **84**, 7413.
51. Strauss, W. M., Dausman, J., Beard, C., Johnson, C., Lawrence, J. B., and Jaenisch, R. (1993). *Science*, **259**, 1904.
52. Choi, T. K., Hollenbach, P. W., Pearson, B. E., Ueda, R. M., Weddell, G. N., Kurahara, C. G., *et al.* (1993). *Nature Genet.*, **4**, 117.
53. Ellis, J. and Bernstein, A. (1989). *Mol. Cell. Biol.*, **9**, 1621.
54. Koller, B. H. and Smithies, O. (1989). *Proc. Natl Acad. Sci. USA*, **86**, 8932.
55. Koller, B. H., Kim, H.-S., Latour, A. M., Brigman, K., Boucher, R. C., Scambler, P., *et al.* (1991). *Proc. Natl Acad. Sci. USA*, **88**, 10730.
56. Chan, S. Y. and Evans, M. J. (1991). *Trends Genet.*, **7**, 76.
57. Miller, S. A., Dykes, D. D., and Polesky, H. F. (1988). *Nucleic Acids Res.*, **16**, 1215.
58. Sambrook, J., Fritsch, E. F., and Maniatis, T. (ed.) (1989). *Molecular cloning, a laboratory manual.* Cold Spring Harbor Press, Cold Spring Harbor, NY.
59. Asubel, F. M., Brent, R., Kingston, R. E., Moore, D. D., Seidman, J. G., Smith, J. A., *et al.* (1988). *Short protocols in molecular biology.* Greene Publishing Associates and Wiley-Interscience, New York.
60. Chomczynski, P. and Sacchi, N. (1987). *Anal. Biochem.*, **162**, 156.
61. Lehrach, H., Diamond, D., Wozney, J. M., and Boedtker, H. (1977). *Biochemistry*, **16**, 4743.

6

Production of transgenic rodents by microinjection of cloned DNA into fertilized one-cell eggs

SARAH JANE WALLER, MEI-YIN HO, and DAVID MURPHY

1. Introduction

Since the pioneering demonstration in the early 1980s that the introduction of recombinant growth hormone DNA into the pronuclei of single cell mouse embryos could result in inheritable alterations in the growth of these mice (1), the techniques of mammalian transgenesis have led to significant advances in our understanding of the regulation of numerous complex biological processes. The power and appeal of transgenesis lies in its ability to allow the study of almost any protein within a truly physiological environment, that of the whole organism. This permits detailed studies of the developmental and physiological regulation of the protein as well as its function in context with the normal processes occurring in the animal.

Once recombinant DNA has been incorporated into the germline, the 'transgene' will be transmitted to the descendants of the founder animal in a classic Mendelian fashion. Appropriate tissue-specific and physiologically regulated expression of the transgene is a prerequisite for many transgenic experiments, and can be obtained only by including the appropriate genetic elements within its structure. Indeed, many transgenic studies are specifically designed to localize and identify such regulatory elements. Once appropriate expression of transgenes has been obtained, either gain-of-function (overexpression of the protein of interest) or loss-of-function (cell-specific expression of cytotoxic proteins) approaches can be employed to derive pertinent information about almost any gene or cell type. Studies of the nervous and immune systems, and the process of oncogenesis are just some examples in which transgenic animals have made major contributions to our knowledge (2–4). Such work has led to the development of transgenic animals as suitable models for a large number of human diseases (5), which should allow the development of novel therapies.

In this chapter we describe the procedures necessary for producing transgenic mice and rats by microinjection of recombinant DNA into the pronuclei of fertilized one-cell eggs. This technique demands precise technical skill and expensive equipment, but its speed and reliability establish it as the most efficient method for the production of transgenic mice, and it is currently the only route to transgenic rats. The methodologies for producing transgenic mice by infection of pre-implantation embryos with recombinant retroviruses or by homologous recombination in embryonic stem cells are described in other chapters in this series.

The mouse has generally been the preferred species for transgenesis, due largely to its well-characterized genetics and embryology, but we will also describe the utilization of the microinjection technique for the production of transgenic rats, a species whose size makes it more accessible to many physiological and neurobiological studies. The procedures required for the generation of a transgenic animal from either species are essentially the same and, unless otherwise stated, any given protocol can be used for rats and mice. In Chapter 7, we present protocols that allow the experimenter to analyse the resulting transgenic rodents, from their identification, through examination of transgene expression at the mRNA and protein level, to post-mortem examination of the whole animal.

2. Summary of microinjection method

The process of making transgenic rodents by microinjection is summarized in flow diagram form in *Figure 1*. The method is outlined below, with reference to the appropriate sections of the rest of this chapter.

(a) Donor females are superovulated (section 5.1) and mated with stud males.

(b) About 12 hours post-coitum (p.c.) the oviducts of the donor females are removed and the fertilized one-cell eggs are collected (section 5.4) and placed into culture.

(c) The eggs are then microinjected (section 6.4) with purified cloned DNA fragments (section 6.2).

(d) Eggs that survive injection are returned by oviduct transfer (section 7.2) to the natural environment provided by a pseudopregnant recipient surrogate mother (section 7.1). Pseudopregnant recipients are produced by mating sexually mature females with vasectomized males (sections 4.1.3 and 4.2.3).

(e) A proportion of the transplanted eggs will survive to term and will either be delivered naturally or by Caesarean section (section 8.1).

(f) Transgenic animals in a litter are identified by analysis of genomic DNA isolated from tail tissue (Chapter 7, section 2).

(g) Transgenic animals are then bred to produce a line (section 8.2) and analysed for transgene expression (Chapter 7, section 3).

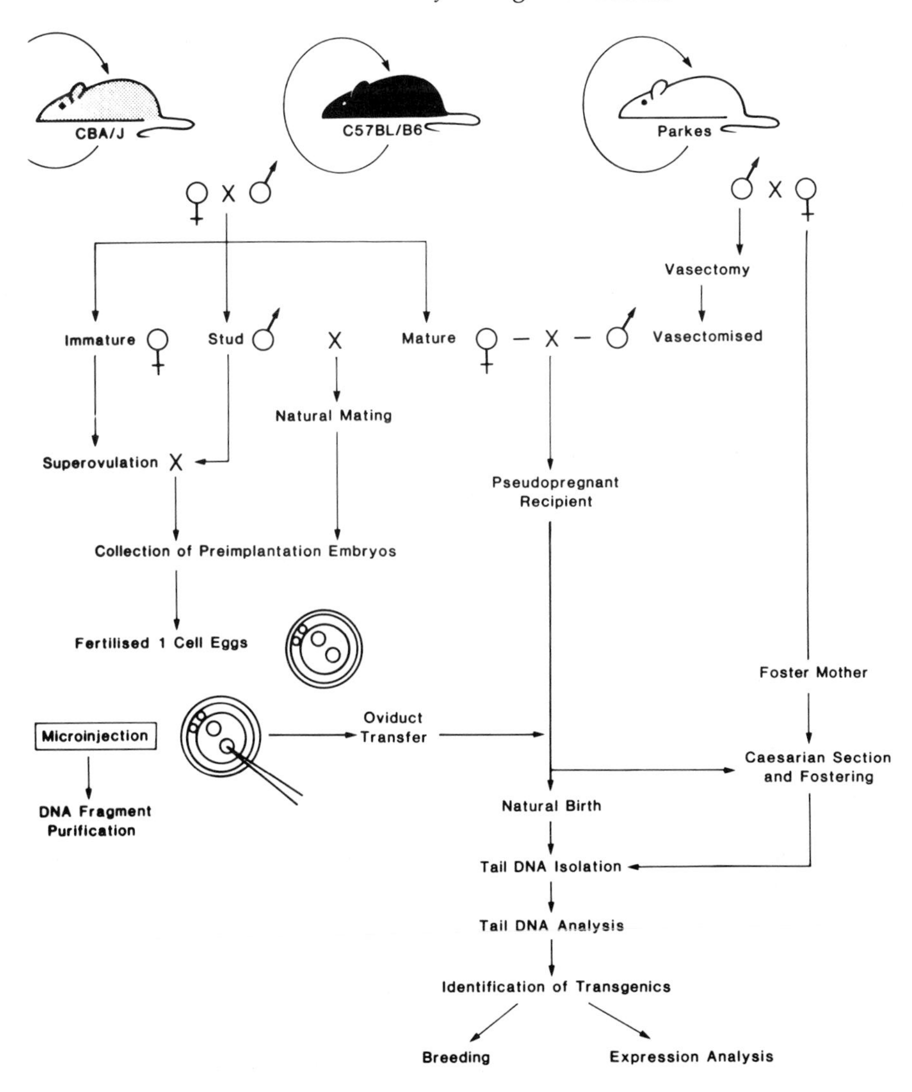

Figure 1. Flow diagram of the process of making transgenic mice by the microinjection of fertilized one-cell eggs with cloned DNA. The process is similar for making transgenic rats, except that a single strain of rats, Sprague Dawley, can be used throughout.

The efficiency of producing transgenic rats or mice varies considerably between experiments. Mice have been used to a much greater extent in transgenic studies, and the efficiencies of each stage of the procedure are well recorded. Under optimal conditions, 60–80% of eggs survive injection. Of these, 10–30% implant in the pseudopregnant recipient, proceed through normal development, and are born. Typically, 10–30% of pups born are transgenic. Fewer data are available for the rat, but in our hands we find that 50–75% of eggs survive injection, 10–30% implant and are born as pups, and 10–33% of these pups are transgenic. Some of the variables governing the efficiency are somewhat intangible, for example experimental dexterity. However, both the condition of the DNA used to inject the eggs (section 6.2) and the choice of mouse strain can have a profound effect on the overall efficiency of the technique, and both must be carefully controlled.

The mechanism by which injected DNA integrates into host chromosomes is unknown, but some information about the nature of the process have been inferred from a study of the state and organization of the inserts found in transgenic mice (6). About 70% of transgenic mice carry exogenous DNA in all of their somatic and germ cells, indicating that integration usually occurs prior to the first round of DNA replication. The remaining 30% of transgenic mice show some degree of mosaicism, and in these animals integration must have occurred at some stage after the first round of DNA replication. The copy number of the transgene can vary considerably; from one to several hundred. However with a particular founder animal, there is usually only one integration site. Multiple copies of a transgene are usually arranged in a head-to-tail tandem array within a single locus. Integration events have been observed on many different autosomes (7), on the X chromosome (8), and on the Y chromosome (D. Murphy, unpublished observations).

3. Animal welfare

The conduct of scientific experiments involving animals is strictly regulated by most governments and institutions. Scientists must consult with the relevant governmental or institutional authorities before commencing upon any series of experiments utilizing live animals and/or genetic engineering. Legal requirements aside, researchers must always abide by the basic principle that animals in their care should not experience any avoidable distress. Guidelines for such care are published by the National Institutes of Health (9).

The techniques using live animals described in this and the following chapter should be referred to only as a guide to learning by the novice. All manipulations involving animals should be taught directly to the novice by a skilled and experienced operative. For the surgical procedures (vasectomy and oviduct transfer) it is recommended that the novice gains skill and confidence on cadavers before moving on to live anaesthetized subjects. If a live animal appears to be in any distress or discomfort, then the operator (novice or skilled) should immediately kill the animal by cervical dislocation or decapitation (see below).

3.1 Sacrifice of rodents

The recommended method for killing mice is to break the neck (cervical dislocation). This is quick and causes the animal the minimum of distress.

Pick the animal up by the tail and place it on top of the cage. Still holding it by the tail, allow the mouse to run away such that it is stretched out with its hindlegs almost in the air and its forelimbs gripping the cage bars firmly. Apply firm pressure at the base of the skull with a blunt instrument, e.g. a spatula blade. At the same time pull on the tail. It is this stretching action that breaks the neck, the pressure at the base of the spine defining the point of dislocation.

Cervical dislocation or decapitation with a rodent guillotine are used to kill rats. However, the greater size and strength of rats usually requires that they are stunned before either of these procedures. Hold the animal by its tail, then quickly swing it over, hitting the back of the animal's head firmly on a hard surface, e.g. a table top. While the animal is still stunned, dislocate its neck by applying pressure to the base of the skull with a large pair of scissors while simultaneously pulling the animal's tail. Alternatively, a guillotine (Harvard Apparatus) can be used following stunning if decapitation is preferred.

3.2 Anaesthetizing rodents

We have found that the anaesthetics described in *Protocol 1* produce a reliable depth of anaesthesia with a minimal degree of mortality for the duration of the operations described in this chapter.

Protocol 1. Anaesthesia of rodents

Equipment and reagents

For mice:

- 100% Avertin stock. Prepare by mixing 10 g of tribromoethyl alcohol with 10 ml of *tertiary* amyl alcohol. Store at 4°C, protected from light. To make the working solution, dilute the stock solution to 2.5% (v/v) in sterile water. Store at 4°C, protected from light.
- Sterile 1 ml disposable syringe fitted with 0.5 × 16 mm needle

For rats:

- CRC anaesthetic. Mix one part Dormicum™ (5 mg/ml midazolam, Hoffmann-La Roche), one part Hypnorm™ (10 mg/ml Fluanisone and 0.2 mg/ml fentanyl, Janssen Pharmaceuticals) and two parts sterile water. Once made up, this solution is stable at room temperature for one week.
- Sterile 1 ml disposable syringe fitted with 0.5 × 16 mm needle

A. *Anaesthetizing mice*

1. Weigh the animal and determine the dose of Avertin required; use 15–17 μl per gram of body weight.
2. Fill a sterile 1 ml disposable syringe fitted with a 0.5 × 16 mm needle with the required dose. Exclude any air bubbles.
3. Restrain the animal as shown in *Figure 2* and introduce the needle into the abdomen of the mouse, avoiding the bladder and the diaphragm. Inject the anaesthetic, then wait momentarily before withdrawing the needle. Accidental subcutaneous injection is revealed by leakage of drops of Avertin through the skin.

Protocol 1. *Continued*

4. The animal will remain fully anaesthetized for 30–60 min; sufficient time to perform the surgical task. Following the operation, leave the animal undisturbed to recover in a quiet, warm place.

B. *Anaesthetizing rats*

1. Weigh the animal and determine the dose of CRC anaesthetic required; use 0.275 ml per 100 g of rat.
2. Introduce the anaesthetic intraperitoneally as described for mice above. If the rat is too large to be restrained with one hand, obtain the help of an assistant to restrain the animal with both hands, or use one of the many commercially available plastic animal restrainers.
3. The rat will stay fully anaesthetized for 20–30 min. Following the operation, allow the rat to recover in a quiet, warm place.

4. Animal stocks

The production and maintenance of transgenic rats and mice necessitates that the investigator is responsible for the care of hundreds and maybe thousands of rodents. As such, only scientists working in institutions with the necessary facilities and experience can carry out transgenic experiments. A suitable animal facility must be able to provide:

- spacious caging, with a regular change of clean, comfortable bedding, and cage washing (and preferably sterilization)
- a food and water supply
- environmental control (i.e. temperature, light, humidity, and ventilation)
- access to veterinary care

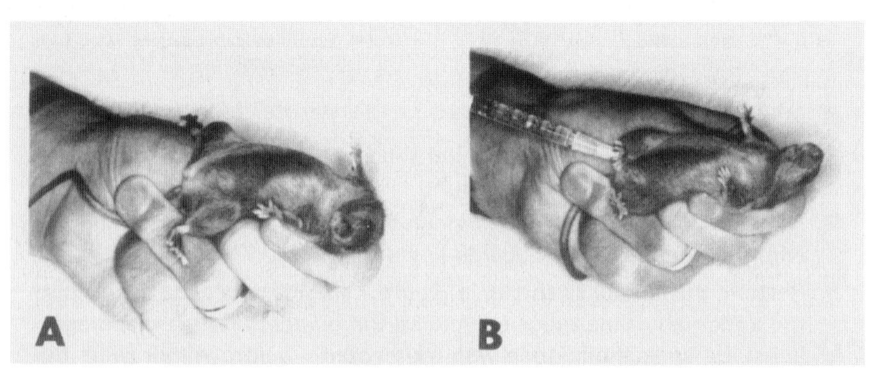

Figure 2. Demonstration of the correct way to restrain (A) and intraperitoneally inject (B) a mouse.

As can be seen from *Figure 1*, a number of different types of animal are required for producing fertilized eggs for microinjection. In addition to these, many more animals will be produced in the process of generating and maintaining transgenic lines. Before embarking on any transgenic study, investigators must therefore ensure that they and their staff will be able to devote considerable time to the maintenance and analysis of these animals.

4.1 Mice stocks required

4.1.1 Mice needed to produce fertilized one-cell eggs

Fertilized one-cell eggs for microinjection are produced as a consequence of a mating between a donor female and a stud male. The choice of the strain of the mice is very important. Brinster *et al.*(10) compared the efficiency of transgenic mouse production following manipulation of C57BL/6J inbred mouse eggs and C57BL/6J × CBA/J hybrid eggs. A number of parameters were shown to be strain-dependent, including yield of eggs from the donor female and the survival of eggs following injection. Overall, the experiments on the hybrid eggs were eightfold more efficient than those on the inbred eggs. Inbred zygotes should be used only when the genetic background of the host animal needs to be carefully controlled.

The microinjection of F2 zygotes resulting from matings between CBA/J × C57BL/6J or /10 F1 hybrid males and superovulated females results in excellent efficiencies. Such a protocol demands the maintenance of the following animals:

(a) At least 20 F1 hybrid stud males caged individually. Sexually mature males will act as good studs for between eight months and a year. An individual male should be presented with a female only on every alternate day. A record of the individual animal's plugging (section 4.1) record should be kept and the animal replaced if it suffers from repeated poor performance.
(b) A supply of immature (four- to five-weeks-old; 12.5–14 g) F1 hybrid females. These are superovulated, mated with the F1 hybrid males, and sacrificed to supply the fertilized one-cell eggs. The number of females required will depend on the number of eggs that are to be injected; ten immature females will supply at least 250 injectable eggs.
(c) Large colonies of CBA/J male mice and C57BL/6J or /10 female mice need to be maintained in order to supply the F1 hybrid donor females by interstrain crosses. To provide 30 F1 females per week, 50 breeding pairs (male CBA/J and female C57BL/6J) will need to be maintained. Alternatively, it is often easier and cheaper to purchase immature F1 hybrid females from a reputable commercial or institutional supplier.

4.1.2 Mice needed to act as recipients for microinjected eggs

Pseudopregnant recipient females (0.5 day p.c.) are used as surrogate mothers to nurture surviving microinjected eggs to birth. Females can be of any strain with good maternal characteristics, but need to be sexually mature and greater than 19 g in weight. We use C57BL/6J × CBA/J F1 females. Oestrous females are made pseudopregnant by mating them with vasectomized males.

4.1.3 Vasectomized male mice

Sexually mature mice are sterilized by vasectomy (*Protocol 2*), and used to engender pseudopregnancy in sexually mature females. Any strain of mouse can be used, but Parkes or Swiss males are particularly suitable because of their good performance; 20–30 vasectomized males, mated on alternate days with females analysed for their stage in the oestrous cycle, should be able to provide at least five pseudopregnant recipients per day.

Protocol 2. Vasectomizing male mice and rats

Equipment and reagents

- Dissection instruments (e.g. from Arnold Horwell): dissection scissors (large and small), fine blunt forceps (one pair curved, one pair straight), sharp curved forceps, watchmaker's forceps (size 5), curved surgical needle (size 10, triangular, pointed), surgical silk sutures (size 5), autoclips and applicator
- Male mice (about two-months-old) or male rats (five-weeks-old)
- Anaesthetics: Avertin for mice, CRC for rats (see *Protocol 1*)
- Bunsen burner
- 70% ethanol in a squeeze bottle

A. *Vasectomizing mice*

The vasectomy operation should be performed on young, healthy, sexually mature (around two-months-old) male mice.

1. Anaesthetize the mouse as described in *Protocol 1*. Place the animal abdomen side up on the lid of a 9 cm glass or plastic Petri dish.
2. Spray the lower abdomen with 70% (v/v) ethanol. Comb the hair away from the incision site (level with the top of the hindlimbs) with a pair of fine forceps.
3. Lift the skin away from the body wall at the incision point with a large pair of dissecting scissors (*Figure 3A*). Stretch the incision with the outer edges of a pair of scissors to prevent bleeding.
4. Cut the body wall. Stretch the incision to prevent bleeding.
5. Introduce a single stitch through the body wall to one side of the incision and leave the silk suture in place (*Figure 3B*).
6. Pull out the fat pad on one side of the animal and with it the testis, epididymis, and vas deferens (*Figure 3C*).
7. Identify the vas deferens located underneath the testis and free it of the membranes holding it (*Figure 3D*).

8. Hold the vas deferens in a loop with one pair of forceps. Heat a large pair of blunt forceps in the flame of a Bunsen burner until glowing red. Grip the vas deferens loop with the red-hot forceps tips. This will burn away the vas deferens and cauterize the ends (*Figure 3E*).
9. Separate the cauterized ends (*Figure 3F*).
10. Return the organs to the inside of the body wall using a blunt pair of forceps.
11. Repeat steps 6–10 on the other side of the animal.
12. Sew up the body wall with at least two stitches (*Figure 3G*).
13. Clip the skin together with autoclips (*Figure 3H*).
14. Cage the vasectomized males individually. Wait for a few weeks after the operation before using the animal to produce pseudopregnant surrogate mothers.

B. *Vasectomizing rats*

When performing vasectomies on rats, it is necessary to remove the testis with attached vas deferens and epididymis individually from each scrotal sac. Use young (five-week-old) male rats.

1. Anaesthetize the rat with CRC cocktail as described in *Protocol 1*.
2. Place the animal abdomen side up and swab the scrotal area with 70% (v/v) ethanol.
3. Lift the scrotal skin covering one testis and make a longitudinal cut of 1.5 cm.
4. Lift the body wall, which is very thin, and carefully make another longitudinal incision of about 1 cm. Introduce a single stitch into the body wall at one end of the incision and leave it in place.
5. With a blunt pair of forceps, gently pull out the fat pad and with it the testis, epididymis, and vas deferens.
6. Identify the vas deferens located underneath the testis and free it of the membranes holding it. Tie a piece of cotton thread tightly around each end of the vas deferens, and then sever the vas deferens with a pair of scissors between the two cotton ties.
7. Return the organs to the inside of the body using a blunt pair of forceps. It is sometimes difficult to relocate the thin body wall and to push the testis back inside—the single stitch introduced at the start of the surgery should help. Stitch up the body wall with two to three stitches.
8. Stitch up the skin, and repeat steps 3–7 on the other side of the animal.
9. Allow the animal to recover in a quiet, warm place, and then cage individually. Wait for three to four weeks before mating with recipient females.

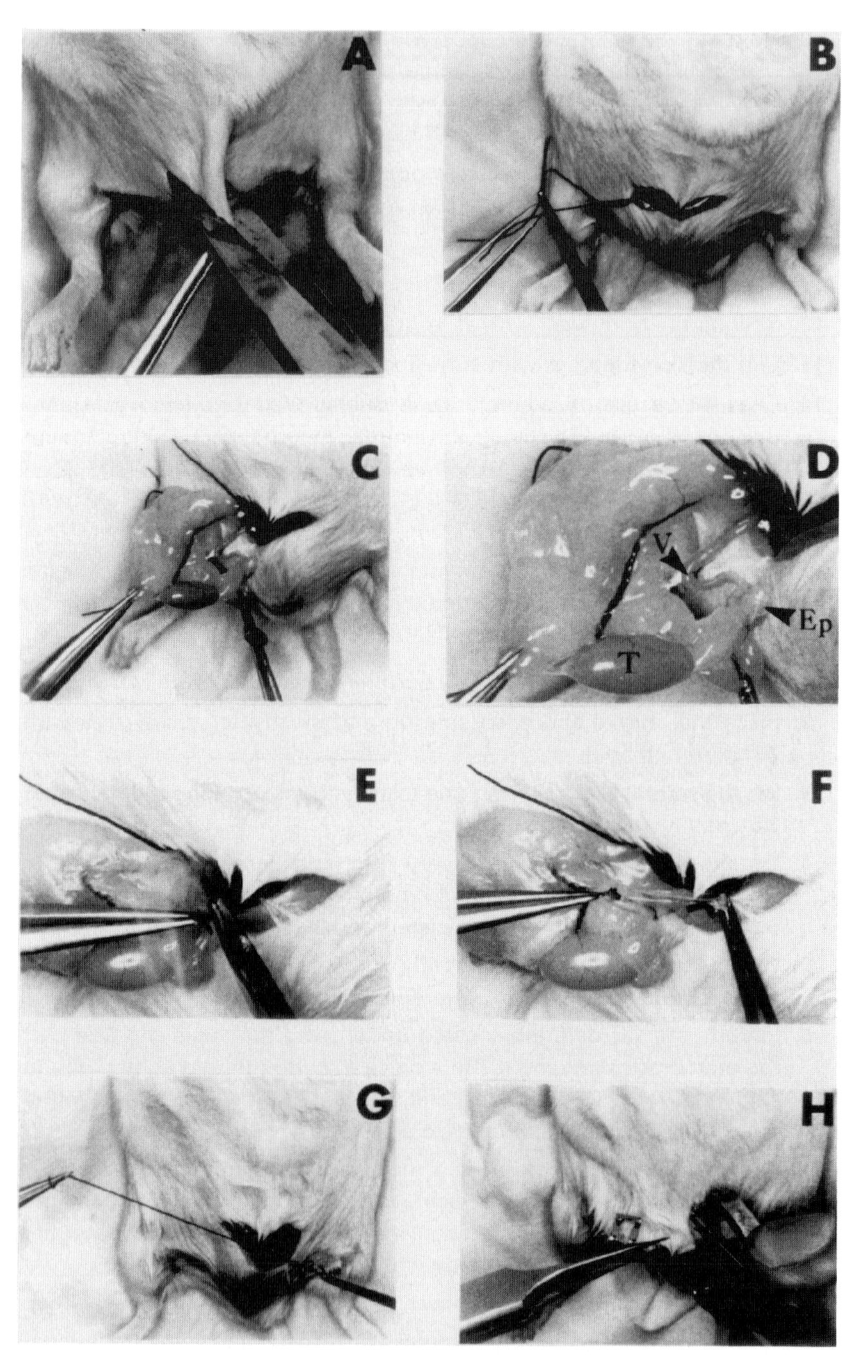

Figure 3. Vasectomizing a male mouse. Refer to *Protocol 2* for details. Key to symbols used in (D): T, testis; Ep, epididymis; V, vas deferens.

4.1.4 Mice needed to act as foster mothers

A larger animal unit may be able to supply suitable foster mothers to receive any pups delivered by Caesarean section, or born to a mother that dies unexpectedly (section 8.1). Whilst such a facility is desirable, it is not an absolute requirement. Foster mothers need to have pups of their own of about the same age as the pups to be fostered, and should be of a strain with good maternal characteristics (e.g. Swiss). We use C57BL/6J × CBA/J F1 females.

4.2 Rat stocks required

The generation of transgenic rats is a more recent development than the procedure for making transgenic mice, and as such, less research has been done on the efficiency of the technique using different rat strains. We use Sprague Dawley rats throughout, although other workers have reported the successful use of fertilized eggs produced from mating Sprague Dawley females with WKY males (11).

4.2.1 Rats required to produce fertilized one-cell eggs

For microinjection, we use fertilized eggs produced from mating super-ovulated immature Sprague Dawley females with Sprague Dawley stud males. To supply 30 immature females per week, ten breeding pairs should be sufficient.

4.2.2 Rats required to act as recipient mothers

We have found that females who have already given birth to one litter (exbreeders) make better surrogate mothers than virgins. Although virgins tend to produce larger litters than exbreeders following the oviduct transfer, they often eat the resulting pups. Oestrous exbreeder females (section 7.1) are mated with vasectomized males the night before the microinjected eggs are to be transferred to the pseudopregnant exbreeder.

4.2.3 Vasectomized male rats

Vasectomized male Sprague Dawley rats are needed to engender pseudo-pregnancy in the exbreeder female rats. *Protocol 2* details the vasectomy operation. Vasectomized male rats quickly grow to a large size and should be replaced every six to eight months.

4.2.4 Rats needed to act as foster mothers

As with mice, it is occasionally necessary to deliver pups by Caesarean section and to foster them to a mother that has a litter of her own of about the same age as the pups to be fostered. We use Sprague Dawley females for this purpose, which are suitably maternal.

5. Collection of fertilized one-cell eggs for microinjection

5.1 Superovulation of females

For most purposes, for example maintaining stocks of normal or transgenic strains, rats and mice can be allowed to mate and breed with the minimum of intervention. Breeding is controlled by the careful regulation of the transfer of males and females between cages. However, 'timed matings', where the timing of ovulation, copulation, and fertilization are controlled, are required to provide a supply of fertilized one-cell eggs for microinjection. The administration of gonadotrophins to immature females at precise times provides an accurate method of 'timing' matings to produce fertilized one-cell eggs. This superovulation (*Protocol 3*) also results in a much greater yield of eggs per female than natural matings (around 30–40 per animal compared to 10–15).

The gonadotrophins used to superovulate both rats and mice are the same, although the doses and timings are different for the two species. About 80–100% of female mice should be impregnated following this procedure, although it is less efficient with rats, resulting in successful impregnation in only 60–80% of females. Other laboratories (11) have reported the use of continual 'minipump' delivery of porcine follicle stimulating hormone (FSH) to rats instead of intraperitoneal injection of FSH. Although this method can result in a very large yield of eggs per rat (60–100), we have found that only a small percentage of the treated rats ovulate and mate.

Protocol 3. Superovulation of immature rats and mice

Reagents

- 50 IU/ml follicle stimulating hormone (FSH; Sigma 4877) in sterile water or 0.9% (w/v) NaCl (store in 1 ml aliquots at −70°C)
- 50 IU/ml human chorionic gonadotrophin (hCG; Sigma CG-2) in sterile water or 0.9% (w/v) NaCl (store in 1 ml aliquots at −70°C)

A. *Superovulation of mice*

1. To immature F1 hybrid female mice (12.5–14 g, four- to five-weeks-old), administer 0.1 ml of FSH intraperitoneally at 10.30 a.m.
2. Two days later, intraperitoneally inject each mouse with 0.1 ml of hCG at 10.30 a.m. Immediately place each female with a stud male.
3. The following morning, check each female for the presence of a copulatory plug. This is evidence that a successful mating has occurred and consists of a white mass of coagulated protein blocking the vagina. In most instances the plug is easily seen, but it sometimes lies deep in

the vagina, and the animal should be examined carefully with the aid of a smooth blunt probe.

B. *Superovulation of rats*

1. Inject four- to five-week-old female rats intraperitoneally with 0.4 ml FSH at 9.00 a.m.
2. Two days later inject each rat with 0.2 ml of hCG at 2.00 p.m. Immediately place each female with a stud male.
3. The following morning, check each female for the presence of a copulatory plug. With rats, the plug has often fallen out by the morning. A vaginal smear should be taken from any animal in which it can not be detected (see *Protocol 9*), and the contents examined under × 400 magnification for the presence of sperm.

5.2 Culture of fertilized one-cell eggs

For the purpose of making transgenic rats or mice, fertilized one-cell eggs need to be maintained outside of their natural environment for between 3 and 36 hours. Fertilized eggs are collected at approximately 12 hours post-coitus (coitus will usually occur around midnight) and are injected at some point during the following 12 hours. The eggs are then returned to the natural environment of a pseudopregnant surrogate mother soon after injection, while the eggs are still at the one-cell stage. Alternatively, the eggs can be cultured overnight and transferred at the two-cell stage.

Eggs at the one- and two-cell stage are maintained in microdrop cultures in a 37°C tissue culture incubator gassed with 5% CO_2. Microdrops are made up of small drops (40 μl) of M16 medium (*Table 1*) arranged in an array in a 35 mm culture dish and covered with liquid paraffin to prevent evaporation. When eggs are being handled outside the incubator, for example when they are being collected or microinjected, they are maintained in M2 medium (*Table 1*). The pH of M2 medium is maintained by the addition of Hepes buffer. Both M2 and M16 media contain bovine serum albumin (BSA). BSA reduces the stickiness of the eggs and adsorbs low level poisons. However, some batches of BSA will not sustain egg development. Each new stock of BSA should therefore be tested for its ability to sustain eggs through several cleavage divisions before being used to culture microinjected eggs. BSA from Sigma (Cat. No. A4161), is the only product that we have found to give survival of eggs to the blastocyst stage.

5.3 Egg transfer pipettes

Two types of egg transfer pipettes need to be prepared in advance; general transfer pipettes (*Figure 4Bi*) and oviduct transfer pipettes (*Figure 4Bii*).

Table 1. M2 and M16 media

Concentrated component stocks

Use Sigma tissue culture grade chemicals throughout. Make up all stocks using sterile disposable plastic containers and pipettes; washed glass items can be contaminated with detergents that are toxic to eggs. Filter all concentrated stocks through 0.45 μm Millipore filters into sterile plastic tubes. Store frozen at −20°C.

- 10 × A stock. Weigh out the following reagents into a 50 ml sterile plastic tube: 2.767 g NaCl (Sigma S5886), 0.178 g KCl (Sigma P5405), 0.081 g KH_2PO_4 (Sigma P5655), 0.1465 g $MgSO_4.7H_2O$ (Sigma M2643), 0.5 g glucose (Sigma G6138), 0.03 g penicillin (Sigma P3032), 0.025 g streptomycin (Sigma S9137). Weigh out 1.305 g sodium lactate (Sigma L4263) into a microcentrifuge tube and add this to the 50 ml tube. Rinse the microcentrifuge tube with double distilled water and use this to make 10 × A stock to 50 ml final volume.
- 10 × B stock. Dissolve 1.0505 g $NaHCO_3$ (Life Sciences 895–1810 IN) and 0.005 g phenol red (Sigma P5530) in water and make to 50 ml final volume.
- 100 × C stock. 0.18 g sodium pyruvate (Sigma P5280) in 50 ml water.
- 100 × D stock. 1.26 g $CaCl_2.2H_2O$ (Sigma C7902) in 50 ml water.
- 10 × E stock. Weigh out 2.979 g Hepes (Sigma H9136) and 0.005 g phenol red (Sigma P5530) into a sterile 50 ml tube and dissolve in 25 ml of double distilled water. Adjust to pH 7.4 with 0.2 M NaOH, then make to 50 ml final volume.

Preparation of M2 and M16 media from concentrated stock(s)

Prepare either M2 or M16 medium by mixing the stock solutions as described below. Measure the double distilled water into a sterile 50 ml plastic tube. Aliquot the concentrated stocks into the water, then carefully rinse the pipette by sucking the liquid up and down. Add the BSA and mix gently until dissolved. Pass through a 0.45 μm Millipore filter using a large sterile disposable syringe, aliquoting into sterile containers. Store at 4°C. Prepare fresh every two weeks.

Stock	M2	M16
10 × A	5.0 ml	5.0 ml
10 × B	0.8 ml	5.0 ml
100 × C	0.5 ml	0.5 ml
100 × D	0.5 ml	0.5 ml
10 × E	4.2 ml	–
Double distilled water	39.0 ml	39.0 ml
BSA (Sigma A4161)	0.2 g	0.2 g
Total volume	50 ml	50 ml

Both are made from hard glass capillaries (BDH Ltd.) and are assembled into a mouth-operated system made up of a mouthpiece, a rubber tube, and a pipette holder (components available from Arnold Horwell). Methods for preparing the pipettes are detailed in *Protocol 4*, and illustrated in *Figure 4*.

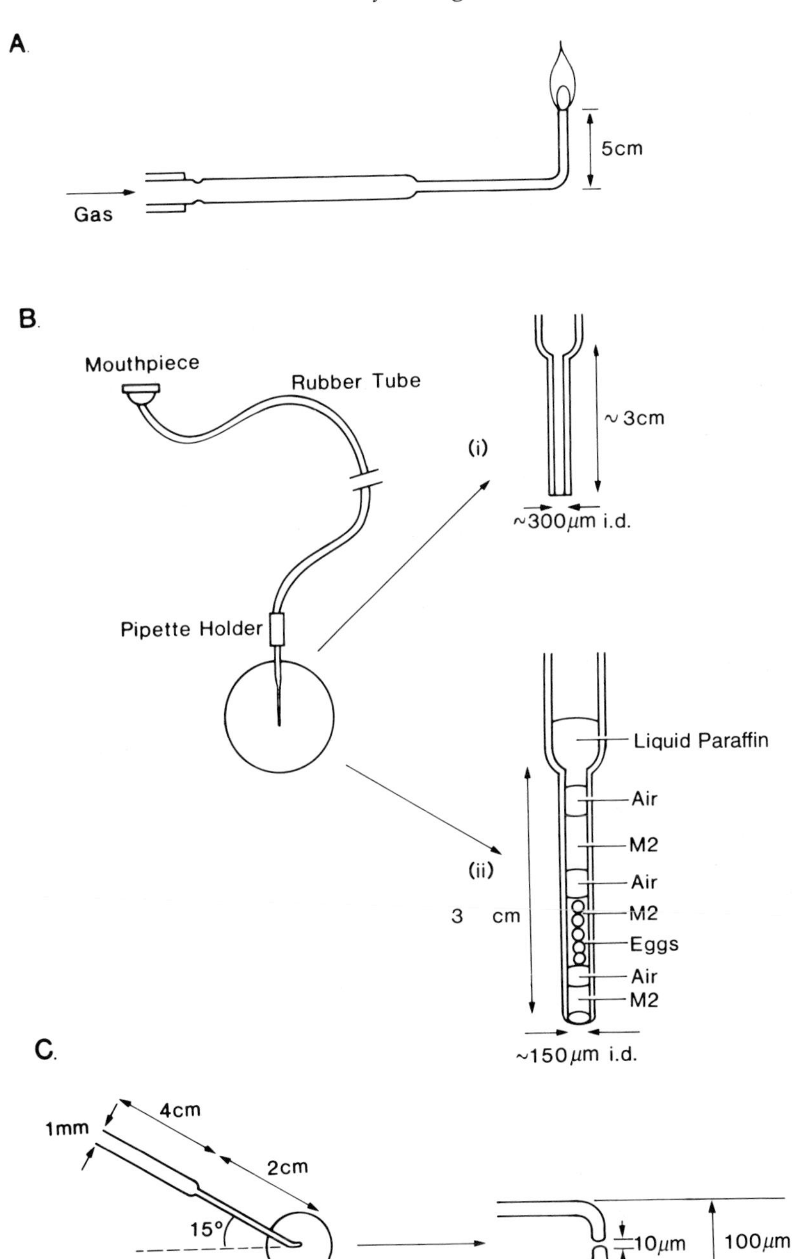

Figure 4. (A) Structure of a gas microflame constructed out of a Pasteur pipette. (B) Design and assembly of a mouth-operated pipette holder containing (i) a general egg transfer pipette and (ii) an oviduct transfer pipette. (C) Design of an egg-holding pipette. Refer to *Protocols 4* and *7* for details.

Protocol 4. Preparation of egg transfer pipettes

Equipment and reagents

- Hard glass capillaries, 1.5 mm o.d. (BDH Ltd.)
- Diamond pencil
- Microforge (e.g. Narishige MF-9)

A. *Preparation of general transfer pipettes (Figure 4Bi)*

1. Set-up a glass microflame burner made from a Pasteur pipette as shown in *Figure 4A*. The flame should be 1 cm high.
2. Soften the middle of a BDH hard glass capillary in the microflame. As soon as the glass begins to soften, withdraw from the heat and simultaneously pull sharply on the ends, but not until the two halves snap.
3. Score the capillary with a diamond pencil and gently snap it into two. If the pipettes are too long (> 5 cm) or too narrow (< 200 μm), score and snap again. The internal diameter of the pipette should be around 300 μm (*Figure 4Bi*) and the end should be flush with no jagged edges.

B. *Preparation of oviduct transfer pipettes (Figure 4Bii)*

1. Follow the procedure for making general transfer pipettes described above, except aim to make the pipettes with internal diameters of around 150 μm.
2. Flame polish the end in a microforge. Ensure that the opening of the pipette is large enough to accommodate a mouse or rat egg (> 120 μm).

5.4 Collection of fertilized one-cell eggs

Fertilized one-cell eggs should be collected from females which have been successfully mated in the morning of the day of microinjection. The procedure is described in *Protocol 5* and illustrated in *Figure 5*, and is the same for mice and rats.

Paraffin oil used to prevent evaporation of the media is frequently a source of toxins that may kill the eggs. We use paraffin oil from Fluka (Cat. No. 76235), which allows more than 50% of eggs to develop to the blastocyst stage in culture.

Protocol 5. Collection of fertilized one-cell mouse eggs and rat eggs

Equipment and reagents

- M2 and M16 egg culture media (*Table 1*)
- 35 mm sterile tissue culture dishes
- Liquid paraffin (Fluka Cat. No. 76235)
- 10 mg/ml hyaluronidase (Sigma Cat. No. H1136) in M2 medium (store at −20°C in 1 ml aliquots)
- 37°C incubator gassed with 5% CO_2
- Dissection instruments as in *Protocol 2*
- Stereo dissecting microscope (e.g. Nikon SMZ-10TD)
- Egg transfer pipettes and mouthpiece (*Protocol 4*)

Method

1. At least 1 h before collecting the eggs, set-up four 35 mm tissue culture dishes containing 2–3 ml M16 medium and two 35 mm culture dishes containing M16 microdrops (40 μl) covered with liquid paraffin. Allow these to equilibrate in a 37°C incubator gassed with 5% CO_2. At the same time prepare five 35 mm tissue culture dishes containing 2–3 ml M2 medium and leave at room temperature.
2. Kill the plugged donor female as described in section 3.1.
3. Lay the animal on its back and soak the abdomen with 70% (v/v) ethanol from a squeeze bottle.
4. Skin the lower half of the animal, cut the body wall, and enter the abdominal cavity to reveal one arm of the reproductive tract as shown in *Figure 5A*, which is associated with a fat pad. Identify the coiled oviduct, which is to be found between the ovary and the uterus.
5. Gripping the uterus with a pair of forceps, pull the reproductive tract away from the rest of the animal. Use a pair of fine forceps to puncture the membrane (the mesometrium) that joins the reproductive tract to the body wall. Trim the mesometrium away from the oviduct (*Figure 5A*).
6. Cut between the ovary and the oviduct (*Figure 5B*).
7. Carefully hold the oviduct with a pair of sharp watchmaker's forceps, then cut between the uterus and the oviduct (*Figure 5C*).
8. Place the oviduct into one of the dishes of M2 medium prepared earlier.
9. Dissect out the oviduct from the other horn of the reproductive tract and then from the rest of the donor females. Place all the oviducts into the same dish of M2 medium.
10. View the oviducts under × 10–20 magnification using a stereo dissecting microscope. Identify the swollen ampulla containing the cumulus mass (the fertilized eggs surrounded by cumulus cells; *Figure 5D*).
11. Using two pairs of sharp watchmaker's forceps, tear the ampulla. The cumulus mass should spill out of the hole (*Figure 5E*). Sometimes it is necessary to tease the eggs out of the ampulla with forceps. Remove the emptied oviduct from the culture dish. Repeat with the rest of the oviducts.
12. Mix the cumulus masses with around 50 μl of 10 mg/ml hyaluronidase. Digestion with hyaluronidase for a few minutes will release the eggs from the cumulus cells. Gently pipetting the eggs up and down will speed the process.
13. Transfer the eggs to a fresh dish of M2 medium to wash away traces

Protocol 5. *Continued*

of the enzyme. Repeat in another fresh dish of M2 medium. Each time the eggs are washed, try to leave behind the cumulus cells.

14. Wash the eggs twice in two of the pre-warmed dishes of M16 medium, then transfer them to the microdrops (prepared in step 1).
15. Incubate at 37 °C in 5% CO_2 until required for microinjection. It is best to leave them for at least an hour before microinjection.

6. Microinjection

Microinjection of DNA into one-cell eggs is a technique that requires both patience and practice to learn. However, once perfected, the technique provides an extremely quick and efficient method for producing transgenic animals. Much time in repeating failed experiments can be saved by using DNA that has been carefully prepared and purified.

6.1 Transgene design

The generation and analysis of transgenic animal lines is a lengthy process, and thus the careful design of the transgene is of great importance. No one wants to spend months generating sufficient transgenic animals for analysis only to discover that the transgene needs to be redesigned. Of course, for the majority of genes very little is known about the location or identity of all the regulatory elements required for efficient expression, and as such a certain amount of serendipity is required in achieving the desired expression pattern for a transgene. However, certain general principles have been determined that can be applied to the construct of most transgenes:

(a) The presence of contiguous vector-derived prokaryotic DNA sequences in a fragment of injected DNA can severely inhibit the expression of some eukaryotic transgenes (12). Bacterial coding sequences such as chloramphenicol acetyl transferase (CAT) and β-galactosidase are often incorporated into transgenes and used as reporters of expression directed by eukaryotic promoter elements. Unlike some vector sequences, these sequences do not seem to inhibit the expression of eukaryotic genes, and any inhibitory effect of prokaryote-derived DNA may therefore be specific to sequences contained within the commonly used lambda and pBR322-derived vectors. As a general rule, it is therefore best to remove all vector sequences prior to injecting cloned transgenes. Linear DNA integrates fivefold more efficiently than supercoiled DNA, and the structure of the fragment ends created by different restriction enzymes has little effect (8).

(b) For the majority of transgenic experiments, appropriate tissue-specific and physiologically regulated expression of the transgene is required.

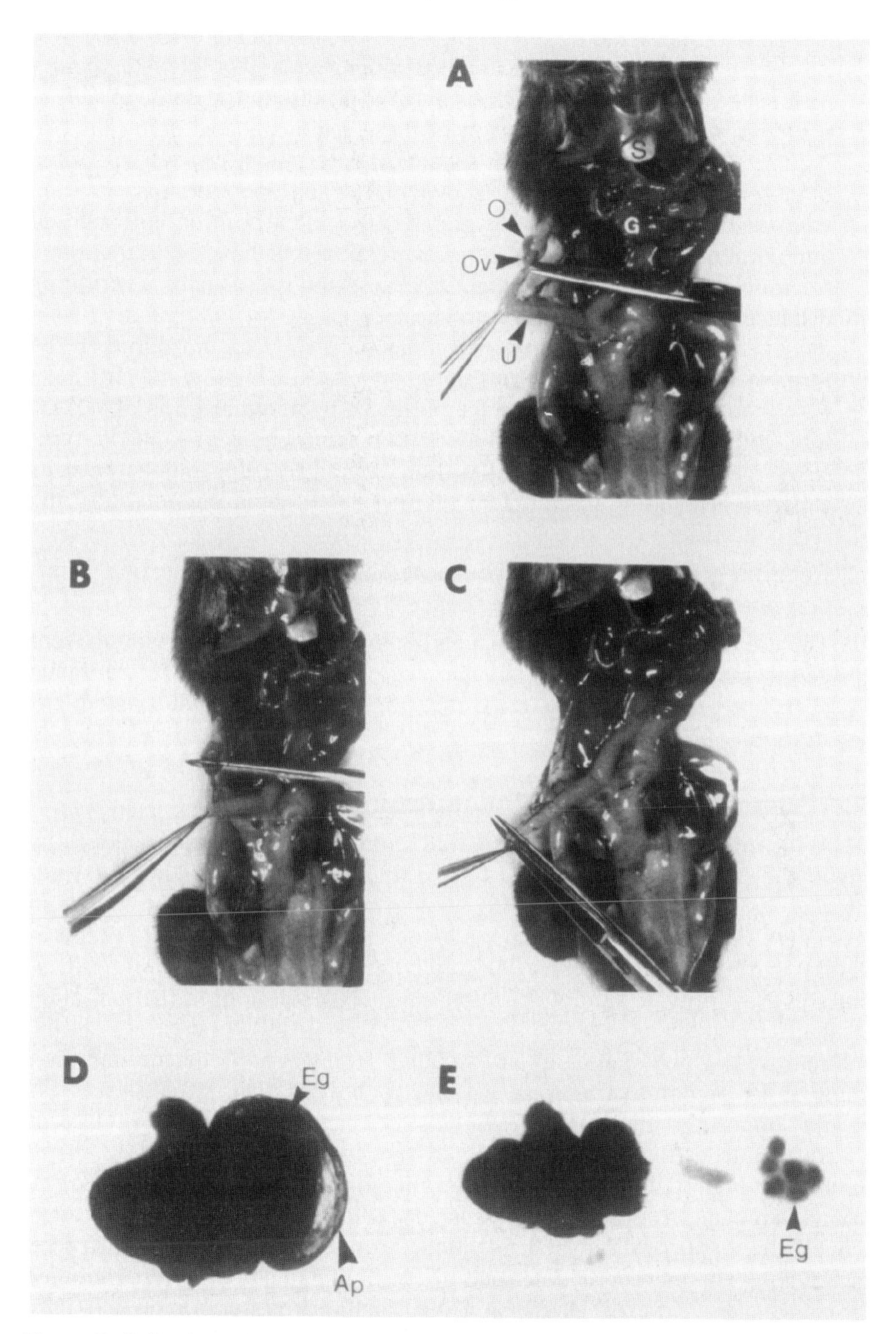

Figure 5. Collection of fertilized one-cell eggs. Refer to *Protocol 5* for details. Key: S, sternum; G, gut; O, ovary; Ov, oviduct; U, uterus; Eg, eggs; Ap, ampulla.

Since the DNA elements generating such specificities are often unknown, it is usually best to use as large a piece of the gene as possible. Regulatory elements can be located in introns and 3′ untranslated regions, as well as upstream sequences, so inclusion of these regions is advisable (13, 14). Adjacent genes may also contain sequences determining correct expression of their neighbours, as has been shown for the neuropeptides oxytocin and vasopressin (15). Locus control regions (LCRs) allow position-independent and copy number-dependent gene expression, but have been identified for only a very small percentage of genes, and are usually a very long distance from the coding region. The technology for manipulating fragments of DNA up to one megabase has recently been developed in the form of yeast artificial chromosomes, or YACs. Schedl *et al.* (16) have described the first transgenic mice produced by microinjection of a YAC into fertilized one-cell eggs. Their 250 kb transgene contained the tyrosinase gene, and was inserted without major rearrangements, with its expression resulting in complete rescue of the albino phenotype of the recipient mice.

(c) Reporter genes should ideally be located in the first or last exons. Insertion into an internal exon may reduce splicing efficiency.

(d) The use of chimeric transgenes, combining pieces of DNA from different genes, may result in unpredictable ectopic expression of the transgene (17). It is best to keep the transgene simple by using regulatory elements from only one gene.

6.2 Preparation of DNA for microinjection

DNA for microinjection can be isolated from cosmid or plasmid clones by any of the standard techniques of lysozyme/Triton X-100 lysis or lysozyme/alkaline lysis followed by banding of supercoiled molecules in ethidium bromide/CsCl gradients. The many commercially available plasmid preparation kits, utilizing alkaline lysis followed by anion-exchange column purification of DNA, provide suitably pure plasmid DNA solutions in the minimum of time.

Vector-free DNA fragments are isolated by restriction enzyme digestion followed by preparative agarose gel electrophoresis. The latter step is extremely important since DNA fragments for microinjection *must* be free of all contaminants that may be toxic to the eggs and *must* be free of all particulate matter that could clog the injection pipette. Isolation of DNA from agarose gels by binding to glass beads (18), followed by passage through a Sephadex G-50 column, and filtration through a 0.45 μm filter provides such DNA. The method is described in *Protocol 6*, including preparation of the glass beads. However, glass bead DNA isolation kits are also commercially available (e.g. Geneclean II from Bio 101).

Purified DNA for microinjection is dissolved in a buffer of 10 mM Tris–

HCl pH 7.4, 0.1–0.25 mM EDTA. Higher concentrations of EDTA and low concentrations of $MgCl_2$ are toxic to eggs. There is no correlation between the concentration of the microinjected DNA and the resulting copy number of the transgene. Excessively high concentrations of DNA are however toxic to the egg. It is estimated that around 1–2 pl of DNA solution is injected into each pronucleus by microinjection. Most investigators use DNA concentrations of 1–5 μg/ml and are therefore microinjecting about 500 copies of the DNA fragment, depending on its size (10). For each different DNA fragment, the DNA concentration that results in maximum integration efficiency with the greatest egg survival will vary. Therefore, some workers make dilutions of the injection DNA stock (for example 1, 2.5, and 5 μg/ml) and rotate between these solutions during an injection session as the microinjection pipettes are changed.

Protocol 6. Purification of DNA for microinjection from agarose gels

Equipment and reagents

- Glass beads. These can be purchased[a] or prepared in the laboratory as described below, but the preparation procedure involves using boiling nitric acid, so *take full safety precautions.* Mix 250 ml of powdered glass flint (available from glass supply companies) with 500 ml sterile water. Allow to settle for 1 h. Discard the settled glass and recover the fines by centrifugation (1000 *g*, 5 min). Resuspend the fines in 200 ml of sterile water. Add 200 ml concentrated nitric acid and bring to the boil. Carry out this step in a fume-hood. Allow to cool, then centrifuge the mixture to recover the glass beads (1000 *g*, 5 min). Discard the nitric acid carefully and safely. Wash away residual acid by resuspending the glass pellet in water and then centrifuge to recover the glass. Repeat until the water has a neutral pH. Store the glass at 4°C as a 50% (v/v) slurry in sterile water.
- DNA digested with appropriate restriction enzymes according to the supplier's instructions
- Gel electrophoresis equipment
- 50 × TAE buffer stock: dissolve 242 g Tris base, 57.1 ml glacial acetic acid, and 100 ml of 0.5 M EDTA in 1 litre of water and autoclave
- Agarose gel prepared in 1 × TAE buffer containing 50 μg/ml ethidium bromide[b]
- 6 M sodium iodide: dissolve 90.8 g of sodium iodide and 0.5 g Na_2SO_3 in water to a final volume of 100 ml, filter through a 0.45 μm Nalgene filter. Add a few Na_2SO_3 crystals to the filtrate, which will not dissolve fully. Store at 4°C protected from light.
- Ethanol wash solution: 50% (v/v) ethanol, 0.1 M NaCl, 10 mM Tris–HCl pH 7.5, 1 mM EDTA (store at −20°C)
- Microinjection TE buffer: 10 mM Tris–HCl pH 7.4, 0.2 mM EDTA
- Sephadex G-50 column in 1 ml syringe, equilibrated in microinjection TE buffer
- 0.45 μm Millipore filter attached to a 1 ml syringe
- Long wave UV transilluminator

Method

1. Fractionate the restricted DNA in an agarose gel[b] prepared in 1 × TAE buffer containing 50 μg/ml ethidium bromide. Use 1 × TAE as the electrophoresis buffer. Visualize the DNA under long wave UV light and excise the desired fragment of agarose gel into a pre-weighed 1.5 ml microcentrifuge tube in as small a volume as possible.
2. Weigh the gel and, assuming a density of 1.0, estimate the volume.

Protocol 6. *Continued*

Add 2 vol. of 6 M sodium iodide. Incubate at 37 °C, mixing occasionally, until the gel dissolves.

3. Resuspend the glass beads by vortexing. To the solubilized agarose gel, add 1 μl of glass suspension for every 2 μg of DNA present. Chill on ice for 1 h with occasional agitation.
4. Centrifuge the mixture; to ensure a loose pellet, spin at low speed for a brief period (1–2 sec in a microcentrifuge).
5. Discard the supernatant. Disperse the glass pellet in 0.5 original gel volumes (see step 2) of 6 M sodium iodide.
6. Pellet the glass beads again as in step 4.
7. Disperse the pellet in 0.5 original gel volumes of ethanol wash solution. Centrifuge to pellet the glass.
8. Repeat step 7 twice.
9. Remove as much of the supernatant as possible without allowing the pellet to dry. Add > 50 μl microinjection TE buffer immediately. Elute the DNA by incubating at 37 °C for 15 min.
10. Pellet the glass beads by centrifugation and transfer the supernatant containing the DNA to a fresh tube. Subsequent to this point, all tubes and tips should be rinsed with filtered water prior to use.
11. To further remove impurities, pass the DNA through a Sephadex G-50 column equilibrated in microinjection TE buffer. Then filter the eluate through a small 0.45 μm Millipore filter attached to a 1 ml syringe.
12. Determine the DNA concentration by spectrophotometry at 260 nm or comparative ethidium bromide staining. Dilute to a concentration of 1–5 μg/ml with microinjection TE buffer.

[a] Glass bead DNA isolation kits are available (e.g. Geneclean II from Bio 101).
[b] The concentration of agarose gel required will depend on upon the size of the DNA fragments to be fractionated (19).

6.3 Microinjection equipment

The equipment required for microinjection is illustrated in *Figure 6* and described in detail in the following sections.

6.3.1 The microscope

The inverted microscope (e.g. Nikon Diaphot) used for microinjection should have the following features:

- image erected optics
- a fixed stage (i.e. the objective lens moves rather than the stage when focusing)

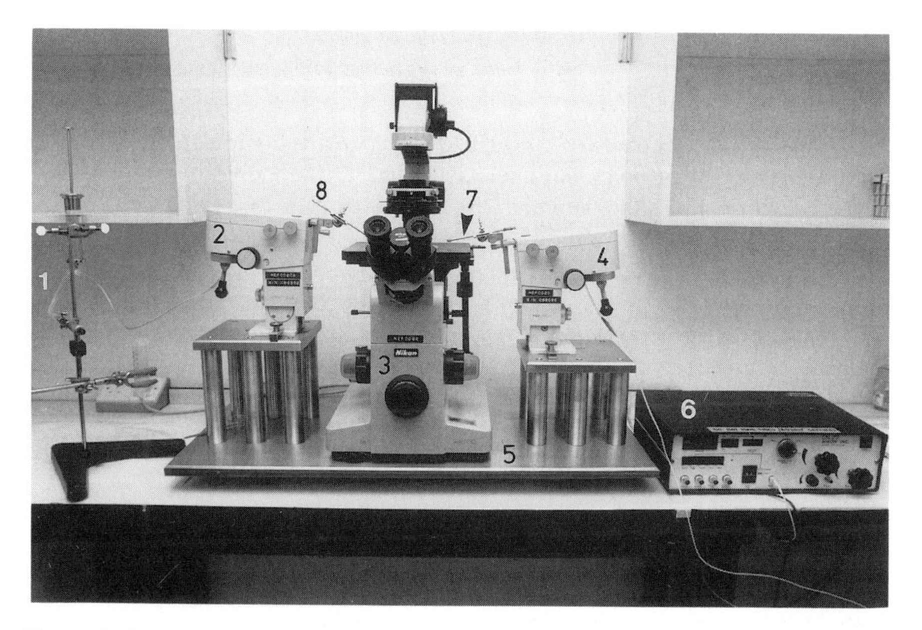

Figure 6. Arrangement of a Nikon-Leitz microinjection apparatus. Key: 1, Agla micrometer syringe linked by liquid paraffin-filled plastic tubing to the egg-holding pipette; 2, left-hand Leitz micromanipulator controlling the egg-holding pipette; 3, Nikon Diaphot inverted microscope; 4, right-hand Leitz micromanipulator controlling the microinjection pipette; 5, baseplate; 6, Narishige automatic Pico-Injector linked by air-filled plastic tubing to the microinjection pipette, and to the compressed air supply (hidden from view); 7, right-hand Leitz instrument tube holding the injection pipette; 8, left-hand Leitz instrument tube holding the egg-holding pipette.

- a condensor with a long working distance
- × 10 magnification eyepieces
- × 4 objective for low magnification work
- × 40 objective for microinjection
- suitable optics. The best optics for visualizing internal structure of one-cell eggs are the Nomarski differential interference contrast (DIC) microscopy system. However, DIC optics are very expensive and glass injection chambers are obligatory. Hoffman modulation contrast optics do not give the same resolution as DIC, but are cheaper and more compatible with plastic injection chambers. Bright field optics can be used if Nomarski or Hoffman optics are unavailable and are preferential to phase contrast microscopy.

6.3.2 Micromanipulation

Microinjection is achieved by the micromanipulation of fertilized eggs using two pipette types:

- the holding pipette, which holds the egg in place
- the injection pipette, which is loaded with the DNA solution and actually pierces the egg

Both of these pipettes are controlled using micromanipulators, the most commonly used type being supplied by Leitz. These manipulators have the advantage of having a convenient joystick control for fine horizontal movement in two planes. One micromanipulator is positioned on either side of the microscope. The left-hand micromanipulator controls the holding pipette, whilst the right-hand one controls the microinjection pipette. These micromanipulators must be positioned on the baseplate with the microscope such that these three components are conveniently positioned relative to each other (see *Figure 6*). The baseplate also ensures that there is no movement between the manipulators and the microscope and reduces vibration.

6.3.3 The injection system

Injection of DNA into the pronucleus of fertilized one-cell eggs can be performed by either one of two systems. Manual injection involves the use of an air-filled tube connected to a 50 ml syringe to which the microinjection needle is attached. Manual squeezing on the syringe forces DNA out of the needle and into the egg. Alternatively, an automatic injection system can be used. A number of these are available which use compressed air to expel the DNA, triggered by a foot-operated pedal. We use the Narishige Pico-Injector PLI-188, with compressed air provided from the Hitachi (Tokico) Package Oilfree Bebicon PO-O 75PSB compressor (maximum output 121 p.s.i.). Although automatic injection systems are more expensive than manually-operated ones, they do possess several advantages over manual injection systems:

- The use of a foot-pedal to trigger DNA injection leaves the hands free to operate the micromanipulators, speeding up the process of microinjection.
- A low constant pressure can be applied to produce continual outflow of the DNA solution, preventing back-flow of M2 medium into the micro-injection pipette and clogging of the pipettes. Pipettes therefore do not have to be changed as often as when using a manual injection system.

The assembly and testing of an automatic injection system, particularly the high pressure gas supply, should be carried out by a trained professional. The injection pressure must be determined empirically. We have found that the starting pressure will be relatively constant if the same parameters are always used to pull the microinjection pipettes, but the injection pressure may need to be altered during the microinjection session.

6.3.4 Injection chamber

Depression slide injection chambers are compatible with inverted microscopes fitted with any optical system. Siliconize the glass slide by rinsing it in a 3% (v/v) solution of dichloromethyl silane in chloroform. Wash the slide thoroughly in water and with a standard household detergent. Then rinse the slide with ethanol and wipe dry with a paper tissue. Place a flat drop of M2 medium into the centre of the depression. The M2 drop should be no more than 1 cm across. Cover the M2 drop with light paraffin oil to prevent evaporation. Place the injection chamber on the stage of microscope and using a × 4 objective lens, focus on the bottom of the M2 drop.

6.3.5 Holding pipettes

A holding pipette is shown in *Figure 3C*. One holding pipette should last for a whole microinjection session. Holding pipettes should not be reused because they tend to get dirty and clogged with egg debris and liquid paraffin. Large numbers of holding pipettes can be prepared in advance (*Protocol 7*) and stored loose in a Petri dish.

Protocol 7. Preparation of holding pipettes

Equipment and reagents

- Leitz glass capillaries (1 mm o.d.)
- Diamond pencil
- Microforge (e.g. Narishige MF-9)
- Leitz single instrument holder with tube
- Agla micrometer syringe MS01 (available from Wellcome Reagents Ltd.)
- One metre of Tygon tubing (3/32 inch i.d., 5/32 inch o.d.)
- Fluorinert electronic liquid (FC77 from 3M Company)
- Light paraffin oil (Fluka Cat. No. 76235)
- Injection chamber (see section 6.3.4)
- Hypodermic needle (5 cm long, 26 gauge) fitted to a 1 ml disposable syringe
- M2 medium (*Table 1*)

Method

1. Draw a Leitz hard glass capillary in a microflame, as described for egg transfer pipettes (*Protocol 4*).
2. Using a diamond pencil score the drawn section around 2 cm away from the shoulder and break the capillary at this point.
3. Mount the capillary in the vertical position in a microforge. Using a × 4 objective, focus on the tip of the capillary. The end should be perfectly flush with no jagged edges and the capillary should be perfectly straight. The external diameter should be no less than 80 μm and no more than 150 μm. Bring the filament of the microforge close up to but not touching the capillary tip. Heat the filament and observe the melting of the tip. When the internal diameter of the tip reaches 10–15 μm, turn off the current to the filament.
4. Carefully move the capillary to the horizontal position. Locate the filament below the capillary 1–2 mm from the tip. Move the filament up to the capillary again, heat the filament, and allow the capillary to bend. Turn off the current to the filament when the capillary has bent by 15° to the horizontal.

Protocol 7. *Continued*

5. The holding pipette is now ready for assembly into the micromanipulation system. Connect a Leitz instrument tube to an Agla micrometer syringe held in a clamp stand via one metre of Tygon tubing. Fill the whole system with light paraffin oil. Ensure that all air bubbles have been excluded.
6. Fill the holding pipette with Flourinert electronic fluid using a needle (5 cm long, 26 gauge) connected to a 1 ml disposable syringe. Insert the holding pipette into the oil-filled Leitz instrument tube and tighten the ring to hold it in place. Carefully clamp the instrument tube into the instrument tube holder of the left-hand micromanipulator. Adjust the Agla syringe until fluid stops flowing out of the holding pipette, but do not allow air to flow into the pipette.
7. By adjusting the instrument tube and the controls of the instrument tube holder, position the tip of the holding pipette above the injection chamber such that it is horizontal. Using the fine controls of the micromanipulator, lower the holding pipette into the injection chamber. Ensure that the holding pipette is free to move in the horizontal plane throughout the field of view when controlled by the joystick, and does not catch on the bottom of the chamber.
8. Draw M2 medium into the holding pipette by adjusting the Agla syringe until the meniscus between the Fluorinert and the M2 medium is near the shoulder of the holding pipette.

6.3.6 Preparation of microinjection pipettes

Microinjection pipettes are prepared using thin walled glass tubing with an outside diameter of 1 mm. Capillaries with internal filaments (e.g. from Clark Electromedical Instruments, Cat. No. GC 100TF-15) are particularly useful because they allow the injection pipette to be filled with DNA from the end distal to the point by capillary action. Pipettes are made on a mechanical pipette puller (e.g. the Vertical Pipette Puller Model 720 available from David Kopf). The appropriate temperature and mechanical pull settings have to be worked out by trial and error. A good pipette will have an opening of around 1 μm. Smaller openings will tend to clog easily, whilst larger openings are more likely to burst the egg. Pipettes should be prepared as required.

The microinjection pipette is operated by the right-hand micromanipulator. Connect a Leitz instrument tube by a one metre length of Tygon tubing to either a 50 ml syringe with a ground glass plunger (manual injection) or to an automatic injection system (section 6.4.3). Load the injection pipette with DNA solution by capillary action. As soon as liquid can be seen at the tip then the pipette is loaded sufficiently. Avoid contaminating the DNA stock solution with, for example, talcum powder dust from disposable gloves or enzymes from the exposed hand. Assemble the pipette into the instrument tube. Clamp the instrument tube on to the instrument tube holder of the micromanipulator. Using the instrument tube holder controls, position the tip of the injection pipette above the injection chamber. Then, using the fine vertical control of the micromanipulator, lower the pipette tip into the drop of M2 medium, all the time monitoring the position of the tip by viewing down the microscope with a × 4 objective. Ensure that the injection pipette can be freely moved in the horizontal plane when operated by the micromanipulator joystick.

6.4 Microinjection of fertilized one-cell mouse or rat eggs

The procedure for microinjecting fertilized one-cell mouse or rat eggs is described in *Protocol 8*, and pictures of an egg before and after injection are shown in *Figure 7*. Although the procedure is the same for both species, it is often harder to see the pronuclei of rat eggs due to the more granular appearance of the rat egg cytoplasm. The membrane of rat eggs is also more elastic than that of mouse eggs, making it more difficult to pierce the egg. As a result, the number of rat eggs that can be injected in a session is fewer than for mouse eggs.

Protocol 8. Microinjection of fertilized one-cell rodent eggs

Equipment and reagents

- Microinjection apparatus (see section 6.3)
- Egg transfer pipette and mouthpiece (*Protocol 4*)
- M2 and M16 egg culture media (*Table 1*)
- Fertilized eggs (*Protocol 5*)

Method

1. Assemble the microinjection apparatus as described in section 6.3.
2. Remove about 20 fertilized eggs from storage in M16 medium at 37 °C using a general egg transfer pipette.
3. Wash the eggs twice in M2 medium, then load them in as small a volume as possible into a transfer pipette and discharge them into the injection chamber. Observe the entry of the eggs into the injection chamber using a × 4 objective. Try to keep the eggs in a group positioned below the holding and injection pipettes. Ensure that no air bubbles are released into the chamber with the eggs since they will completely disrupt the chamber; not only is it likely that the eggs will be lost, but also the chamber may have to be reassembled.
4. Readjust the vertical position of the holding and injection pipettes such that they are in the same plane as the eggs and are consequently able to readily manipulate the eggs. Bring the tip of the holding

Protocol 8. *Continued*

pipette up to an egg and, by adjusting the Agla syringe, apply light suction such that the egg is held on to the end of the pipette.

5. Bring the egg to the centre of the field of view. Switch to the ×40 DIC objective and focus on the egg. By focusing up and down, locate the position of the larger of the two pronuclei (usually the male). By adjusting the Agla syringe, the holding pipette and the injection needle, move the egg until the larger pronucleus is in a position central to, but away from, the holding pipette (*Figure 7A*). Ensure that the egg is tightly held by the holding pipette by applying slightly more suction. However, although the zona pellucida may be deformed, do not allow the egg itself to be pulled into the holding pipette.
6. Focus on the pronucleus to be injected. Bring the tip of the injection pipette up to the zona pellucida of the egg. Adjust the fine vertical micromanipulator control to place the tip of the injection pipette in the same focal plane as the targeted pronucleus.
7. Inject the egg. The zona pellucida is easily pierced by the injection needle. When the injection needle appears to be inside the nucleus, squeeze on the injection syringe. One of three things will then happen:
   (a) The egg nucleus may swell. This is indicative of a successful injection. Continue to apply pressure until the nucleus has reached roughly twice its normal volume (*Figure 7B*) then quickly withdraw the injection pipette.
   (b) A small, clear bubble may appear at the end of the injection needle and the perivitelline space may swell. This indicates that the egg membrane, which can be very elastic (particularly with rat eggs), has not yet been pierced. To penetrate the membrane, continue to push the pipette through to the far side of the nucleus, pull back until the tip is again in the nucleus and then squeeze on the injection syringe again.
   (c) Nothing. This is probably due to the injection needle being blocked. Change the needle and, if particulate contamination in the DNA stock is suspected, change the DNA stock also.
8. Following injection, cytoplasmic granules may be seen to flow out into the perivitelline space. This is indicative of egg lysis. If eggs lyse on two or three successive injections, then change the injection needle. The injection needle should also be changed if it appears to be getting dirty or clogged.
9. Following injection, return to the × 4 objective and place the egg above the holding and injection pipettes, well away from the uninjected eggs. The post-injection eggs should be divided into two groups; those that have survived injection and those that have not.

10. Eggs should be maintained in M2 medium for a maximum of 15 min. A skilled operative should be able to inject between 15 and 40 eggs in such a time. Once all the eggs in a batch have been injected, remove those that have survived from the injection chamber, wash them twice in M16 medium, and return them to microdrop culture. Maintain them there until transfer to a 0.5 day post-coitus pseudopregnant female (section 7). Eggs can be transferred to a pseudopregnant recipient that same day, or following culture overnight by which time the eggs have reached the two-cell stage (*Figure 7C*).[a]

[a] Transfer of rat eggs to a pseudopregnant recipient on the same day as microinjection results in a greater percentage of eggs developing to term. However, the day of transfer of microinjected mouse eggs has no effect on the number developing to term.

7. Transfer of microinjected eggs to the oviducts of pseudopregnant recipient females

7.1 Generation of pseudopregnancy in recipient female mice and rats

Pseudopregnant recipient females are used as surrogate mothers to nurture surviving microinjected eggs to birth. Recipient females are made pseudopregnant by mating them with vasectomized males. To increase the chances of a successful mating, only females that are in oestrous are paired with the vasectomized males. Different criteria are used to determine the stage of the oestrous cycle for mice and rats; for mice, one can simply examine the state of the vagina (*Table 2*), whereas for rats, a smear of cells from the vagina needs to be taken and the contents examined under high magnification ($\times$ 400) (*Protocol 9*). Each female that is determined to be in oestrous is placed with a vasectomized male in the afternoon before the day on which the microinjected eggs are to be transferred to the pseudopregnant female. The following morning the female mice are examined for the presence of a copulatory plug, evidence that a successful mating has occurred. Between

Table 2 Identification of ovulating mice

Stage of oestrous cycle	Vaginal characteristics
Dioestrous	Opening small, tissues blue and moist
Pro-oestrous	Opening gaping, tissues red-pink and moist, dorsal and ventral surface folds
Oestrous (ovulating)	Opening gaping, tissues pink and moist, pronounced folds
Metoestrous	Tissues pale and dry, white cell debris

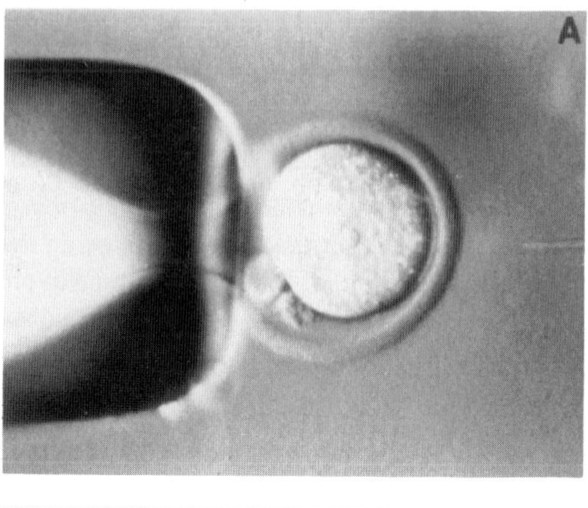

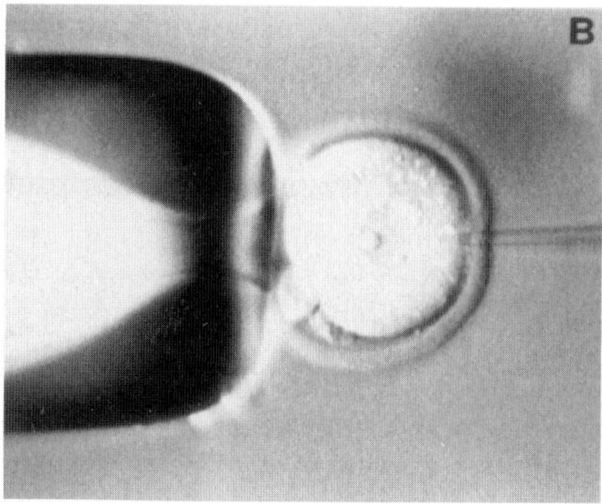

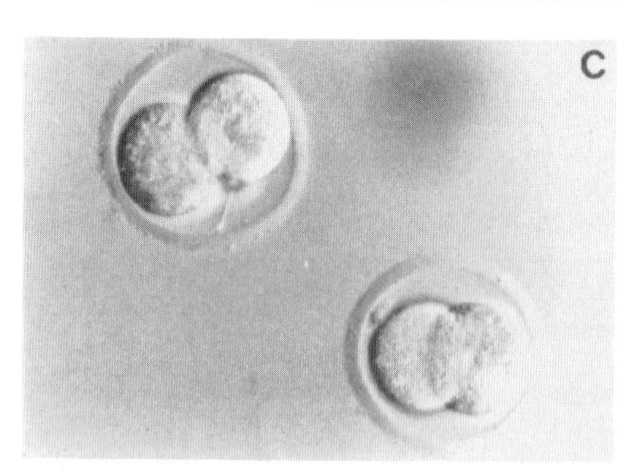

Figure 7. (A) An egg ready for injection, viewed at high power magnification (× 400) using DIC optics. The holding pipette is on the left, gripping the zona pellucida of the egg. The micro-injection needle is on the right. The large male pronucleus, containing several nucleoli, is clearly visible in the middle of the egg. (B) Microinjection and consequent swelling of the male pronucleus. (C) Two-cell stage eggs viewed under DIC at × 400 magnification.

50% and 80% of females that are visually determined to be in oestrous will be impregnated. However, vasectomized male rats do not produce copulatory plugs, and thus there is no reliable way to ensure that mating has occurred and that pseudopregnancy has been engendered. Since vaginal stimulation can induce pseudopregnancy in rats, female rats are gently prodded in the vagina with a smooth blunt probe for one minute before being placed with the vasectomized males, as a back up procedure.

Protocol 9. Identification of ovulating rats

Equipment and reagents

- Phosphate-buffered saline (PBS): 20 mM sodium phosphate buffer pH 7.4 containing 0.8% (w/v) NaCl, or 0.9% (w/v) NaCl
- 1 ml syringe (without needle)
- Petri dish or microscope slide
- Light microscope with × 40 magnification (and × 400 magnification if checking for the presence of sperm)

Method

1. Use female rats that have given birth to one litter, and been weaned for at least one week. Identify them by marking the tails.
2. Fill a 1 ml syringe with 0.2–0.3 ml PBS or 0.9% (w/v) saline.
3. Lift up the rat's tail to expose the opening to the vagina.
4. With the other hand, gently insert the syringe into the opening until resistance is felt. Expel the contents of the syringe into the vagina, and then draw the fluid back into the syringe. This will collect cells from the vagina into the fluid.
5. Transfer the contents of the syringe to a labelled Petri dish or microscope slide. Repeat with several other rats.
6. Examine the cells under a light microscope with × 40 magnification and identify ovulating rats from the following characteristics:

Stage of oestrous cycle	Cell characteristics
Dioestrous	Few epithelial cells and leucocytes
Pro-estrous (pair rats at this stage)	Many nucleated epithelial cells, yellow appearance
Oestrous	Many large cornified epithelial cells, brownish appearance
Metoestrous	Fewer cornified epithelial cells, many leucocytes

7. If checking for the presence of sperm, examine the vaginal smear under × 400 magnification.

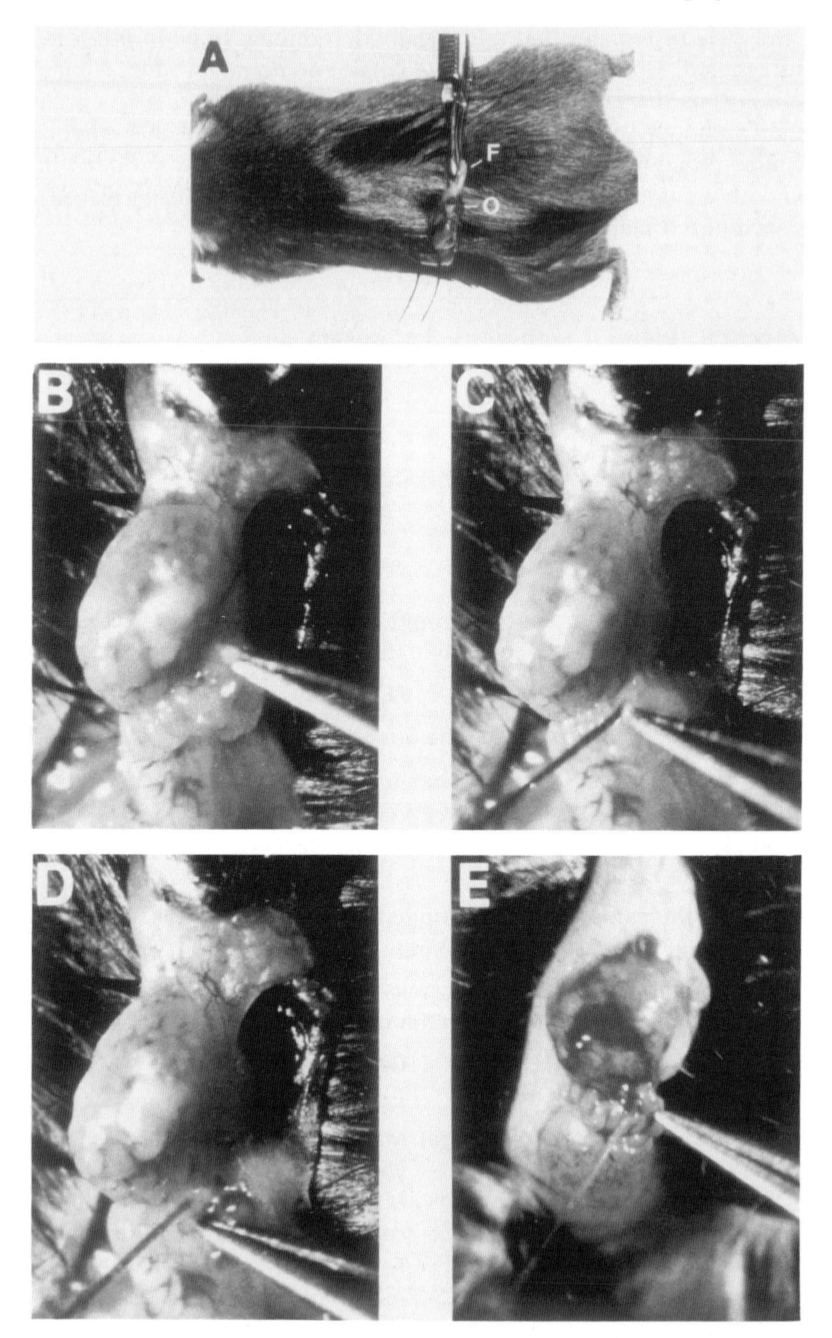

Figure 8. Microsurgical introduction of material into the oviduct of a recipient female mouse (oviduct transfer). Key to symbols used in (A): F, fat pad; O, ovary. Refer to *Protocol 10* for details.

7.2 Delivery of microinjected eggs to surrogate mothers by oviduct transfer

This procedure requires considerable skill, and is best practised first on cadavers. A dye can be used instead of eggs while practising the expelling of the contents of the oviduct transfer pipette into the ampulla (*Figure 8B–D*). An ideal litter size resulting from the oviduct transfer would be five to ten pups. The number of eggs required to achieve this will vary with the competence of the operator; 15–20 eggs delivered to each oviduct should be suitable. The procedure is described in *Protocol 10*, and illustrated in *Figure 8*.

Protocol 10. Oviduct transfer

Equipment and reagents

- Microinjected eggs in microdrop culture (from *Protocol 8*)
- M2 medium (*Table 1*)
- Pseudopregnant recipient female mouse or rat (see section 7.1)
- CRC (for rats) or Avertin (for mice) anaesthetic (see *Protocol 1*)
- Dissection instruments (as in *Protocol 2*)
- Artery clip (1.5 in) (e.g. available from Arnold Horwell)
- Fibre optic illumination (e.g. available from Nikon)
- Surgical microscope with optional assistant's viewing head (e.g. Carl Zeiss Jena OPM 212T with head model 050)
- Oviduct transfer pipettes and mouthpiece (*Protocol 4*)
- Epinephrine: prepare 0.5% (w/v) epinephrine in 0.1 M HCl and dilute to 0.1% (v/v) in PBS for a working solution
- Ampicillin (Binotal; from Bayer)
- Paraffin oil (Fluka Cat. No. 76235)
- 70% ethanol in squeeze bottle

Method

1. Anaesthetize a 0.5 p.c. pseudopregnant recipient female as described in *Protocol 1*.
2. Place the animal abdomen side down on the lid of a 9 cm Petri dish if operating on a mouse, or straight on to the stage of the stereo dissecting microscope if using a rat. Spray the back with 70% (v/v) ethanol.
3. Comb away the hair from the incision site using a fine pair of forceps. Cut the skin transversely 1 cm down from the spine at a point level with the last rib.
4. Make a 1 cm (2 cm for rats) cut through the skin with a large, sharp pair of scissors. Stretch the incision to prevent bleeding.
5. Locate the orange-coloured ovary beneath the body wall (not usually visible in rats). Make a 3–5 mm cut through the body wall with a fine, sharp pair of scissors at a point a few millimeters away from the ovary. Stretch the incision to prevent bleeding.
6. Introduce a single stitch through the body wall on one side of the incision and leave the silk suture in place.

Protocol 10. *Continued*

7. Pull out the fat pad joined to the ovary with a fine but blunt pair of forceps. The ovary, oviduct, and uterus will be pulled out with the fat pad (*Figure 8A*).
8. Attach an artery clip to the fat pad, carefully avoiding the ovary. The reproductive tract is held in position over the back of the animal by the artery clip with the coils of the oviduct uppermost and the ovary towards the rear of the animal (*Figure 8A*).
9. Transport the mouse on the Petri dish lid to the stage of a stereo dissecting microscope. Illuminate the oviduct with a fibre optic light source and view under $\times$ 10–20 magnification (*Figure 8B*).
10. The opening of the oviduct, the infundibulum, is located in a cavity below the ovary and behind the coils of the oviduct. Gently move the oviduct coils down to reveal the cavity. Sometimes the infundibulum can be seen behind the transparent membrane that covers the ovary, the oviduct, and the cavity. Identify an area of the membrane, preferably above the infundibulum, that is free of capillaries. Tear the membrane at this point using watchmaker's forceps. Any local bleeding can be halted by applying 0.1% (w/v) epinephrine to the area through a syringe fitted with a 26 gauge needle.
11. Prepare the eggs for transfer. Remove a maximum of 20 microinjected eggs from microdrop culture and wash them in M2 medium. Load the eggs into an oviduct transfer pipette (*Figure 4Bii*). First fill the pipette as far as the shoulder with liquid paraffin oil. The viscosity of the oil affords a greater degree of control when transporting eggs. Then take up a small amount of air, then a small amount of M2, then more air. Then introduce the eggs into the pipette in M2 medium. Draw up the eggs in a minimum volume of medium such that they are stacked in a rank. Then introduce another small air bubble and a final column of M2.
12. Return to the animal. Use a paper tissue to mop up any body fluids, blood, or blood clots around the oviduct. *Figures 8B–D* are photographs of a demonstration operation performed on a cadaver and consequently the tissue is unnaturally dry. *Figure 8E* is a picture of an actual operation being performed on a live mouse. Note the blood and fluid which, unless removed obscures the oviduct.
13. Grip the tip of the infundibulum with a sharp pair of watchmaker's forceps. Pull out the infundibulum from behind the oviduct coils and through the punctured membrane such that the opening can be accessed by the oviduct transfer pipette (*Figure 8B*).
14. Push the tip of the oviduct transfer pipette into the opening of the infundibulum. The opening can not be seen and is located by gentle

prodding with the tip of the pipette. The tissue will 'give' at the opening and the pipette will enter the infundibulum (*Figure 8B*).

15. Expel the contents of the pipette and monitor by the appearance of bubbles of air in the ampulla. When three bubbles have appeared, one can be sure that the eggs have been successfully deposited into the ampulla. In *Figures 8B, C,* and *D,* blue dye has been used in order to more clearly demonstrate the emptying of the pipette into the ampulla.
16. Withdraw the oviduct transfer pipette. Remove the artery clip and, gripping the fat pad with a blunt pair of forceps, return the reproductive tract to the inside of the body wall. Sew up the body wall with one or two stitches then clip the skin together with an autoclip (two autoclips are required for rats).
17. Repeat with the other side of the reproductive tract if the availability of microinjected eggs allows, and if the animal is sufficiently anaesthetized.
18. If the recipient is a rat, inject it with 50 mg ampicillin (Binotal) intraperitoneally. This prevents low grade post-surgery infections, and can optimize the number of pups subsequently delivered. This step is not necessary with mice.
19. Leave the animal to recover in a quiet, warm place, and then transfer it to an individual cage.
20. Examine the animals regularly during the days after the operation to ensure that they do not pick up any infections. Mouse pups should be delivered 19–20 days after the operation, whereas the gestation period for rats is 21–22 days.

8. Maintaining transgenic offspring

Pups born from a successful oviduct transfer need to be analysed to determine which ones are transgenic. This is covered in detail in the next chapter. However, there are several procedures for maintaining identified transgenic animals as 'lines' that will be covered in this section.

8.1 Caesarean section and fostering

Often, the number of microinjected eggs developing to term in a surrogate mother is very low. Under such circumstances the embryos are 'overnourished' and grow very large compared to the embryos of a normal sized litter. If the embryos grow too large to be delivered down the birth canal, the pregnancy will proceed beyond the normal gestation period and there is a risk that the pups will die in the uterus. It is therefore sometimes necessary to rescue pups by Caesarean section and fostering. It may also be necessary to foster valuable unweaned pups born normally to a mother that dies unexpectedly, or pups in a small litter who are not being fed sufficiently by the mother. Foster mothers need to have pups of their own of about the same age as the pups to be fostered and should be of a strain with good maternal characteristics, for example Swiss mice or Sprague Dawley rats. The procedures for Caesarean section and fostering are given in *Protocol 11*.

Protocol 11. Procedures for Caesarean section and fostering

Equipment and reagents

- Dissection instruments (see *Protocol 2*)
- 70% (v/v) ethanol

A. *Caesarean sectioning*

1. Kill the mother (section 3.1).
2. Soak the abdomen in 70% (v/v) ethanol, skin the lower half of the animal, and cut away the body wall, revealing the embryos in the uterus. Cut out the uterus.
3. Cut the uterus open then gently squeeze the embryos free of uterine tissue. Dissect away the membranes surrounding the pups. Cut the umbilical cord.
4. Wipe away any fluid from the mouth and nose area with a paper tissue and gently squeeze the chest with blunt forceps to stimulate breathing.
5. Place the pups on a damp tissue and keep warm until fostered.

B. *Fostering*

1. Remove the foster mother from her own pups.
2. Quickly mix in the foster pups with the natural pups.
3. Try to make the foster mother urinate on the pups. This can usually be induced by picking the animal up as shown in *Figure 4A*. Although this may not be possible with a larger, stressed rat, the Sprague Dawley foster mothers that we use often foster the pups without this stimulus.
4. Return the mother to the cage then leave undisturbed for an hour or two.
5. Remove some of the natural pups to leave a litter of 12 pups at the most. The foster mother will be unable to care for a larger litter.

8.2 Breeding transgenic lines

As soon as a founder animal has been identified as being transgenic (see Chapter 7), it is necessary to mate the animal in order to establish a line. The line must of course be established before the founder is sacrificed for analysis. This is easier if the founder is male since a male founder can be very quickly mated with a large number of females, whereas it is necessary for a female founder to produce and wean at least one litter before she can be sacrificed. It is best to positively identify first generation transgenic rodents by hybridization analysis before the founder is sacrificed.

8.2.1 Natural matings

Mating transgenic animals to generate a line requires the minimum of intervention. The animals should be kept on a constant 12 hour light–dark cycle. Females maintained on such a cycle will ovulate every four to five days, five hours after the onset of darkness, at around midnight. Males will copulate with ovulating females in the middle of the dark period (around 00.30) and fertilization will take place between 30 minutes and two hours later. Several female animals can be put with one transgenic male, left for a week to ensure mating has occurred, and then separated until delivery. Alternatively, females can be pre-selected for their stage in the oestrous cycle (*Protocol 9*) and checked for a copulatory plug the following morning. Only one male should be paired with a transgenic female.

8.2.2 *In vitro* fertilization

Transgenic male founder animals are occasionally found to have impaired fertility. With mice, *in vitro* fertilization (IVF; ref. 20) can be used to generate offspring and establish a line. The technique involves superovulating wild-type female mice and then fertilizing the eggs with sperm taken from the dissected epididymis of males. IVF embryos are then surgically transferred to a 0.5 day p.c. pseudopregnant surrogate. Mouse strains used for the IVF

Table 3 Fertilization medium (FM)

All chemicals used should be tissue culture grade. Prepare the following concentrated stocks and filter each through a 0.45 μm Millipore filter into a sterile plastic tube. Store frozen at −20 °C.

- 10 × A stock. Dissolve 7.013 g NaCl (Sigma S5886), 1.0 g glucose (Sigma G6138), 0.201 g KCl (Sigma P5405), 21.3 mg Na_2HPO_4 (Sigma S5136), 0.102 g $MgCl_2.6H_2O$ (Sigma M2393) in double distilled or Milli-Q water, and bring to 100 ml final volume.
- 10 × B stock. Dissolve 2.106 g $NaHCO_3$ (Life Technologies 895-1810IP), 1.0 g phenol red (Sigma P5530) in water, and bring to 100 ml final volume.
- 100 × C stock. Dissolve 55 mg sodium pyruvate (Sigma P5280) in 10 ml water.
- 100 × D stock. Dissolve 0.264 g $CaCl_2.2H_2O$ (Sigma C7902) in 10 ml water.

To prepare fertilization medium, mix 10 ml of 10 × A, 10 ml of 10 × B, 1 ml of 100 × C, and 1 ml of 100 × D. Make to 100 ml by adding 78 ml double distilled water. Add 3.0 g BSA (Sigma A4161) and mix gently until dissolved. Pass through a 0.45 μm Millipore filter using a large sterile disposable syringe, and store in sterile plastic containers at 4 °C.

procedure should be the same as those used for original transgenic mouse production, usually C57BL/6J × CBA/J F1 hybrids. *Protocol 12* describes the procedures using mice maintained on a 12 hour light–dark cycle (06.30 lights on and 18.30 lights off). The fertilization medium required is listed in *Table 3*.

Protocol 12. *In vitro* fertilization of mouse eggs

Equipment and reagents

- 100 ml fertilization medium (*Table 3*)
- M16 egg culture medium (*Table 1*)
- 50 IU/ml FSH (see *Protocol 3*)
- 50 IU/ml hCG (see *Protocol 3*)
- 37°C tissue culture incubator gassed with 5% CO_2
- Paraffin oil (Fluka 76235)
- 35 mm sterile tissue culture dishes
- Dissection instruments (as for *Protocol 2*)
- Egg transfer pipette and mouthpiece (*Protocol 4*)

Method

1. Start the superovulation of the donor females on day 1 (IVF will be performed on day 4). Inject 5 IU FSH intraperitoneally to each of ten four-week-old immature female mice at 20.00 h.
2. On day 3 at 20.00 h, inject the female mice intraperitoneally with 5 IU hCG.
3. In the evening of day 3, pre-incubate one 35 mm sterile tissue culture dish containing 1 ml FM medium, one dish containing 0.5 ml FM medium under paraffin oil, and five dishes containing 0.5 ml FM medium under paraffin oil at 37°C with 5% CO_2.
4. Early in the morning of day 4, prepare three 35 mm tissue culture dishes containing 2–3 ml M16 medium, and one 35 mm dish containing M16 microdrop cultures. Incubate at 37°C with 5% CO_2.
5. At 06.45 h on day 4, kill the donor male[a] by cervical dislocation and place it abdomen side up. Soak the lower abdomen with 70% (v/v) ethanol and open up the body cavity. Gently pull out the testes with a pair of watchmaker's forceps. Dissect out the epididymis (a white mass of coils at the base of the testes) with a pair of fine scissors, and immediately transfer it to the pre-gassed 35 mm culture dish containing 0.5 ml FM medium covered with paraffin oil. Repeat the procedure for the other epididymis.
6. Quickly, with a pair of fine forceps, tease away the membrane and gently squeeze the sperm out of the epididymis. The sperm should exit in a continuous stream. Do this rapidly since the sperm should be at room temperature for as short a time as possible.
7. Incubate the sperm for 30 min at 37°C with 5% CO_2.

8. To each pre-incubated culture dish containing 0.5 ml FM medium under paraffin oil (from step 3), add 50 μl of sperm. The final concentration of sperm should be approx. $1–2 \times 10^6$ sperm/ml.
9. Incubate the diluted sperm mixture for 2 h at 37°C with 5% CO_2.
10. At 09.40–10.00 h, kill the superovulated females by neck dislocation and dissect out the oviducts from both sides (*Protocol 5*). Collect all of the oviducts into the pre-gassed dish containing 1 ml of FM medium without oil.
11. Tease away the ampullae membrane to release the cumulus mass from the oviducts. Transfer the cumulus masses from four oviducts to each pre-gassed dish containing 0.5 ml FM plus sperm (from step 8).
12. Incubate the sperm/egg mixture at 37°C under 5% CO_2 for 5–6 h.
13. At 15.00 h, wash the eggs three times in pre-gassed M16 medium to remove excess sperm. Transfer the eggs to the M16 microdrop and culture overnight.
14. On the next day, surgically transfer the two-cell embryos to the oviducts of 0.5 day post-coitum pseudopregnant surrogate mothers (*Protocol 10*).

[a] If one does not wish to sacrifice the transgenic male, anaesthetize it with 2.5% (v/v) Avertin (*Protocol 1*) and, following the removal of the sperm (step 6), cauterize the severed ends of the epididymis near the testis and vas deferens with red-hot forceps. Return the testes to the body, stitch up the body wall, and clip the skin with autoclips, as described in *Protocol 2*.

8.3 Cryopreservation of transgenic rodent lines

Valuable transgenic lines can be frozen and preserved indefinitely in freezing straws in liquid nitrogen. Cryopreservation not only protects the transgenic lines from environmental catastrophes, but is also an economical and labour-saving method of maintaining lines for future detailed analysis. Mouse blastocysts can be generated by culturing fertilized one-cell embryos in M16 medium (two and a half to three days post-hCG; 20, 21). However, rat eggs do not survive well in culture, and blastocysts are best obtained directly from the animal, at five days post-hCG. The blastocysts will have to be flushed from the rat oviducts.

There are two general methods for cryopreservation of mouse embryos (22, 23). One is an equilibrium method (slow cooling; *Protocol 13*), which utilizes a programmable freezing machine. The second is a fast cooling method (non-equilibrium; *Protocol 14*) which needs only access to a −70°C freezer. The major difference between these two types of cryopreservation is the rate of cooling, and whether the cooling rate allows for osmotic equilibrium to become established between the extracellular and intracellular compartments with respect to cryoprotectants:

(a) In equilibrium cooling, embryos are exposed to moderate cryoprotectant concentration and are cooled slowly (0.3–2°C/min). Embryos are dehydrated during this slow cooling process.

(b) In non-equilibrium cooling, embryos are exposed to high molar concentrations of cryoprotectant and are cooled rapidly (10°C–50°C/min). Embryos are dehydrated rapidly, exposure time to cryoprotectants is reduced, and vitrification occurs soon after cooling begins. This method is however more sensitive to minor variations in protocol, especially time.

One should estimate the number of embryos to freeze by taking into account the number of embryos needed to produce adequate numbers of male and female mice to rescue the line.

For example:

Frozen embryos	*Recovered from straws*	*Freeze–thaw survival*	*Pups*	*Transgenic animals*
100%	88%	77%	7.7%	4%

Therefore, one should freeze at least 400 embryos in at least two different sessions per line to be sure of obtaining 15–16 transgenic animals when thawed. Individual experimenters should also optimize the procedure for their laboratories using wild-type embryos before attempting the cryopreservation with valuable transgenic lines.

Protocol 13. Slow cooling (equilibration) method of freezing mouse embryos

Equipment and reagents

- PBS with Ca^{2+} and Mg^{2+}. Prepare PBS-A by dissolving 10 g NaCl (Sigma S5885), 0.25 g KCl (Sigma P5405), 1.44 g Na_2HPO_4 (Sigma S5136), 0.25 g KH_2PO_4 (Sigma P5655) in 1 litre of water. Autoclave. To 160 ml of PBS-A, add 20 ml of autoclaved 6.8 mM $CaCl_2.2H_2O$ (Sigma C7902) and 20 ml of autoclaved 4.9 mM $MgCl_2.6H_2O$ (Sigma M 2393).
- Fetal calf serum (FCS) (from Hyclone Laboratories, Cat. No. A-1111-L)
- Glycerol (Life Technologies 5514VA)
- Freezing solution 1: PBS with Ca^{2+} and Mg^{2+} containing 10% (v/v) FCS
- Freezing solution 2: PBS with Ca^{2+} and Mg^{2+} containing 10% (v/v) FCS and 5% (v/v) glycerol
- Freezing solution 3: PBS with Ca^{2+} and Mg^{2+} containing 10% (v/v) FCS and 10% (v/v) glycerol
- Sucrose (Sigma S1888)
- Freezing solution 4: PBS with Ca^{2+} and Mg^{2+} containing 10% (v/v) FCS, 10% (v/v) glycerol, and 0.6 M sucrose
- Egg culture medium M16 (*Table 1*)
- Freezing straws and plugs (available from Biotechnologies International, Cat. No. ZA475 and ZA511 respectively)
- Forceps
- Programmable freezing machine (e.g. the Kryo 10 series II controlled rate freezer, available from Planer Products Ltd.)
- Liquid nitrogen storage tank
- 37°C tissue culture incubator gassed with 5% CO_2
- Stereo microscope
- 35 mm tissue culture dishes
- Egg transfer pipettes and mouthpiece (*Protocol 4*)

A. *Freezing embryos*

1. Prepare two 35 mm tissue culture dishes containing freezing

solution 1, one dish containing freezing solution 2, and one dish containing freezing solution 3. Incubate at room temperature.

2. Transfer the mouse embryos (eight-cell to early blastocyst stage) from M16 culture medium to freezing solution 1. Rinse briefly.

3. Transfer the embryos to the second dish of freezing solution 1, and rinse.

4. Place the embryos in the dish containing freezing solution 2 and incubate at room temperature for 10 min.

5. Transfer the embryos to freezing solution 3 and incubate at room temperature for 10 min. During this incubation period, label the straws, load them with the embryos, and seal the straws.

6. Begin cooling within 10–30 min from when the embryos are first exposed to freezing solution 3. Cool the straws in the programmable freezing machine from room temperature to −6°C at 2°C/min and hold at −6°C for 5 min.

7. Remove the straws from the freezing machine. Hold the straws in a vertical position and manually seed them by grasping the straws at a point above the embryos with a pair of cold forceps dipped in liquid nitrogen. Hold for a few seconds until the liquid at the point of contact freezes.

8. Return the straws to the freezing machine and maintain at −6°C for 10 min.

9. Cool the straws at 0.5°C/min to −30°C.

10. Finally, plunge the straws into liquid nitrogen quickly and store.

B. *Thawing frozen embryos*

1. Prepare one 35 mm tissue culture dish containing freezing solution 4, two dishes containing freezing solution 1, and two dishes containing M16 egg culture medium. Incubate at room temperature.

2. Remove the straws quickly from liquid nitrogen and hold at room temperature for 30 sec before placing in a 37°C water-bath until the ice melts.

3. Expel the embryos into the dish of freezing solution 4 and incubate at room temperature for *exactly* 10 min.

4. Rinse the embryos twice in the two dishes of freezing solution 1.

5. Transfer the embryos into M16 culture medium and rinse twice.

6. Culture the embryos or transfer them into recipients (*Protocol 10*).

Protocol 14. Fast cooling (non-equilibration method of freezing mouse embryos)

Equipment and reagents

- As for *Protocol 13*
- Freezing solution 5: PBS with Ca^{2+} and Mg^{2+} containing 10% (v/v) FCS, 3.25 M glycerol, and 0.5 M sucrose
- Freezing solution 6: PBS with Ca^{2+} and Mg^{2+} containing 10% (v/v) FCS and 0.5 M sucrose
- −70°C freezer

A. *Freezing embryos*

1. Prepare two 35 mm tissue culture dishes containing freezing solution 1 and one dish containing freezing solution 5. Incubate at room temperature.
2. Transfer the mouse embryos (eight-cell to morula stage or two and a half days post-hCG) from M16 culture medium to one of the dishes containing freezing solution 1.
3. Rinse the embryos twice in the second dish of freezing solution 1.
4. Place the embryos into the dish of freezing solution 5 and incubate at room temperature for 20 min.
5. Load the embryos into pre-labelled straws and seal.
6. Keep the embryos at room temperature so that the total exposure to freezing solution 5 is 20 min.
7. Place the straws horizontally on to the bottom of −60°C to −80°C freezer.
8. Cool the straws in the freezer for 5–15 min.
9. Plunge the straws into liquid nitrogen using cooled forceps and store.

B. *Thawing embryos*

1. Prepare two 35 mm tissue culture dishes containing M16 egg culture medium and warm in a 37°C incubator gassed with 5% CO_2. Also prepare two dishes containing freezing solution 1 and one dish containing freezing solution 6. Incubate at room temperature.
2. Quickly remove the straws from liquid nitrogen and place into a 25°C water-bath until the ice melts.
3. Expel the embryos into the dish containing freezing solution 6 and incubate at room temperature for *exactly* 12 min.
4. Transfer the embryos into a dish of freezing solution 1 and rinse twice in the second dish of freezing solution 1.
5. Transfer the embryos into a dish of pre-warmed M16 and rinse twice.
6. Culture the embryos or transfer them into recipients (*Protocol 10*).

Cryopreservation of rat embryos has been less well researched. *Protocol 15* describes a method for freezing rat blastocysts by direct plunging into liquid nitrogen (24).

Protocol 15. Cryopreservation of rat embryos

Equipment and reagents

- As for *Protocol 13* (although with different freezing media; see below)
- Freezing medium VS1: PBS with Ca^{2+} and Mg^{2+} containing 0.4% BSA (Sigma A4161), 20% (v/v) DMSO (Sigma D2650), 15.5% (w/v) acetamide (Sigma A0500), 10% (v/v) propylene glycol (Sigma 1009), 6% (w/v) polyethylene glycol 8000 (Sigma P2139)
- Freezing medium VS2: PBS with Ca^{2+} and Mg^{2+} containing 0.4% BSA and 12.5% (v/v) freezing medium VS1
- Freezing medium VS3: PBS with Ca^{2+} and Mg^{2+} containing 0.4% BSA and 25% (v/v) freezing medium VS1
- Freezing medium VS4: PBS with Ca^{2+} and Mg^{2+} containing 0.4% BSA and 50% (v/v) freezing medium VS1
- 0.4% BSA in PBS

A. *Freezing embryos*

1. Prepare one 35 mm tissue culture dish containing freezing medium VS1 and chill at 4°C. Prepare another 35 mm tissue culture dish containing freezing medium VS2 and one containing freezing medium VS3. Incubate at room temperature.
2. Place the rat blastocysts (five days post-hCG, see *Protocol 3*) in the dish containing freezing medium VS2 and incubate for 5 min at room temperature.
3. Transfer the embryos to the dish containing freezing medium VS3 and incubate at room temperature for 5 min.
4. Wash the embryos in the chilled (4°C) dish of freezing medium VS1. Transfer the embryos in a volume of 40 μl to a pre-cooled (4°C) polypropylene tube or freezing straw. Chill at 4°C for 15 min.
5. Plunge the straws into liquid nitrogen and store.

B. *Thawing embryos*

1. Chill freezing media VS4 and VS3 to 4°C.
2. Prepare one 35 mm tissue culture dish containing freezing medium VS2 and two dishes containing 0.4% BSA in PBS with Ca^{2+} and Mg^{2+}. Incubate all three dishes at room temperature. Prepare two 35 mm tissue culture dishes containing egg culture medium M16 and incubate at 37°C with 5% CO_2.
3. Thaw the tubes on ice.
4. Add 200 μl of chilled freezing medium VS4 to the thawed tube and incubate at 4°C for 10 min.
5. Add 400 μl of chilled freezing medium VS3 and leave at 4°C for 10 min.

Protocol 15. *Continued*

6. Transfer the embryos to the dish containing freezing medium VS2 and maintain at room temperature for 5 min.
7. Wash the embryos twice in the dishes of 0.4% BSA in PBS with Ca^{2+} and Mg^{2+}.
8. Transfer the embryos into pseudopregnant recipient mothers (see *Protocol 10*).

Acknowledgements

We would like to thank Sandra Jones and her staff for their excellent care of our transgenic animals, and Francis Leong for photography.

References

1. Palmiter, R. D., Brinster, R. L., Hammer, R. E., Trumbauer, M. E., Rosenfeld, M. G., Birnberg, N. C., *et al.* (1982). *Nature*, **300**, 611.
2. Carter, D. A. (1993). In *Transgenesis techniques* (ed. D. Murphy and D. A. Carter), pp. 7–22. Humana Press, New Jersey.
3. Hui, K. M. (1993). In *Transgenesis techniques* (ed. D. Murphy and D. A. Carter), pp. 37–52. Humana Press, New Jersey.
4. Murphy, D. (1993). In *Transgenesis techniques* (ed. D. Murphy and D. A. Carter), pp. 23–36. Humana Press, New Jersey.
5. Lathe, R. and Mullins, J. J. (1993). *Transgenic Res.*, **2**, 286.
6. Palmiter, R. D. and Brinster, R. L. (1986). *Annu. Rev. Genet.*, **20**, 465.
7. Lacy, E., Roberts, S., Evans, E. P., Burtenshaw, M. D., and Constantini, F. (1983). *Cell*, **34**, 343.
8. Krumlauf, R., Chapman, V. M., Hammer, R. E., Brinster, R. L., and Tilghman, S. M. (1985). *Nature*, **319**, 224.
9. *NIH guide for the care and use of laboratory animals*. (1985). National Institutes of Health Publication No. 85–23, Washington, DC.
10. Brinster, R. L., Chen, H. Y., Trunbauer, M. E., Yagle, M. K., and Palmiter, R. D. (1985). *Proc. Natl Acad. Sci. USA*, **82**, 4438.
11. Mullins, J. J., Peters, J., and Ganten, D. (1990). *Nature*, **344**, 541.
12. Chada, K., Magram, J., Raphael, K., Radice, G., Lacy, E., and Constantini, F. (1985). *Nature*, **314**, 377.
13. Oberdick, J., Smeyne, R. J., Mann, J. R., Zackson, S., and Morgan, J. I. (1990). *Science*, **248**, 223.
14. Brinster, R. L., Allen, J. M., Behringer, R. R., Gelinas, R. E., and Palmiter, R. D. (1988). *Proc. Natl Acad. Sci. USA*, **85**, 836.
15. Young III, W. S., Reynolds, K., Shepard, E. A., Gainer, H., and Castel, M. (1990). *J. Neuroendocrinol.*, **2**, 917.
16. Schedl, A., Montoliu, L., Kelsey, G., and Schütz, G. (1993). *Nature*, **362**, 258.
17. Swanson, L. W., Simmons, D. M., Azzira, J., Hammer, R., and Brinster, R. (1985). *Nature*, **317**, 363.

18. Vogelstein, B. and Gillespie, D. (1979). *Proc. Natl Acad. Sci. USA*, **82**, 4438.
19. Sambrook, J., Fritsch, E. F., and Maniatis, T. (ed.) (1989). *Molecular cloning: a laboratory manual*, Vol. 2. Cold Spring Harbor Press, Cold Spring Harbor, NY.
20. Hogan, B., Constantini, F., and Lacy, E. (ed.) (1986). *Manipulating the mouse embryo—a laboratory manual*. Cold Spring Harbor Laboratory, Cold Spring Harbor, NY.
21. Whittingham, D. G. (1971). *J. Reprod. Fert. Suppl.*, **14**, 7.
22. Pomeroy, K. O. (1991). *GATA*, **8**, 95.
23. Kasai, M., Komi, J. H., Takakamo, A., Tsudura, H., Sakurai, T., and Machada, T. (1990). *J. Reprod. Fert.*, **89**, 91.
24. Kono, T., Suzuki, O., and Tsunoda, Y. (1988). *Cryobiology*, **25**, 170.

7

Genomic and expression analysis of transgenic animals

SARAH JANE WALLER, JUDITH McNIFF FUNKHOUSER, KUM-FAI CHOOI, and DAVID MURPHY

1. Introduction

Analysis of transgene expression is a cardinal part of any transgenic study. Transgenic pups born from microinjected eggs need to be identified (section 2), and then bred to produce a line (see Chapter 6, section 8.2). Expression of the transgene in the transgenic animals can then be analysed at the RNA (section 3) and protein (section 4) levels. It is very important to examine for transgene expression in as many tissues as possible, including those in which one does not expect to find its expression. Most transgenic animals that express their transgenes do so in a manner that is appropriate to the regulatory elements present. However, a number of transgenes are expressed ectopically, due either to influences of the adjacent cellular DNA or to unforeseen interactions between *cis*-acting elements in fusion genes. Such ectopic expression may produce unexpected and inappropriate results, including ill health, which should be taken into account when interpreting the results of the study. A thorough post-mortem examination is therefore also recommended for all animals expressing new transgenes (section 5).

Many transgenic rodent lines do not express their transgenes. This is thought to be due either to the presence of inhibitory sequences within the transgene (for example prokaryotic sequences, see Chapter 6, section 6.1) or to the site of integration of the transgene. The exogenous DNA may, for example, be integrated into a chromosomal location that is transcriptionally inactive. Furthermore, the presence of transgene RNA does not necessarily imply that the transcript will be correctly translated and processed, underlining the importance of protein expression analysis.

2. Identification of transgenic rodents

Transgenic animals are identified by hybridization analysis of high molecular weight genomic DNA isolated from tail tissue. The tissue can be taken from rat pups when they are two-weeks-old, although mice pups should be weaned (at least three-weeks-old) before they are tailed. A method for the isolation of tail DNA is given in *Protocol 1*.

Protocol 1. Isolation of genomic DNA from rodent tails

Equipment and reagents

- Proteinase K buffer: 50 mM Tris–HCl pH 8.0, 100 mM EDTA, 100 mM NaCl, 1% SDS, 100 μg/ml Proteinase K (Sigma P-0390)
- TE-equilibrated phenol. Melt 100 g of molecular biology grade phenol (Gibco-BRL) at 65 °C. Equilibrate overnight with 30 ml Tris–HCl pH 7.4, 24 ml 0.5 M EDTA, 7.5 ml 10 M NaOH, and 243.5 ml distilled autoclaved water. Remove the aqueous phase, add 100 mg 8-hydroxyquinoline and 200 μl 2-mercaptoethanol to the phenol, and store frozen in a bottle protected from light.
- TE buffer: 10 mM Tris–HCl pH 7.4, 0.1 mM EDTA
- Phenol/chloroform: mix TE-equilibrated phenol with an equal volume of chloroform and a 1/50th volume of isoamyl alcohol (store at 4 °C in a bottle protected from light)
- 1 μg/ml RNase A in TE buffer (incubate in a boiling water-bath for 10 min before use to destroy any contaminating DNase)
- Electronic identification system (e.g. Lab-Track from Stoelting) (optional)

Method

NB Genomic DNA is sensitive to shearing, and should not be vortexed or subjected to repeated freeze–thaw cycles.

1. Whilst restraining the animal with one hand, use a sharp pair of scissors to cut off around 1 cm of tail. Place the piece of tail in 0.7 ml of Proteinase K buffer. Before releasing the animal, mark it permanently by clipping an ear or toe, or with an electronic identification system so that it can be identified later.
2. Mince the recovered tail tissue using a pair of sharp, fine scissors whilst still in the Proteinase K buffer. Incubate at 50°C overnight, preferably with gentle agitation.
3. In the morning add 5 μl of 1 mg/ml RNase A and incubate at 37 °C for at least 1 h.
4. Add 0.7 ml of TE-equilibrated phenol. Gently mix with the digested tail solution until homogeneous (this could take up to 20 min).
5. Separate the phases by centrifugation at 7000 *g* in a microcentrifuge for 10 min. Transfer the viscous aqueous phase (top layer) and the interphase to a fresh tube.
6. Add 0.7 ml of phenol/chloroform and gently mix the phases. Separate again by centrifugation at 7000 *g* in a microcentrifuge. Transfer the upper (aqueous) layer to a fresh tube. This time avoid any interphase.
7. Add 0.8–1.0 ml of 100% ethanol and gently mix. A large white stringy precipitate of genomic DNA should appear. Pellet this by centrifugation in a microcentrifuge for 2 min.

8. Discard the supernatant and rinse the pellet with 1 ml of cold 70% (v/v) ethanol. Re-pellet the DNA by centrifugation for 1 min. Remove as much of the supernatant as possible and dry the pellet briefly under vacuum.
9. Resuspend the DNA by adding 250 μl of TE buffer. Incubate at 37°C for several hours to allow the DNA to dissolve. Store at 4°C.
10. Assay the DNA by spectrophotometry at 260 nm; 50 μg/ml of double-stranded DNA has an $A_{260} = 1.0$.

The method used to analyse tail DNA depends upon the nature of the transgene. If the transgene has a homologue within the rodent genome, then Southern hybridization analysis (ref. 1; *Protocol 2*) must be undertaken. If the transgene is wholly or partially non-homologous to any part of the rodent genome then slot blot analysis (ref. 2; *Protocol 4*) or PCR (ref. 3; *Protocol 5*) can be used. However, for initial identification of founder animals it is best to use Southern analysis, regardless of the nature of the transgene. Slot blots and PCR, unlike Southern analysis, can not provide information on transgene copy number and integrity, although these two techniques are useful for the rapid screening of subsequent generations.

2.1 Southern analysis of tail DNA

Southern analysis of genomic DNA involves several steps. DNA is first digested with the chosen restriction enzyme and the resulting fragments are separated in an agarose gel. The DNA is then transferred to a membrane filter and this is hybridized with a labelled probe. A foreign gene integrated in the transgenic animal's genome can be distinguished from an endogenous homologue in two ways:

(a) By differing restriction endonuclease cleavage patterns. Even if the transgene is identical to the endogenous gene it will be flanked by different restriction enzyme sites and can be distinguished accordingly. Alternatively, if the transgene is derived from another species it may be possible to distinguish it from the homologous rodent counterpart by virtue of different internal restriction enzyme sites.
(b) By using transgene-specific probes which have little homology to the host genomic DNA, for example probes specific for reporter genes.

Protocol 2 describes Southern blotting of tail DNA and *Protocol 3* describes probe preparation and the hybridization of the labelled probe with the Southern blot. Probes are usually radioactively labelled pieces of DNA, and, for safety, must be hybridized to and washed from the filter in tightly sealed vessels (e.g. strong thermally sealable plastic bags or tightly capped bottles). Simple and safe hybridization systems are available from several companies (e.g. Hybaid).

Protocol 2. Southern blotting of tail DNA[a]

Equipment and reagents

- Tail DNA (from *Protocol 1*)
- Appropriate restriction enzyme, with buffer containing 1 mM spermidine (Sigma Cat. No. S0266)
- 50 mM sodium phosphate buffer pH 7.0
- 50 × TAE buffer: mix 242 g Tris base, 57.1 ml glacial acetic acid, 100 ml 0.5 M EDTA, pH 8.0, make up to 1 litre and autoclave
- 0.8% agarose gel. Prepare this as follows. Using a hot plate or microwave oven, dissolve 2 g of molecular biology grade agarose in 250 ml water containing 5 ml of 50 × TAE buffer. Cool to 50°C and add ethidium bromide solution to give a final concentration of 0.5 μg/ml. Pour into a 20 cm × 20 cm gel mould and allow to set. CAUTION: ethidium bromide is a potent carcinogen.
- Gel electrophoresis equipment
- 10 × loading dye: 50% (v/v) glycerol, 10 × TAE, 0.25% (w/v) bromophenol blue, 0.25% (w/v) xylene cyanol
- DNA size marker (e.g. 1 kb ladder from Life Technologies)
- Transfer buffer: 1.5 M NaCl, 0.25 M NaOH
- Nylon hybridization membrane (e.g. Hybond-N from Amersham)[b]
- Capillary transfer system[a] filled with transfer buffer. Place a platform in a reservoir of transfer buffer and cover it by a wick made of four thicknesses of 3MM paper cut to the width of the platform, soaked in transfer buffer, and dipping into the reservoir.
- 0.25 M HCl
- 1.5 M NaCl, 0.5 M NaOH
- UV transilluminator (312 nm)
- Oven at 50°C

Method

1. Digest 10–15 μg of tail DNA with the chosen restriction enzyme for several hours.
2. Add 1/10 vol. of loading dye to the digested tail DNA, and load on to the 0.8% agarose gel. Also load a positive control, such as the DNA used to microinject the eggs, and a DNA size marker. Run the gel in 1 × TAE buffer until the bromophenol blue dye has reached the end.
3. Photograph the gel on a UV transilluminator (312 nm) with a ruler placed alongside the marker DNA lane to enable migration distances of individual DNA fragments to be determined.
4. Depurinate the separated DNA fragments by gently agitating the gel in several volumes of 0.25 M HCl for 30 min at room temperature.
5. Denature the DNA by gentle agitation in several volumes of 1.5 M NaCl, 0.5 M NaOH for 30 min.
6. Equilibrate the gel by gentle agitation in transfer buffer for 15–30 min.
7. Place the gel on the platform of the capillary transfer system. Wet a 20 cm × 20 cm piece of hybridization membrane in distilled water and then transfer buffer, and place this on the gel, avoiding any air bubbles. Cover the membrane with six 20 cm × 20 cm sheets of 3MM paper, wetting the first two in transfer buffer. Then stack several paper towels on top, followed by a 1 kg weight, and allow the transfer to proceed for at least 12 h.
8. Before removing the membrane from the gel, mark on it the positions

of the tracks on the filter. Rinse the membrane in 50 mM sodium phosphate buffer pH 7.0, and then bake at 80°C for 1 h.

9. Place the membrane, DNA side down, on a piece of Saran wrap, and covalently cross-link the DNA to the matrix by exposure to a UV transilluminator set at 312 for 2 min.
10. Rinse the membrane in 50 mM sodium phosphate buffer pH 7.0.
11. Hybridize the membrane to a probe (*Protocol 3*), or store wrapped in plastic at −20°C for later analysis.

[a] Southern blotting is covered in detail in a companion volume to this book (4).
[b] A variety of different transfer membranes are available commercially. The composition of the transfer buffer required to yield good Southern blots varies with the type of membrane used and hence the supplier's recommendations should be followed.

Protocol 3. Labelling of probes and hybridization to Southern blots[a]

Equipment and reagents

For random primed labelling:

- 200 ng DNA fragment for use as probe
- OLB buffer. Prepare the following stock solutions. Solution A: 1 ml of 1.25 M Tris–HCl pH 8.0, 0.125 M $MgCl_2$, 18 μl 2-mercaptoethanol, 5 μl 0.1 M dATP, 5 μl 0.1 M dGTP, 5 μl 0.1 M dTTP. Solution B: 2 M Hepes pH 6.6 (store at 4°C). Solution C: random hexadeoxyribonucleotides (Pharmacia Cat. No. 27-2166-01) in TE buffer at 90 OD A_{260} U/ml (store at −20°C). Mix solutions A:B:C in the ratio of 100:250:150 and store −20°C.
- 10 mg/ml BSA
- [α-^{32}P]dCTP (3000 Ci/mmol, Amersham International)
- 2 U/μl *E. coli* DNA polymerase I, Klenow fragment (e.g. from Boehringer Mannheim)
- Sephadex G-50 column. To prepare Sephadex G-50 slurry, mix 30 g of Sephadex G-50 with 250 ml TE buffer pH 8.0 and autoclave. A Sephadex G-50 column consists of 1 ml Sephadex G-50 slurry in a 1 ml disposable plastic syringe plugged with autoclaved glass woo! (BDH Ltd.)
- OLB-C stop solution: 20 mM Tris–HCl pH 7.5, 20 mM NaCl, 2 mM EDTA, 0.25% SDS, 1 mM dCTP

For 5′ end labelling:

- 200 ng synthetic oligonucleotide to be used as probe
- 5 × polynucleotide kinase buffer: 0.3 M Tris–HCl pH 7.8, 50 mM $MgCl_2$, 25 mM DTT, 0.5 mM spermidine, 0.5 mM EDTA
- 10 U/μl polynucleotide kinase (e.g. from Boehringer Mannheim)
- TE buffer: 10 mM Tris–HCl pH 7.4, 0.1 mM EDTA
- Sephadex G-50 column (see above for details)
- [γ-^{32}P]dATP (> 5000 Ci/mmol, Amersham International)

For hybridization:[a]

- Southern blot (from *Protocol 2*)
- Hybridization buffer: 0.5 M sodium phosphate buffer pH 6.8, 7% SDS, 1 mM EDTA, 15% (v/v) deionized formamide (Fluka Cat. No. 47670)
- 0.1% (w/v) SDS, 50 mM sodium phosphate buffer pH 6.8
- X-ray film, casettes, and developing chemicals

A. *Random primed labelling (for DNA fragments)*

1. Adjust 200 ng of probe DNA to 32 μl final volume with water. Boil for 3 min, and then rapidly quench on ice.
2. Add 10 μl OLB buffer, 2 μl 10 mg/ml BSA, 5 μl [α-^{32}P]dCTP, and 1 μl Klenow enzyme. Incubate at 25°C for 4–16 h.

Protocol 3. *Continued*

3. Stop the reaction by adding 50 μl OLB-C stop solution.
4. Purify the probe through a Sephadex G-50 column. To do this, place the Sephadex G-50 column into a 15 ml sterile plastic container and centrifuge at 1000 *g* for 4 min. Remove the cap from a microcentrifuge tube and place the tube into the 15 ml container, underneath the G-50 column. Apply the labelled probe to the top of the G-50 column, and collect it into the microcentrifuge tube by centrifuging at 1000 *g* for 4 min.
5. Denature the purified DNA probe by boiling for 3 min, then quench on ice.

B. *5′ end labelling (for oligonucleotide probes)*

1. Mix 200 ng oligonucleotide with 2 μl 5 × polynucleotide kinase buffer, 5 μl [γ-^{32}P]dATP, 1 μl polynucleotide kinase in a total volume of 10 μl. Incubate at 37 °C for 1 h.
2. Add 90 μl TE buffer and purify through a Sephadex G-50 column (see above for details). Denaturation is not required.

C. *Hybridization*

1. Pre-hybridize the Southern blot by incubation in hybridization buffer at the required temperature[b] for at least 5 min.
2. Hybridize the labelled probe to the membrane filter at a concentration of 25 ng/ml at the hybridization temperature[b] for at least 12 h.
3. Carefully remove the excess probe and discard it safely.
4. Wash the membrane at the hybridization temperature[b] several times in 0.1% (w/v) SDS, 50 mM sodium phosphate buffer pH 6.8.
5. Wrap the membrane filter in Cling film and expose it to X-ray film. Develop the film according to the manufacturer's instructions.

[a] Carry out the hybridization in leak-proof containers, e.g. heat sealed plastic bags or using one of the hybridization systems available commercially (e.g. Hybaid).
[b] The temperature at which the probe should be hybridized is usually 15 °C lower than the melting temperature of the probe (5). Oligonucleotide probes over 40 bases in length and random primed DNA probes (6) which have a perfect match to the target sequence, should be hybridized and washed at 65 °C.

An example of such an analysis is shown in *Figure 1*; 11 pups have been born to a surrogate mother impregnated with eggs microinjected with a fragment of the human *c-fos* gene (D. Murphy, unpublished data). Southern hybridization of *Pst*I cleaved tail DNA reveals that the exogenous DNA, which has a different pattern of internal *Pst*I sites than the endogenous gene,

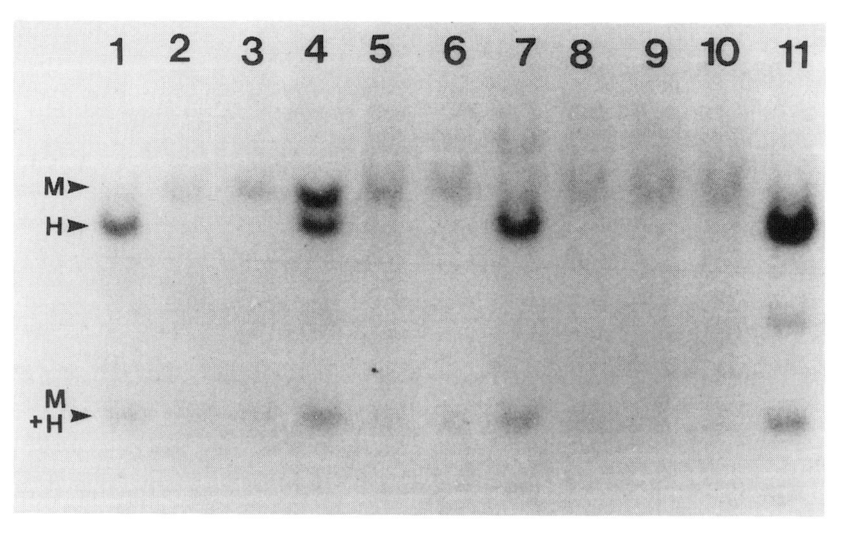

Figure 1. A Southern hybridization analysis of high molecular weight DNA isolated from mouse tail tissue. Genomic DNA was isolated from tail biopsies of 11 mice resulting from the injection of fertilized one-cell eggs with a fragment of the human *c-fos* gene. The DNA was cleaved with the *PstI* restriction endonuclease, fractionated through a 0.8% agarose gel, transferred to a filter, and hybridized with a *c-fos* probe. Both mouse (M) and human (H) *fos* sequences are detected.

is present only in mice numbers 1, 4, 7, and 11. The endogenous bands provide controls for the quantity and quality of the tail DNA preparation.

2.2 Slot or dot blotting of tail DNA

Slot blotting or dot blotting is a quick and easy method of analysing tail DNA, but it can only be applied to transgenic animals bearing a transgene with no homology to rodent DNA (for example a hybrid gene with a viral or prokaryotic reporter element). A suitable method is described in *Protocol 4*.

Slot or dot blot analyses should be performed in duplicate with one filter being analysed with a transgene probe and the other with an endogenous rodent gene probe. This is because tail DNA is difficult to quantitate due to contamination with RNA. Hybridization to an endogenous rodent probe ensures that false negatives are avoided and affords a degree of comparative quantitation. Inclusion of negative controls (i.e. non-transgenic DNA) is also essential.

Protocol 4. Analysis of tail DNA using dot blots or slot blots

Equipment and reagents

- Nylon hybridization membrane (e.g. Hybond-N from Amersham)
- 20 × SSPE: 3.6 M NaCl, 200 mM sodium phosphate buffer pH 6.8, 20 mM EDTA
- 5 M ammonium acetate
- Slot or dot blot apparatus (e.g. from Schleicher and Schuell)

Protocol 4. *Continued*

Method

1. Adjust 10 μg of tail DNA to 0.3 M with NaOH in 200 μl final volume.
2. Incubate in a boiling water-bath for 10 min. Quench on ice.
3. Stand the DNA at 25°C for 5 min, then add 0.4 ml of 5 M ammonium acetate.
4. Soak a pre-cut filter (e.g. Hybond-N) for at least 5 min in 20 × SSPE.
5. Assemble the slot or dot blot apparatus. Apply the DNA to the apparatus using gentle vacuum.
6. Remove the membrane filter from the apparatus and allow it to dry at room temperature.
7. Bake the filter at 80°C for 1 h.
8. Place the filter, DNA side down, on a piece of Saran wrap. Covalently cross-link the DNA to the matrix by exposure to a UV transilluminator set at 312 nm, for 2 min.
9. Hybridize a labelled probe to the filter as described in *Protocol 3*.

2.3 Analysis of tail DNA using PCR

PCR (3) is a rapid way of identifying transgenic offspring, and in addition it is extremely sensitive. This procedure is described in *Protocol 5*. Careful design of the primers is very important to allow specific amplification of transgene DNA:

(a) If the transgene contains a reporter gene that does not have a mammalian homologue, (e.g. encoding bacterial chloramphenicol transferase), the primers can be designed to detect the sequence of the reporter.

(b) The transgene may contain a homologous gene from another species. In this case, the primers should be designed to detect the region of least homology (e.g. from intron or 3′ untranslated region sequences).

(c) In cases where the transgene is from the same species as the transgenic animal, it may be possible to design primers that give different sized products for the transgene and its endogenous counterpart (e.g. by amplifying across an intron).

One of the most serious considerations when using PCR is the absolute need to avoid even trace contamination with extraneous DNA since this will be amplified and generate a signal. Therefore, the experimenter *must* avoid contamination of the PCR reagents with PCR products, which will give false positives. Inclusion of negative controls in all series of PCR reactions to check for contaminations is essential.

Protocol 5. Amplification of tail DNA using PCR

Equipment and reagents

- Thermocycler (e.g. the Trio-Thermoblock TB1, available from Biometra)
- Forward and reverse primers[a]
- 10 × PCR buffer: 200 mM Tris–HCl pH 8.3, 20 mM $MgCl_2$, 250 mM KCl, 0.5% (v/v) Tween-20, 1 mg/ml BSA
- 2 mM solutions of dATP, dCTP, dGTP, and dTTP (Pharmacia ultrapure dNTP set)
- 5 U/μl *Taq* DNA polymerase (e.g. Boehringer Mannheim Cat. No. 1146 173)
- Mineral oil (Sigma M 3516)
- UV transilluminator

Method

1. In a 0.5 ml microcentrifuge tube, mix:
   - 1 μg denatured tail DNA
   - 50 pmol each primer
   - 10 μl each of 2 mM dATP, dCTP, dGTP, and dTTP
   - 10 μl 10 × PCR buffer
   - 1 U *Taq* DNA polymerase
   - water to 100 μl
2. Overlay with 75 μl mineral oil to prevent evaporation.
3. Start the reaction. The conditions will depend on the melting temperature of the primer oligonucleotides,[a] but a typical reaction would use 95°C for 1 min (denaturation), 60°C for 2 min (annealing), 72°C for 2.5 min (extension), for 30 cycles.
4. Load 1/10th of the reaction mixture on to a 1% agarose gel and carry out electrophoresis (see *Protocol 2*). Check that the product is of the correct size by visualizing under UV light.

[a] Primers are usually designed to be 18–28 nt in length with 50–60% GC content. The annealing temperature of the PCR reaction should be 5°C below the calculated melting temperature (T_m) of the primers. A general rule of thumb calculation of the T_m for oligonucleotides is 2°C for each A or T plus 4°C for each G or C.

3. Analysis of transgene expression

Transgene RNA can be detected in individual tissues using Northern analysis (section 3.1) or at the cellular level using *in situ* hybridization (section 3.2). Both of these methods require that all equipment and solutions are RNase-free. Diethylpyrocarbonate (DEPC) is a potent (but highly toxic) inhibitor of RNase, and should be added to most solutions to a final concentration of 0.1% (v/v). The solution is then left to stand at room temperature for several hours and autoclaved, which degrades the DEPC to ethanol and carbon dioxide. Tris buffers react chemically with DEPC and can not be treated with this reagent, although they can be made up with DEPC-treated water. Plasticware should be washed in 0.1% DEPC and then autoclaved. Sterile disposable plasticware is generally regarded as RNase-free as supplied.

3.1 Northern analysis of RNA

RNA from the tissues of choice is fractionated on a denaturing agarose gel and is then transferred to a filter, a process known as Northern blotting. The filter is subsequently incubated with a specific labelled probe to allow hybridization to occur (7). Isolation of RNA from animal tissues requires the inhibition or inactivation of RNases present in the tissue extract and inactivation of RNases in reagents and equipment. The two methods described here use guanadinium thiocyanate as a lytic agent which also inactivates RNases. The first (miniprep) method (*Protocol 6*) can process tissues the size of a mouse thymus, two mouse ovaries, or a single rat hypothalamus. The second (maxiprep) method (*Protocol 7*) should be chosen when larger amounts of tissue are to be processed, such as rodent hearts, whole brain, or testis.

Protocol 6. Miniprep method for isolation of RNA from tissues of transgenic animals

Equipment and reagents

- Guanadinium thiocyanate solution A (GTC-A). Prepare 4 M guanadinium thyocyanate containing 25 mM sodium citrate pH 7.0, and 0.5% (v/v) Sarkosyl. Filter through a 0.45 μm Nalgene filter. Add 36 μl of 2-mercaptoethanol per 5 ml immediately prior to use.
- 100% ethanol and 95% ethanol
- Chloroform/isoamyl alcohol (49 : 1)
- 1 ml syringe fitted with a 25 gauge needle
- Aqueous phenol. Melt 100 g molecular biology grade phenol (Life Technologies) at 65°C, and equilibrate overnight with an equal volume of DEPC-treated water. Remove the aqueous phase, add 100 mg 8-hydroxyquinoline and 200 μl 2-mercaptoethanol, and store frozen.
- DEPC-treated 2 M sodium acetate pH 5.5
- UV spectrophotometer

Method

1. Collect tissue samples into DEPC-treated microcentrifuge tubes and freeze immediately on dry ice. The tissue can be stored at −70°C or processed immediately.
2. Add 0.5 ml ice-cold GTC-A to the sample and allow the tissue to thaw briefly on ice. Disrupt the tissue by passing it several times through a 25 gauge needle attached to a 1 ml syringe.
3. Add 50 μl of 2 M DEPC-treated sodium acetate pH 5.5, 0.5 ml aqueous phenol, 100 μl chloroform/isoamyl alcohol (49 : 1). Shake vigorously for 20 sec, then incubate on ice for 10 min.
4. Centrifuge at 10 000 *g* in a microcentrifuge at 4°C for 10 min.
5. Very carefully remove the aqueous phase, avoiding the interphase

which contains genomic DNA. Precipitate the RNA by adding 2 vol. of 100% ethanol to the aqueous phase and keeping at −70°C for 1 h.

6. Pellet the RNA by centrifugation at 10 000 *g* at 4°C for 10 min.
7. Resuspend the pellet in 250 μl of GTC-A. Re-precipitate the RNA with ethanol at −70°C for 1 h.
8. Pellet the RNA by centrifugation at 10 000 *g* at 4°C for 10 min. Wash the pellet with 95% (v/v) ethanol, dry briefly under vacuum, and resuspend the RNA in 12 μl DEPC-treated water. This often requires heating the sample to 65°C combined with vigorous vortexing.
9. Dilute 2 μl of the RNA in 0.5 ml DEPC-treated water and read its A_{260} by UV light spectrophotometry. An A_{260} = 1.0 is equivalent to 40 μg/ml RNA.

Protocol 7. Maxiprep method for isolation of RNA from tissues of transgenic animals

Equipment and reagents

- Guanadinium thiocyanate solution A (GTC-A); see *Protocol 6*
- Guanadinium thyocyanate solution B (GTC-B). Prepare 5 M guanadinium thyocyanate containing 10 mM EDTA and 50 mM Tris–HCl pH 7.5. Filter through a 0.45 μm Nalgene filter. Add 36 μl of 2-mercaptoethanol per 5 ml immediately prior to use.
- DEPC-treated 2 M sodium acetate pH 5.5
- Homogenizer (e.g. Ultra-Turax T25, Janke and Kunkel GmbH)
- Aqueous phenol (see *Protocol 6*)
- Chloroform/isoamyl alcohol (49:1)
- 100% ethanol and 95% ethanol
- DEPC-treated 4 M LiCl (autoclave and filter through a 0.45 μm Nalgene filter)
- UV spectrophotometer

Method

1. Collect the tissues, either into ice-cold GTC-A, or into a dry tube and immediately freeze on dry ice.
2. Using 1 ml of ice-cold GTC-A per 50 mg tissue, homogenize the samples using an Ultra-Turax homogenizer.
3. Add 0.1 ml 2 M DEPC-treated sodium acetate pH 5.5, 1 ml aqueous phenol, and 0.2 ml chloroform/isoamyl alcohol (49:1) per ml of GTC-A used in step 2. Mix vigorously for 20 sec (do not vortex) and stand on ice for 15 min.
4. Centrifuge at 10 000 *g* at 4°C for 20 min. Carefully transfer the aqueous phase to a fresh tube, avoiding the interphase. Precipitate the RNA by adding 2 vol. of 100% ethanol to the aqueous phase and keeping at −70°C for 1 h.
5. Pellet the RNA by centrifugation at 10 000 *g* at 4°C for 30 min.

Protocol 7. *Continued*

6. Resuspend the pellet in 0.3 ml of GTC-B per ml of GTC-A added in step 2. Add 6 vol. of 4 M LiCl, mix, and allow the RNA to precipitate at 4°C overnight.
7. Pellet the RNA by centrifugation at 10 000 *g* at 4°C for 90 min.
8. Resuspend the pellet in 5 ml GTC-A. Re-precipitate with 2 vol. of ethanol at −70°C for 1 h.
9. Centrifuge at 10 000 *g* at 4°C for 45 min. Wash the RNA pellet with 70% ethanol, dry it briefly, and resuspend in DEPC-treated water. This often requires heating the sample to 65°C combined with vigorous vortexing.
10. Dilute 2 μl of the RNA in 0.5 μl DEPC-treated water and determine the RNA concentration by UV light spectrophotometry. An A_{260} = 1.0 is equivalent to 40 μg/ml RNA.

Once isolated, RNA from transgenic animals is fractionated through a denaturing agarose gel, blotted on to a membrane, and hybridized to a specific labelled probe, usually radiolabelled (*Protocol 8*). After washing away excess probe, RNA bands which hybridized to the probe are detected by autoradiography. Northern blot analysis allows the experimenter to determine in which tissues the transgene is expressed, and to analyse changes in that expression in response to physiological stimuli. The probe used must be able to distinguish between endogenous and transgene RNA. This can be achieved by including reporter genes in the transgene or by designing the transgene such that the RNA produced will be a different size to its endogenous homologue and hence distinguishable by Northern blot analysis.

Protocol 8. Northern blotting of transgene RNA[a]

Equipment and reagents

- 20 × MAE buffer: 0.4 M MOPS pH 7.0, 0.1 M sodium acetate, 20 mM EDTA
- 20 × SSPE buffer: 3.6 M NaCl, 0.2 M sodium phosphate buffer pH 6.8, 20 mM EDTA
- Loading buffer: 5 ml deionized formamide (Fluka Cat. No. 47670), 0.5 ml 20 × MAE buffer, 1.7 ml formaldehyde, 0.3 ml DEPC-treated water
- 10 × Loading dye: 50% (v/v) glycerol, 1 × MAE buffer, 0.25% (w/v) bromophenol blue, 0.25% (w/v) xylene cyanol, 10 mg/ml ethidium bromide (N.B. take care in handling ethidium bromide which is a potent carcinogen)
- Nylon hybridization membrane (e.g. Hybond-N from Amersham)[b]
- Gel electrophoresis equipment
- Capillary transfer system[a] filled with 20 × SSPE buffer (see *Protocol 2* for description of this apparatus)
- UV transilluminator (312 nm)
- Oven at 80°C
- 1% formaldehyde agarose gel. The volumes of reagent needed will depend on the gel dimensions. As an example, the following is to prepare a 20 cm × 20 cm gel. Using a hot plate or microwave, dissolve 3 g of molecular biology grade agarose in 235 ml DEPC-treated water. Cool to 65°C and add 51 ml formaldehyde and 15 ml of 20 × MAE buffer. Pour into a 20 cm × 20 cm gel mould in a fume-hood and allow the gel to set.

Method

1. Mix 3 vol. of loading buffer with each RNA sample, and heat at 65°C for 15 min to denature the RNA. Add 1/10 vol. of 10 × loading dye and mix. Load the samples into the gel slots. Run the gel in 1 × MAE buffer at 4 V/cm until the bromophenol blue has reached the end of the gel.
2. Place the gel on the platform of a capillary transfer system filled with 20 × SSPE, as described in *Protocol 2*. Wet a 20 cm × 20 cm piece of nylon hybridization membrane in distilled water and then in 20 × SSPE buffer, and place this on the gel, avoiding any air bubbles. Cover the membrane with six 20 cm × 20 cm sheets of 3MM paper, wetting the first two in 20 × SSPE buffer. Then stack several paper towels on top, followed by a 1 kg weight, and allow the transfer to proceed for at least 12 h.
3. Before removing the membrane from the gel, mark on it the positions of the tracks. Bake at 80°C for 1 h.
4. Place the membrane, RNA side down, on a piece of Saran wrap, and covalently cross-link the RNA to the matrix by exposure to a UV trans-illuminator set at 312 nm for 2 min. Examine the ethidium bromide stained RNA on the transilluminator, and mark the positions of the 18S and 28S ribosomal RNA bands on the membrane. It is useful to photograph the filter at this stage.
5. Rinse the membrane in 50 mM sodium phosphate buffer pH 7.0 or 2 × SSPE. Either hybridize to a labelled probe immediately and process for autoradiography (*Protocol 3*), or store wrapped in plastic at −20°C for later analysis.

[a] Northern blotting is covered in detail in a companion volume to this book (4).
[b] A variety of different transfer membranes are available commercially. The composition of the transfer buffer required to yield good Northern blots varies with the type of membrane used and hence the supplier's recommendations should be followed.

3.2 *In situ* hybridization

In situ hybridization (ISH) is used to localize transgene expression to a population of cells within the tissues of a transgenic animal. The basic procedure is to expose tissue sections to a specific probe, either radiolabelled or non-radioactively labelled, then wash away excess probe, and dip the sections in liquid film emulsion. Tissues that are expressing the transgene give autoradiographic signals and, in addition, counterstaining with a number of histological stains can be used to visualize tissues in general.

3.2.1 Tissue preparation

Freshly frozen tissue can be used for most ISH applications. Following sectioning (*Protocol 9*), we routinely fix the tissue in 4% (w/v) paraformaldehyde

(*Protocol 10*). Oligonucleotide probes will penetrate even tissues fixed with paraformaldehyde plus 0.2% (v/v) glutaraldehyde. The fixatives mentioned later in the chapter for use with immunohistochemistry (section 4.2) are also suitable for use with ISH.

When the tissue is fragile or contains abundant RNase activity, or when better morphology is required, fixation by transcardial perfusion of the animal is preferred (8). This procedure is described in detail elsewhere (9). Briefly, the animal is given an overdose of anaesthetic, and perfused via the aorta with an initial isotonic wash solution containing heparin for about 1 min followed by a fixative for 10 min. The tissue is then removed and placed in fixative for 2–4 h, then in 30% (w/v) sucrose overnight, and finally frozen and sectioned as in *Protocol 9*.

Protocol 9. Tissue preparation for *in situ* hybridization

Equipment and reagents

- Glass slides. Soak the slides in chromic acid cleaning solution (20 g potassium dichromate, 20 ml sulfuric acid, 200 ml distilled water) for 5 min, then rinse thoroughly in water, and finally in double distilled water. Dip the clean slides in 0.25 M ammonium acetate and oven dry. Finally, place them in 50 μg/ml poly-L-lysine (Sigma P1274) in 10 mM Tris–HCl pH 8.0 for 30 min, then oven dry them.[a]
- Isopentane cooled with a dry ice–acetone mixture
- Cryostat (e.g. the model 2800 Frigocut N, available from Reichert Jung, Cat. No. 0398 18229)
- Clay Adams black slide boxes (Thomas Scientific)

Method

1. Remove the tissue from the animal and immediately freeze it in the cooled isopentane.
2. Cut 10 μm tissue sections with the cryostat and dry these on to the prepared slides.
3. Store at −70°C in Clay Adams black boxes until ready to use.[b]

[a] Alternatively, the more expensive ready-to-use Superfrost/Plus slides (Fisher Scientific, Pittsburgh, PA) can be used.

[b] If the animal has not been perfusion fixed, the tissue sections can be stored frozen for several months before fixation, which is performed just before *in situ* hybridization (*Protocols 10* and *11*).

3.2.2 Probe preparation and *in situ* hybridization

Oligodideoxynucleotides or DNA fragments can be used as probes. Oligonucleotide probes are frequently used because there is no problem with tissue penetration, and they can be designed to distinguish between endogenous mRNAs and transgenic mRNA. The probe can be labelled with a radioactive isotope (*Protocol 10*), which provides the greatest sensitivity, or with a non-radioactive compound, such as digoxigenin (*Protocol 11*). Non-radioactively labelled probes give clearer localization but are less sensitive than radioactive probes. While both types of labelled probe can be hybridized to sections made from freshly frozen tissue (*Protocol 9*), better morphology is achieved with non-radioactive probes if perfusion-fixed tissue is used.

The hybridization temperature, length and temperature of wash, and the concentration of probe must all be optimized for each probe (10). The conditions described here are for use with a 48mer with 62% GC content. Conditions close to these will work for probes of 40–50 bases with 50–65% GC content. As a general rule of thumb, probes with higher GC content require higher hybridization temperatures and longer washes. For example, to hybridize a 48mer probe with 55% GC content, use a hybridization temperature of 37°C, and shorten the 56°C washes in *Protocol 10*, step 6 to 15 min each. For a probe of 35 bases with 72% GC content, a hybridization temperature of 42°C or 45°C with a longer wash time would be better. For a detailed discussion of the variables affecting the results, see Stahl *et al.* (11). For each new probe it is advisable to define the best conditions with a short series of experiments in which one variable is altered at a time.

Controls are very important for interpretation of the results of ISH. Negative controls are essential, and are best performed by hybridizing the probe for the transgene to the same tissue taken from a non-transgenic animal. For positive controls, probes designed to identify an endogenous gene are useful when used with a probe that detects only the transgene.

Protocol 10. *In situ* hybridization using an ^{35}S-labelled oligonucleotide probe

Equipment and reagents

All solutions should be made up in DEPC-treated water to avoid RNase contamination. For probe labelling:

- 200 ng oligonucleotide probe[a] in 2 μl of DEPC-treated water
- 20 × SSC: 3 M NaCl, 0.3 M sodium citrate pH 7.0, prepared in DEPC-treated water and autoclaved
- Hybridization buffer: make 20 ml containing 50% (v/v) deionized formamide, 1 × SSC, 10 mg sheared salmon sperm DNA (Sigma Cat. No. D9156), 2 mg Ficoll, 2 mg polyvinylpyrrolidone, 2 mg BSA, 5 mg tRNA, 2 g dextran sulfate, and 15 000 U of heparin (Sigma Cat. No. H9266) (store at −20°C in 1 ml aliquots)
- 10 × labelling reaction buffer[b]: 140 mM sodium cacodylate pH 7.2, 1 mM cobalt chloride, 0.1 mM DTT
- [α-^{35}S]dATP (1000–1500 Ci/mmol, New England Nuclear NEG 034H)
- 18 U/μl terminal deoxynucleotidyl transferase (TdT)[b]
- TEN buffer: 50 mM Tris–HCl pH 7.4, 10 mM EDTA, 150 mM NaCl
- Sephadex G-50 column, prepared in a 1 ml syringe (see *Protocol 3*)

For hybridization:

- Tissue sections (from *Protocol 9*)
- Fixative: 4% (w/v) paraformaldehyde in 0.1 M sodium phosphate buffer pH 7.4. Make the fixative shortly before use. Heat 4 g paraformaldehyde in 50 ml of water to 60–70°C. Titrate with 10 M NaOH until the solution clears. Cool to 40°C and add 50 ml of 0.2 M sodium phosphate buffer pH 7.4.
- Phosphate-buffered saline (PBS): 20 mM phosphate buffer pH 7.4 containing 0.8% (w/v) NaCl
- 20 × SSC (see above)
- Xylene

Protocol 10. *Continued*

- 0.1 M triethanolamine containing 0.25% (v/v) acetic anhydride and 0.2% (w/v) NaCl (add the acetic anhydride to the buffer just before use)
- Xylene-based mounting medium
- Chloroform
- Dark-room fitted with a safe light equipped with a Kodak #2 red filter
- Kodak NTB2 film emulsion (available from International Biotechnologies Inc.)
- Kodak D-19 developer
- Clay Adams black slide boxes (Thomas Scientific)
- 5% (v/v) sodium thiosulfate
- Histological counterstain (e.g. 1.0% (w/v) neutral red pH 6.5)

A. *End-labelling the oligonucleotide probe*

1. Mix in a sterile microcentrifuge tube:
   - 200 ng probe — 2 μl
   - 10 × labelling reaction buffer — 3 μl
   - [α-^{35}S]dATP — 5 μl
   - DEPC-treated H_2O — 17.5 μl

   Centrifuge briefly to mix and also to remove air bubbles.
2. Add 2.5 μl TdT, mix gently, and incubate at 38°C for 1 h.
3. Place a microcentrifuge tube with the lid removed into a 15 ml sterile disposable tube, followed by the Sephadex G-50 column. Transfer the reaction mixture on to the Sephadex G-50 column. Rinse the microcentrifuge tube that had contained the labelling reaction with 30 μl of TEN buffer and add this to the Sephadex G-50 column. Centrifuge the Sephadex G-50 column at 1000 *g* for 5 min to collect 50 μl of eluate containing the purified labelled probe into the microcentrifuge tube.
4. Add 10 μl of purified labelled probe to 1 ml of hybridization buffer containing 10 μl of 1 M DTT.
5. Count 10 μl of the diluted probe in a liquid scintillation counter. It should have a count of at least 100 000 c.p.m. The probe must be used the same day.

B. *In situ* hybridization

1. For sections prepared from freshly frozen tissue (*Protocol 9*), place the slides into fixative for 15 min.
2. Rinse the sections twice in PBS for 2 min each time.
3. Place the slides in 0.1 M triethanolamine containing 0.25% (v/v) acetic anhydride and 0.2% (w/v) NaCl for 10 min at 25°C. This reduces non-specific binding of the probe.
4. To further permeabilize the cell membranes, dehydrate the sections for 1 min each in 70%, 90%, and 100% ethanol. Delipidate the tissue by placing the sections in chloroform for 5 min, then 100% ethanol, and 90% ethanol for 1 min each. Allow to air dry.

5. Apply 60 μl of the radiolabelled probe (from part A, step 5) to each slide and cover with a coverslip cut from Parafilm or Nescofilm. Place the slides in a humid chamber and incubate overnight at 40°C.[a]
6. Wash the slides four times in 1 × SSC at 56°C, 20 min each time.[a]
7. Wash the slides twice for 1 h each time in 1 × SSC at 25°C. Rinse in sterile water and air dry.
8. In the dark-room, using the safe light, dip the slides in Kodak NTB2 emulsion melted at 45°C and diluted 1:1 with water. Leave the slides to dry (approx. 1 h) in darkness (i.e. no safe light). Pack the slides into black slide boxes containing desiccant and leave at 4°C from three days to one month to expose the emulsion.
9. Develop the slides in Kodak D-19 for 2.5 min, rinse in water for 30 sec, and fix in 5% (v/v) sodium thiosulfate for 6 min. Finally, rinse in running water for 5 min.
10. The slides can also be counterstained with the usual histological stains (e.g. in 1.0% (w/v) neutral red). Place in the stain for 3 min, dehydrate through 70%, 95%, and 100% ethanol for 15 sec each, then clear twice in xylene for 2 min each. Finally, add a coverslip using a xylene-based mount medium.

[a] The conditions in this protocol are for a 48mer oligonucleotide with 62% GC content. The conditions must be optimized for use with other probes (see section 3.2.2).
[b] 3′ end-labelling kits are available. The authors routinely use Du Pont-NEN kit Cat. No. NEN 009Z.

Protocol 11. *In situ* hybridization using a digoxigenin labelled oligonucleotide probe

Equipment and reagents

All solutions should be made up in DEPC-treated water to avoid RNase contamination.
For probe labelling:

- 200 ng oligonucleotide probe[a] in 2 ml of DEPC-treated water
- 10 × labelling reaction buffer[b] (see *Protocol 10*)
- Terminal deoxynucleotidyl transferase (TdT)[b] (see *Protocol 10*)
- TEN buffer (see *Protocol 10*)
- 1 mM ATP (Pharmacia)
- 0.1 mM digoxigenin-11-dUTP (Boehringer Mannheim Cat. No. 109308)

For hybridization:

- Reagents as for *Protocol 10*
- 3% (v/v) hydrogen peroxide in PBS
- 0.1 M acetate buffer pH 6.0: dissolve 27 g sodium acetate in 1 litre of water and titrate to pH 6.0 with diluted acetic acid
- Tissue sections: while non-radioactive probes can be hybridized to freshly frozen tissue (*Protocol 9*), better morphology is achieved using perfusion-fixed tissue, which should then be frozen and cuts as in *Protocol 9*
- PBS containing 1% BSA and 0.1% (v/v) Tween-20

Protocol 11. *Continued*

- GDN (glucose oxidase, 3,3′-diaminobenzidine dihydrochloride (DAB), nickel). First prepare solution A by dissolving 2.5 g nickel ammonium sulfate in 50 ml 0.2 M sodium acetate buffer pH 6.0. Also prepare solution B by dissolving 50 mg 3,3′-diaminobenzidine dihydrochloride (DAB; Sigma Cat. No. D5637) in 50 ml double distilled water. (Note: DAB is a carcinogen—handle with care.) Next mix solutions A and B and add 200 mg β-D-glucose, 40 mg ammonium chloride and 100–200 U of glucose oxidase (Sigma type VII, Cat. No. G2133). This is the final GDN reagent; prepare fresh just before use.
- Anti-digoxigenin POD Fab_2 fragment (Boehringer Mannheim Cat. No. 1207733)
- Xylene
- Xylene-based mounting medium

A. *End-labelling the oligonucleotide probe*

1. Mix in a sterile microcentrifuge tube:
   - 200 ng probe 2 μl
   - 10 × labelling reaction buffer 3 μl
   - 1 mM ATP 1 μl
   - 0.1 mM digoxigenin-11-dUTP 3 μl
   - DEPC-treated H_2O 18.5 μl

   Centrifuge briefly to mix and remove air bubbles.
2. Add 2.5 μl TdT, mix gently, and incubate at 38°C for 1 h.
3. Add 20 μl TEN buffer and place on ice.[c]

B. *In situ hybridization*

1. Place the slides bearing the tissue sections in PBS for 5 min.
2. To block non-specific binding, place the slides in 0.1 M triethanolamine containing 0.25% (v/v) acetic anhydride and 0.2% (w/v) NaCl for 10 min.
3. Dehydrate the tissue sections successively in 70%, 90%, and 100% (v/v) ethanol for 1 min each, then place in chloroform for 5 min to delipidate and permeabilize the membranes. Rehydrate in 100%, 90%, and 70% ethanol (in that order) for 1 min each.
4. Rinse the slides in 1 × SSC for 2 min.
5. Dilute the digoxigenin labelled probe 1:50 with hybridization buffer. Apply 50 μl of diluted probe to each slide and cover with a coverslip cut from Parafilm or Nescofilm. Incubate at 40°C overnight.[a]
6. Wash the slides four times in 1 × SSC at 56°C for 15 min each time.[a]
7. Wash the slides twice for 1 h each time in 1 × SSC at 25°C.
8. Rinse the slides in PBS pH 7.4 for 2 min.

9. To block endogenous peroxidase, treat the tissue sections with 3% (v/v) hydrogen peroxide in PBS for 3 min.
10. Rinse the slides in PBS for 2 min.
11. To block non-specific binding of antibody, place the slides into PBS containing 1% BSA and 0.1% (v/v) Tween-20 for 10 min.
12. Apply 100–200 μl of anti-digoxigenin POD diluted 1:50 in PBS containing 1% BSA and 0.1% (v/v) Tween-20 to each slide. Incubate at 37°C for 1.5–2 h in a moist chamber.
13. Rinse the slides three times in PBS containing 0.1% (v/v) Tween-20 for 5 min each.
14. Rinse the slides in 0.1 M sodium acetate buffer pH 6.0.
15. Incubate the slides in freshly made up GDN reagent.
16. Rinse the slides twice in 0.1 M sodium acetate buffer pH 6.0 for 2 min each time.
17. Dehydrate the tissue sections for 15–30 sec each in 70%, 95%, and then 100% ethanol.
18. Clear the tissue sections twice in xylene for 2 min each time. Finally, add a coverslip using a xylene-based mounting medium.

[a] The conditions in this protocol are for a 48mer oligonucleotide with 62% GC content. The conditions must be optimized for use with other probes (see section 3.2.2).
[b] 3′ end-labelling kits are available. The authors routinely use Du Pont-NEN kit Cat. No. NEN 009Z.
[c] The probe can be stored frozen at −20°C for up to one year if not required for immediate use.

Typical results obtained using *Protocols 10* and *11* are shown in *Figure 2*.

These two protocols can be combined for double ISH (8). The procedure is as follows:

(a) Hybridize with both probes simultaneously and then wash with 1 × SSC.
(b) Block endogenous peroxidase and incubate with the anti-digoxigenin antibody as in *Protocol 11*, steps 9–17.
(c) Air dry and then dip in 2% (v/v) collodion plastic (Electron Microscopy Sciences).
(d) Air dry, and then proceed with *Protocol 10*, steps 8–10.
(e) Develop gently to avoid losing the emulsion. The plastic layer prevents undesirable reactions between the peroxidase product and the emulsion.

4. Analysis of transgene products

4.1 Reporter enzyme assays

The reporter genes contained in many transgenes encode enzymes, the activities of which can be rapidly assayed to detect transgene expression or

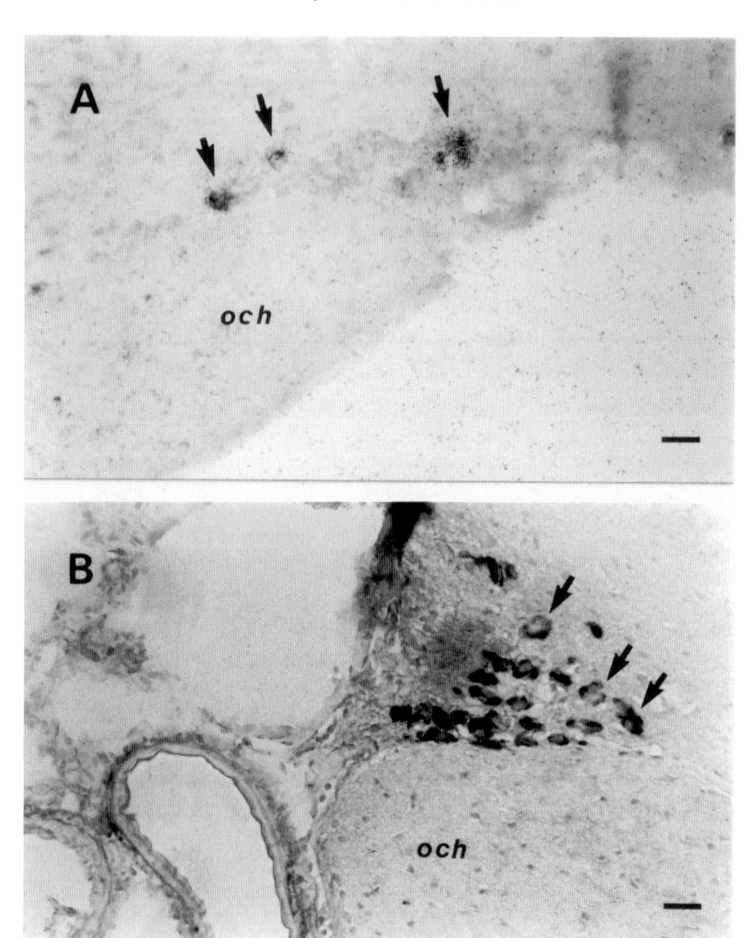

Figure 2. (A) *In situ* hybridization of transgenic bovine vasopressin mRNA in the supraoptic nucleus of a transgenic mouse produced by microinjection of fertilized one-cell eggs with a bovine vasopressin transgene. An oligonucleotide probe specific for bovine vasopressin was labelled with [α-^{35}S]dATP as described in *Protocol 10*. Three labelled cells are indicated by the *arrows*. (B) *In situ* hybridization of endogenous vasopressin mRNA in the rat supraoptic nucleus. An oligonucleotide probe specific for rodent vasopressin was labelled with digoxygenin dATP as described in *Protocol 11*. Labelled cells are indicated by the *arrows*. In both figures the scale bar represents 8 μm. Och: optic chiasma.

to analyse transcriptional activity of linked transgene sequences. The three most commonly used reporter enzymes are chloramphenicol acetyl transferase (CAT), β-galactosidase, and luciferase.

4.1.1 Chloramphenicol acetyl transferase (CAT)

The bacterial enzyme CAT is often used as a reporter gene as it has no mammalian counterpart, and its protein coding region is relatively small (660 bp).

The enzyme catalyses the transfer of acetyl from acetyl CoA to chloramphenicol. Several assays exist for CAT. In the assay described in *Protocol 12*, ^{14}C-labelled acetyl CoA is used as substrate and, following the reaction, the labelled acetylated chloramphenicol is assayed by liquid scintillation spectroscopy (12, 13). Quantification is achieved using a standard curve produced by assaying purified CAT enzyme (Pharmacia Cat. No. 27–0847–01).

Protocol 12. Chloramphenicol acetyl transferase assay

Equipment and reagents

- Liquid nitrogen
- [^{14}C]acetyl CoA solution: dilute an aliquot of [^{14}C]acetyl CoA containing 1 mCi (Amersham) to a concentration of 5 mCi/ml with 0.5 mM unlabelled acetyl CoA (Sigma Cat. No. A2056)
- 8 mM chloramphenicol
- 0.25 M Tris–HCl pH 7.8
- 0.25 M Tris–HCl containing 5 mM EDTA
- Ethyl acetate
- Liquid scintillation vials and fluid
- Liquid scintillation counter

Method

1. Homogenize the tissue of interest in 0.2 ml of 0.25 M Tris–HCl pH 7.8. Freeze–thaw the homogenate three times by immersion in liquid nitrogen for 3 min followed by incubation in a 37°C water-bath for 3 min. Vortex thoroughly after the final thawing step.
2. Centrifuge at 10 000 *g* at 4°C for 10 min, and retain the supernatant.
3. Heat the supernatant at 65°C for 10 min to destroy endogenous CAT inhibitors.
4. Mix in a sterile microcentrifuge tube:
   - tissue extract (from step 3) 30 μl
   - 8 mM chloramphenicol 20 μl
   - diluted [^{14}C]acetyl CoA 20 μl
   - 0.25 M Tris–HCl containing 5 mM EDTA 30 μl

   Incubate at 37°C for 1 h.
5. Transfer the samples to an ice-bath and extract twice with cold ethyl acetate as follows. Add 100 μl of ethyl acetate to each sample, vortex, and separate the phases by centrifugation at 10 000 *g* for 3 min at 4°C. Retain 80 μl of the organic phase and re-extract the remaining mixture with 100 μl of ethyl acetate. This time, retain 100 μl of the organic phase and combine with the first 80 μl.
6. Add the combined organic extracts to 2 ml of scintillation fluid and determine the extent of ^{14}C-acetylation of chloramphenicol by liquid scintillation spectroscopy.

4.1.2 β-Galactosidase assays

The *lacZ* gene from *E. coli* encodes the enzyme β-galactosidase which catalyses the hydrolysis of β-galactosides at an optimum pH of 7.3 (14). Some mammalian tissues, including brain, contain endogenous β-galactosidase activity, but as the optimum pH of the mammalian form is 3.5, false positives can be avoided by performing the assays at pH 7.5 and by including normal tissue as a control. Enzyme assays can be performed *in vitro*, using a simple photometric assay that measures the hydrolysis of the substrate *o*-nitrophenyl β-D-galactopyranoside (ONPG) by β-galactosidase (ref. 14; *Protocol 13*), or *in situ* using 5-bromo-4-chloro-3-indolyl-β-D-galactopyranoside (X-Gal) as the substrate, the cleavage product of which is dark blue and resistant to common solvents (refs 15, 16; *Protocol 14*).

Protocol 13. *In vitro* assay for β-galactosidase

Reagents

- 4 mg/ml *o*-nitrophenyl β-D-galactopyranoside (ONPG; Sigma Cat. No. N1127) in 0.1 M sodium phosphate buffer pH 7.5
- 1 M Na_2CO_3
- 0.1 M $MgCl_2$ containing 4.5 M 2-mercaptoethanol
- β-Galactosidase (600–1200 U/mg protein, Sigma Cat. No. G6512)

Method

1. Prepare the tissue extracts as described in *Protocol 12*, but do not heat the samples to 65°C prior to assay.
2. Mix in a sterile microcentrifuge tube:
   - tissue extract 30 μl
   - 0.1 M $MgCl_2$ containing 4.5 M 2-mercaptoethanol 3 μl
   - ONPG 66 μl
   - 0.1 M sodium phosphate pH 7.5 201 μl
3. Incubate at 37°C for 30 min. A faint yellow colour should appear.
4. Stop the reaction by the addition of 0.5 ml of 1 M Na_2CO_3.
5. Assay the reaction by measuring the absorbance of each sample at 420 nm. Calculate the enzyme activity against a standard curve derived using purified β-galactosidase.

Although large pieces of tissue must be sectioned before *in situ* detection of β-galactosidase, small pieces of tissue (for example mouse embryos; ref 17), can be fixed, rinsed, and reacted as a whole piece. It may then be necessary to add detergents (e.g. 0.2% (v/v) NP-40 and 0.1% (w/v) sodium deoxycholate) to the reaction mixture to aid in permeabilization of the cells. The tissue may be then observed as a whole or embedded in plastic, e.g. LR white (Electron Microscopy Sciences) for sectioning at 0.5–2.0 μm.

Protocol 14. *In situ* assay of β-galactosidase

Reagents

- Reaction mixture: add 1 ml of X-Gal (Sigma Cat. No. B4252; 40 mg/ml in dimethyl sulfoxide) to 39 ml of PBS pH 7.0 containing 5 mM potassium ferricyanide, 5 mM potassium ferrocyanide, and 2 mM $MgCl_2$
- Xylene
- Xylene-based mounting medium

Method

1. Prepare and section the tissues as for ISH (*Protocol 9*). For better preservation of the morphology fix the slides for a short time (15 min to 1h) in 2% (w/v) paraformaldehyde or in 2% (w/v) paraformaldehyde plus 0.2% (v/v) glutaraldehyde, and then rinse in PBS.
2. Place the sections into a Coplin jar containing about 40 ml of the reaction mixture, and incubate at 25°C overnight. If the sections were fixed with glutaraldehyde, it may be necessary to incubate at 30°C or 37°C overnight.
3. Rinse the tissue sections three times in PBS.
4. Dehydrate the sections in 70%, 95%, and then 100% ethanol, 15–30 sec each. Clear twice in xylene for 2 min each.
5. Add a coverslip using a xylene-based mounting medium.
6. Examine the slides with bright field or DIC microscopy. A dark blue reaction product will be seen in cells containing β-galactosidase.

Tissue sections which are assayed *in situ* for β-galactosidase can be observed with bright field microscopy to locate the reaction product and compared to the section seen in phase contrast microscopy. Tissues which have a strong reaction product can be counterstained lightly with haematoxylin and eosin (*Protocol 18*) to better define the tissue areas, but often phase contrast or Nomarski optics (DIC) will provide sufficient information. Control slides should always be reacted along with the tissue of interest. Tissue which does not express the gene, preferably the same tissue from a wild-type animal, should be reacted as a negative control. For tissues which contain endogenous mammalian β-galactosidase, a useful positive control is to react the sections at pH 4.0 to detect a dark blue reaction product from the endogenous enzyme.

Rapid delivery of the fixative to tissues which are dense and difficult to penetrate (e.g. brain) or are fragile and easily degraded (e.g. lung) can be achieved by transcardial perfusion of the fixative (18). This can result in clearer microanatomy and in more consistent results.

4.1.3 Luciferase

The enzyme luciferase, obtained from the firefly *Photinus pyralis*, catalyses the ATP-dependent oxidation of the substrate luciferin, and results in the emission of light (19). The activity of luciferase in extracts of transgenic tissues can be assayed in the presence of ATP and luciferin by measuring the light output using a luminometer (*Protocol 15*). The total light output is rapid and proportional to the amount of luciferase present over a wide range of enzyme concentrations when assayed in the presence of excess substrate. Advantages of this reporter enzyme include its high sensitivity and its use of a non-radioactive substrate.

Protocol 15. Luciferase assay

Equipment and reagents

- Liquid nitrogen
- Homogenization buffer: 100 mM potassium phosphate pH 7.8 containing 1 mM DTT
- Incubation buffer: 25 mM glycylglycine pH 7.8, 15 mM $MgSO_4$, 4 mM EGTA, 15 mM potassium phosphate pH 7.8, 1 mM DTT, 2 mM ATP
- Luciferin solution: 0.2 mM luciferin (Sigma Cat. No. L6882) dissolved in 25 mM glycylglycine pH 7.8, 15 mM $MgSO_4$, 4 mM EGTA, 2 mM DTT
- Luminometer with autoinjection system (e.g. the Lumac Biocounter M2500)

Method

1. Homogenize the tissue in homogenization buffer.
2. Freeze–thaw the homogenate three times as for CAT assays (*Protocol 12*).[a] After the final thawing step, vortex the lysate thoroughly.
3. Centrifuge the lysate at 10 000 *g* for 5 min at 4°C.
4. Recover the supernatant and briefly vortex it just before starting the assay. Do not heat the supernatant to 65°C.
5. Mix 0.1 ml of tissue extract with 0.36 ml of incubation buffer.
6. Using a luminometer fitted with an automatic injection system, initiate the reaction by injecting 0.2 ml of luciferin solution into each sample. Measure the integrated light output for 5 sec after the luciferin injection.

[a] For an alternative procedure, to avoid any loss of luciferase activity during freeze–thawing, homogenize the tissue in 1% (v/v) Triton X-100, 25 mM glycylglycine pH 7.8, 15 mM $MgSO_4$, 4 mM EGTA, and 1 mM DTT, and then proceed from step 3.

4.2 Immunohistochemistry

Protein products of transgenes can be detected by immunohistochemistry (20). This technique can also be used to detect specific proteins which will identify the cell type in which the transgene is expressed. Most protocols use a primary antibody to attach to the antigen of interest, followed by a secondary antibody which is specific to the IgG of the host animal which produced the primary antibody. The secondary antibody can be labelled with an enzyme tag, such as alkaline phosphatase or horse-radish peroxidase (which will react with chromogens to produce a permanent result), or with a fluorescent tag, such as fluorescein or rhodamine. Fluorescent tags are more sensitive but the results are not permanent and must be photographed soon after the reaction. The primary antibody should be a high affinity monoclonal antibody or an affinity purified polyclonal antibody highly specific for the protein of interest. The secondary antibody should be the whole molecule or Fab_2 fragment. Purchased antibodies should be specifically for immunocytochemistry or immunohistochemistry; good sources include Amersham and Dako Ltd.

Gentle fixation of the tissue of interest is best for preserving most antigens. Fixatives like picric acid paraformaldehyde (PAF) (21) and periodic acid lysine paraformaldehyde (PLP) (22) are excellent for immunohistochemistry. Freshly frozen tissue can be sectioned (*Protocol 9*) and the slides placed in fixative for 15 minutes prior to use. Alternatively, the tissue (especially if it contains many degradative enzymes) can be fixed by perfusion before sectioning (8), or by immersing small pieces of tissue in the fixative, followed by cryoprotection in 30% (w/v) sucrose and freezing in isopentane cooled with a dry ice/acetone mixture. This will reduce degradation of the antigen and improve the morphology and localization. In all cases it is best to react the tissues with antibody as soon as possible after sectioning.

Control slides processed in parallel with the experimental slides are necessary to help optimize the reaction and to enable correct interpretation of the results. These should include a negative control performed by omitting the primary antibody or preferably by adsorbing the primary antibody with the antigen prior to its use.

Protocol 16 uses a peroxidase enzyme tag on the secondary antibody, with diaminobenzidine as a substrate for the peroxidase and a glucose oxidase reaction as the source of H_2O_2, plus a nickle enhancement of the reaction product (ref. 22; GDN reaction).

Protocol 16. Immunohistochemistry with a peroxidase labelled secondary antibody

Reagents

- Fixed tissue sections (see *Protocol 9*)
- PBS (see *Protocol 10*)
- 3% (v/v) hydrogen peroxide in PBS containing 0.1% sodium azide
- PBS containing 0.05% (v/v) Tween-20
- PBS containing 1% BSA and 0.05% (v/v) Tween-20
- Blocking serum: pre-immune serum from the animal in which the secondary antibody was produced
- Xylene
- Primary antibody[a] (e.g. from Amersham or Dako Ltd.)
- Secondary peroxidase tagged antibody[a] (e.g. from Amersham or Dako Ltd.)
- 0.1 M acetate buffer pH 6.0: dissolve 27 g sodium acetate in 1 litre of water and titrate to pH 6.0 with diluted acetic acid
- Freshly prepared GDN solution (see *Protocol 11*)
- Xylene-based mounting medium
- 1% (v/v) glutaraldehyde (optional; see step 13)

Protocol 16. *Continued*

Method

1. Rinse the slides holding the fixed sections in PBS pH 7.4 for 2 min.
2. Treat the sections with 3% (v/v) hydrogen peroxide in PBS containing 0.1% sodium azide for 3 min to block endogenous peroxidase activity.
3. Rinse the slides in PBS for 2 min.
4. Incubate the sections with PBS containing 1% BSA, 0.05% (v/v) Tween-20, and 2% (v/v) blocking serum for 10 min at 25°C to block non-specific binding of the antibody.
5. Drain the slides (do not rinse) and apply 50–100 μl of primary antibody, diluted to a range of appropriate concentrations in PBS containing 1% BSA and 0.05% (v/v) Tween-20.[a] To prevent evaporation of the primary antibody, place a coverslip cut from Parafilm on top of the solution. Incubate the sections overnight at 4°C in a moist chamber.
6. Rinse the slides twice in PBS containing 0.05% (v/v) Tween-20 for 5 min each.
7. Incubate the sections with PBS containing 1% BSA, 0.05% (v/v) Tween-20, and 2% (v/v) blocking serum for 10 min at 25°C.
8. Drain the slides, and apply 100 μl of the secondary peroxidase tagged antibody, diluted 1:100 in PBS containing 1% BSA and 0.05% (v/v) Tween-20. To prevent evaporation, place a coverslip cut from Parafilm on top of the solution. Incubate in a moist chamber at 25°C for 1.5–2 h.
9. Rinse the slides three times in PBS for 5 min each.
10. Rinse the slides in 0.1 M acetate buffer pH 6.0.
11. Incubate the sections in freshly-prepared GDN solution for 20 min.
12. Rinse the slides in 0.1 M acetate buffer pH 6.0.
13. Optional: post-fix the sections in 1% (v/v) glutaraldehyde for 10 min.
14. Rinse the slides in PBS for 5 min.
15. Dehydrate the sections through 70%, 95%, and then 100% ethanol, 15–30 sec each.
16. Rinse the slides twice in xylene for 2 min each. Mount a coverslip with a xylene-based mounting medium.
17. View the slides with bright field microscopy or with Nomarski optics. The reaction product will be black against a light grey background.

[a] The appropriate concentration of primary and secondary antibody will require experimental determination. Tenfold dilutions above and below the concentration recommended by the manufacturer should provide a sufficient range from which the optimal dilution can be chosen.

For less abundant antigens the reaction can be enhanced by using a biotinylated secondary antibody followed by a streptavidin–peroxidase complex. This will yield a greater signal by having more than one enzyme molecule per molecule of primary antibody. This procedure replaces *Protocol 16*, step 8 with the following three steps:

(a) Apply 100 μl of secondary biotinylated antibody (available from Amersham) diluted 1:100 with PBS containing 1% BSA and 0.05% (v/v) Tween-20. Incubate the sections in a moist chamber for 2 h at 25°C.

(b) Wash the slides three times in PBS, 5 min each.

(c) Apply 50–100 μl of streptavidin–peroxidase complex (Sigma S 5512). Incubate in a moist chamber for 45–60 min.

Then proceed with the rest of *Protocol 16* from step 9.

If a fluorescent label is used, an antiquench agent, e.g. *p*-phenylenediamine (Sigma Cat. No. P 6001), should be added to the section after the PBS rinses of *Protocol 16*, step 9, which follow the application of the fluorescent tagged secondary antibody. 10–20 μl of a 1 mg/ml solution of *p*-phenylenediamine in PBS containing 80% (v/v) glycerol should be applied to each section, following which the section should be coverslipped using an aqueous mounting medium (e.g. Geltol, available from Shandon Scientific Ltd, Cat. No. 484940) and sealed with nail polish. The slide must be examined using a microscope equipped with appropriate filters for the chosen fluorescent tag.

5. Post-mortem examination of transgenic rodents

5.1 Post-mortem procedure

A complete necropsy on freshly killed animals should be performed when a new transgene is being studied. *Protocol 17* describes a systematic approach to detect gross pathological changes in transgenic rats and mice. Abnormal tissues should be placed in fixative for histological examination (section 5.2, *Protocol 18*) as soon as possible to minimize post-mortem changes. Adjacent unaffected tissue should be included with these samples. Do not assume that abnormalities seen in the same organ are identical. Most organs can be fixed using 10% (v/v) buffered formalin, except eyes and testis which are better preserved in Bouin's solution. The brain and eye should be fixed whole, while the thickness of other samples should be 5 mm for best penetration of fixative. If whole organs need to be fixed, small incisions will help the penetration of fixative. In general, the tissue should occupy no more than 10% of the fixative volume.

Prior knowledge of the anatomy and pathology of the animal species examined is required (23, 24). The post-mortem examination should be performed with a complete history of the case foremost in the investigator's mind. The information should include the strain of animal, age and sex, the transgene introduced, recent or relevant manipulations performed (drug, surgical, or other treatments), and observations on the state of the animal (alert, quiet, uncoordinated, other abnormal functions, or gross abnormalities). Some transgenes by their very nature, or by insertional mutagenesis, affect development and may result in fetal abnormalities or death *in utero*. Post-mortem examination of fetuses from pregnant female mice should then be performed.

Protocol 17. Post-mortem procedure of rodents

Equipment and reagents

- Dissecting board and pins
- Dissecting instruments: scalpel, scissors (small and large), forceps (two pairs, blunt)
- 10% buffered formalin[a]: prepare by mixing 100 ml formalin (40% (v/v) formaldehyde), 4 g NaH_2PO_4, 6.5 g anhydrous Na_2HPO_4. Add water to 1 litre final volume
- Bouin's solution[a]: prepare this fresh by mixing 75 ml picric acid (saturated aqueous solution), 25 ml formalin (40% (v/v) formaldehyde), 5 ml glacial acetic acid
- Labelled containers in which to fix organs
- Disposable plastic or latex gloves to prevent infection with zoonotic diseases

Method

1. Sacrifice the animal by cervical dislocation, overdose of anaesthetic, or by being placed in a CO_2 chamber. Examine the dead animal and note any abnormalities.
2. Stretch out the animal on its back and pin to a dissecting board through each footpad.
3. Soak the skin with 70% ethanol. Open the skin from the lower jaw to the genital area and pin it down with pins.
4. Cut up the midline of the body wall.
5. Break the mandible at the point in between the incisors. Dissect out the tongue and attached oesophagus, trachea, lungs, and heart. Sever the oesophagus at the diaphragm, avoiding spillage of the stomach contents. If the stomach is engorged with feed, tie the distal oesophagus with a short piece of thread before severing.
6. Examine the thoracic organs. Cut into the oesophagus from anterior to posterior and examine the exposed surface. Palpate the lobes of the lung. Cut into the full-length of the trachea and examine the exposed surfaces. Examine the heart while it is still attached to the lungs; cut into both the right and left auricles and ventricles. The semilunar valves and atrio-ventricular valves should also be examined (25).
7. Dissect the liver at its hilus from the alimentary tract. Gently squeeze the gall bladder and check for flow of bile. Incise into and examine the liver parenchyma.

8. Remove the spleen, and examine with random incisions into the parenchyma.
9. Examine the digestive tract separately from the rest of the organs to minimize contamination with abdominal contents. Remove the entire gastro-intestinal tract, from stomach to rectum, and lay aside. Examine the pancreas, the incised greater curvature of the stomach, and representative lengths of large and small intestine at the end of the necroscopy or using another set of instruments.
10. Examine the adrenal glands.
11. Release the kidneys from their attachment to the body wall. Gently dissect out the ureters up to their connection with the bladder. Dissect out the bladder and the urethra to the external orifice. Remove and examine the reproductive organs together with the urinary tract. Make a longitudinal cut in the kidney to separate it into two equal halves, exposing the cortex, medulla, and pelvis. Peel away the renal capsule and examine the cortical surface. Open up the bladder and urethra to examine the internal surfaces.
12. From the ventral aspect, sever the head from the neck at the first cervical vertebra. Remove the skin, then cut away the cranium, starting at the base of the skull, and free it from the underlying meninges. Remove the brain, anterior end first, cutting at the cranial nerves at the base to free the brain. If the brain is to be examined histologically, place the entire organ in 10% buffered formalin. Otherwise, cut it in 0.5 cm transverse sections for inspection.
13. To remove the eye from its socket, cut from the lateral canthus both dorsally and ventrally to meet at the medial canthus. If the eye is to be examined histologically, place it in Bouin's solution (change to alcohol after 24 h to minimize irreversible yellow staining of tissues). Otherwise, separate the organ into two with a longitudinal cut through the midline, including the optic nerve, and examine the internal structures.
14. Once the eyes have been removed, the Harderian glands can be examined. Gently pull out the horseshoe-shaped gland at the back of the socket with a pair of forceps.

[a] These fixatives are those used in the protocol given, but other fixatives of choice can be substituted instead.

5.2 Histological examination of transgenic rodents

Tissue abnormalities observed during a post-mortem examination need to be examined at the microscopic level. The changes may include hyperplasia, hypertrophy, neoplasia, and inflammation; these can be detected by histological examination (*Protocol 18*). Tissues removed from the animal need to undergo several steps of processing before they can be examined under the microscope. The tissues are first fixed to prevent tissue decay and maintain tissue and cellular structure for study. This is followed by dehydration with alcohol to remove water from the tissue for the later step of paraffin embedding. The tissues are then cleared by replacing the alcohol with a wax solvent, and impregnated to support them when they are sectioned. The tissues are finally embedded with molten paraffin wax and sectioned.

The haematoxylin and eosin stain is usually sufficient to give initial information on the nature of the lesion. Staining for tissue components (26) and immunohistochemical staining (*Protocol 16*) can also be performed if necessary.

Protocol 18. Histological examination of transgenic tissues

Equipment and reagents

- Automatic tissue processor (e.g. Hito Kinnett 2000 available from Reichert Jung, Cat. No. 042460686)[a]
- Microtome (to generate 4–5 μm thickness sections, e.g. Reichert Jung 2055 microtome, Cat. No. 90734)
- Tissue embedding equipment (e.g. Reichert Jung tissue embedding centre, Cat. No. 038621438)
- Staining dishes and racks (a range of sizes is needed)
- Water-bath (e.g. Fisher Tissue Prep model 135 Flotation bath, Cat. No. 15–464–12) set to 50°C and containing 0.1% gelatin (Sigma, Cat. No. G1890)
- Warming plate (for warming slides, e.g. Fisher slide warmer, Cat. No. 12–594–1)
- 10% buffered formalin (see *Protocol 17*)
- 1% eosin (1% (w/v) aqueous solution; BDH Ltd. Cat. No. 35084)
- Glass slides and coverslips (clean and grease-free)
- Haematoxylin stain. Prepare by dissolving 50 g aluminium ammonium sulfate in hot distilled water. Also dissolve 2.5 g haematoxylin (Sigma, Cat. No. H3136) in 50 ml 100% ethanol. Mix the two solutions and boil. Remove from the heat and transfer to a fume-hood. Add 1.5 g mercuric oxide (beware of excessive bubbling) and re-heat briefly. Cool the flask in running water. Add 20 ml glacial acetic acid. Filter if too much precipitate is present.
- Slide boxes and slide racks
- Ethanol
- Toluene
- Embedding paraffin (Paraplast plus, available from Thomas Scientific, Cat. No. 6761-E75)
- Flat forceps
- Mounting medium (e.g. Entellen available from Merck, Cat. No. 7961)
- Acid/alcohol (1% HCl in 70% ethanol)
- Xylene

Method

1. Fix the tissues in 10% buffered formalin for at least 24 h. Bone and other calcified tissue can not be processed unless the calcium salts are first removed (25).
2. Trim the tissues and place them in the automatic tissue processor. Agitate the tissues for 2 h in each of the following solutions: 50% ethanol, 75% ethanol, 95% ethanol, 100% ethanol (twice), toluene (twice), Paraplast heated to 60°C (twice).
3. Embed the sample in Paraplast. Techniques vary with different models of tissue embedder; hence follow the instructions provided. During embedding, do not allow the specimen to dry or harden.

4. Cut the embedded tissues using a microtome, 4–5 μm thickness. Good quality sections require practice.
5. Float the sections in the flotation bath containing 0.1% gelatin, and then transfer the sections to clean slides and leave overnight.
6. Remove the paraffin wax by submerging the slides in xylene, twice for 10 min each.
7. Dip the slides in absolute alcohol, followed by 95% ethanol (2 min) and 70% ethanol (2 min).
8. Wash the slides in water for 1 min.
9. Stain the sections in haematoxylin for 1 min.
10. Wash the slides in running tap-water (*not* distilled water or Milli-Q water) until the sections are blue (approx. 1 min).
11. Dip the slides in acid/alcohol one to four times for a few seconds each time.
12. Repeat step 9.
13. Treat the slides with 1% eosin for 2 min.
14. Wash the slides in running water for 2 min.
15. Dip the slides in 70% ethanol and then 90% ethanol, 1 min each time.
16. Wash the slides in absolute ethanol, two changes (2 min each time).
17. Clear the sections in two changes of xylene (2 min each time).
18. Mount the sections in a suitable mounting medium (e.g. Entellen).
19. Examine the sections under a light microscope (nuclei are stained blue, cytoplasm is stained pink).[b]

[a] A series of glass jars will suffice if an automatic tissue processor is not available.
[b] Refs 27 and 28 describe the histology of normal tissues.

References

1. Southern, E. (1974). *J. Mol. Biol.*, **98**, 503.
2. Kafatos, F. C., Jones, W. C., and Efstratiados, A. (1979). *Nucleic Acids Res.*, **7**, 1541.
3. Saiki, R. K., Scharf, S., Faloona, F., Mullis, K. B., Horn, G. T., Erlich, H. A., *et al.* (1985). *Science*, **230**, 1350.
4. Glover, D. M. and Hames, B. D. (ed.) (1995) *DNA cloning 1: a practical approach. Core techniques*. Oxford University Press, Oxford and New York.
5. Murphy, D. (1993). In *Transgenesis techniques* (ed. D. Murphy and D. A. Carter), pp. 449–51. Humana Press, New Jersey.
6. Feinberg, A. P. and Vogelstein, B. (1984). *Anal. Biochem.*, **132**, 6.
7. Thomas, P. S. (1980). *Proc. Natl Acad. Sci. USA*, **77**, 5201.

8. Young III, W. S., Mezey, E., and Seigel, R. E. (1986). *Mol. Brain Res.*, **1**, 231.
9. Funkhouser, J. M. (1993). In *Transgenesis techniques* (ed. D. Murphy and D. Carter), pp. 379–93. Humana Press, New Jersey.
10. Morrell, J. I. (1989). In *Techniques in immunocytochemistry*, Vol. 4 (ed. G. R. Bullock and P. Petrusz), pp. 127–46. Academic Press, London.
11. Stahl, W. I., Eakin, T. J., and Baskin, D. G. (1993). *J. Histochem. Cytochem.*, **41**, 1735.
12. Gorman, C. M., Moffat, L. F., and Howard, B. H. (1982). *Mol. Cell. Biol.*, **2**, 1044.
13. Sleigh, M. J. (1986). *Anal. Biochem.*, **156**, 251.
14. Shimohama, S., Rosenberg, M. B., Fagan, A. M., Wolff, J. A., Short, M. P., Breakefield, X. O., *et al.* (1989). *Mol. Brain Res.* **5**, 271.
15. Sanes, J. R., Rubenstein, J. L. R., and Nicolas, J. F. (1986). *EMBO J.*, **5**, 3133.
16. Goring, D. R., Rossant, J., Clapoff, S., Breitman, M. L., and Tsui, L.-C. (1987). *Science*, **235**, 456.
17. Allen, N. D., Cran, D. G., Barton, S. C., Hettle, S., Reik, W., and Surani, A. (1988). *Nature*, **333**, 852.
18. Oberdick, J., Smeyne, R. J., Mann, J. R., Zackson, S., and Morgan, J. I. (1990). *Science*, **248**, 223.
19. Brasier, A. R., Tate, J. E., and Habener, J. F. (1989). *BioTechniques*, **7**, 1116.
20. Polak, J. M. and Van Noorden, S. (ed.) (1989). *Introduction to immunocytochemistry: current techniques and problems. Royal Microscopical Society Microscopy Handbook 11*. Oxford University Press, Oxford.
21. McClean, I. W. and NaKane, P. K. (1974). *J. Histochem. Cytochem.*, **22**, 1077.
22. Zamboni, L. and DeMartino, C. (1967). *J. Cell. Biol.*, **35**, 148A.
23. Foster, H. L., Small, J. D., and Fox, J. G. (ed.) (1982). *The mouse in biomedical research*, Vols 1, 2, and 3. Academic Press, Orlando.
24. Jubb, K. V. F., Kennedy, P. C., and Palmer, N. C. (ed.) (1985). *Pathology of domestic animals*, vol. I–III. Academic Press, London.
25. King, J. M., Dodd, D. C., and Newson, M. E. (1984). *Aust. Vet. J.*, **61**, Appendix 1:7–18.
26. Culling, C. F. A., Allison, R. T., and Barr, W. T. (ed.) (1985). *Cellular pathology technique*. Butterworths, London.
27. Dellmann, H. D. and Brown, E. M. (ed.) (1976). *Textbook of veterinary histology*. Lea & Febiger, Philadelphia.
28. Gude, W. D., Cosgrove, G. E., and Hirsch, G. P. (ed.) (1982). *Histological atlas of the laboratory mouse*. Plenum Press, New York.

8

Expression using a defective herpes simplex virus (HSV-1) vector system

FILIP LIM, PHILIP STARR, SONG SONG, DEAN HARTLEY, PHUNG LANG, YAMING WANG, and ALFRED I. GELLER

1. Introduction

1.1 Genetic intervention in the brain

Genetic intervention in the brain through the use of vector systems capable of transferring recombinant genes into mammalian neurones is becoming an increasingly useful strategy for investigating the molecular basis of various brain functions ranging from regulating homeostasis to information processing and higher cognitive functions (1, 2). Genes have been isolated and characterized which encode molecules such as signal transduction enzymes, transcription factors, and components of the neurotransmitter release machinery that may play critical roles in regulating neuronal physiology. Relatively little is known about the function of these molecules in mediating alterations in neuronal physiology and the resulting changes in complex behaviours. Genetic intervention in the brain has the theoretical potential to reveal precise information about specific brain functions through the use of mutations in neuronal genes, and then subsequent expression of the altered gene products in their normal cellular milieu.

Genetic intervention may also have gene therapy applications for many of the disorders that affect the nervous system. Some gene therapy strategies are directed towards a specific disease. For example, expression of tyrosine hydroxylase and other catecholamine biosynthetic enzymes in striatal neurones is a therapeutic strategy for Parkinson's disease (3). Expression of neurotrophic factor(s) and/or their receptor(s) in specific groups of neural cells may be applicable to physical injury to an area of the nervous system (4). The molecular pathways underlying normal neuronal function may be exploited to develop gene therapy strategies that are applicable to multiple diseases and a more sophisticated understanding of the molecular details of these pathways may result in new insights into how defects in these pathways cause diseases.

For example, a more complete understanding of the regulation of long-term changes in neurotransmitter release may result in new approaches to diseases involving altered neurotransmitter release, such as the epilepsies.

The brain has not yielded easily to genetic intervention. Because most neurones are post-mitotic, vectors such as retrovirus vectors, that rely on DNA replication for stable maintenance in a cell (5) are of limited utility in the brain. Thus, gene transfer strategies for the brain require a vector system that can persist stably in post-mitotic cells, and which is also capable of altering the genetic composition and consequent physiology of a specific class of neurones during a defined period of time. As a first step towards addressing this problem, we and others have developed a defective herpes simplex virus type one (HSV-1) vector system which allows genes to be introduced directly into neurones, in culture or in a specific area of the adult mammalian brain (1, 2, 6–9), thereby enabling a genetic analysis of the molecular events responsible for complex brain functions. The role of a recombinant protein (expressed from a HSV-1 vector) in neuronal physiology is assessed using the tools of modern neuroscience, including neurochemistry, protein biochemistry, immunohistochemistry, electrophysiology, and behavioural studies.

The defective HSV-1 vector system can direct expression of reporter genes both in cultured neurones and in neurones *in vivo* (1, 2, 6, 7). Furthermore, the vector system can sustain long-term (> two weeks) expression of *E. coli* β-galactosidase in peripheral and central nervous system neurones (6, 7, 10). Over the past few years, a number of studies have established that this HSV-1 vector system can be used to introduce functional gene products into multiple cell types, including both peripheral and CNS neurones. For example, the human nerve growth factor receptor (p75) was expressed in cultured cortical neurones (11), nerve growth factor was expressed in cells in the superior cervical ganglia (12) and in both cultured striatal and basal forebrain cells (13), a glucose transporter gene was expressed in hippocampal cells (14), the gene for a GluR6 receptor subtype was introduced into these same cell types (15), an adenylate cyclase was expressed in cultured sympathetic neurones (16), and tyrosine hydroxylase was expressed in cultured striatal neurones (17). In each case, the recombinant gene product produced a functional response. Furthermore, introduction of an HSV-1 vector expressing tyrosine hydroxylase into the partially denervated striatum in a rat model of Parkinson's disease resulted in persistence of the vector DNA, long-term expression of tyrosine hydroxylase, and both biochemical and behavioural recovery for up to one year (18).

NB: HSV-1 is a dangerous human pathogen that is transmitted by exchange of solutions but is not transmitted as an aerosol. It is essential to obtain the required approvals from the appropriate Biosafety committee before initiating work. In the USA, most procedures with HSV-1 vectors are classified at the Biosafety level two containment level. Before beginning work with the virus, consult the relevant Biosafety committee about the required precautions for working safely with the virus and establish laboratory procedures to decontaminate waste material and any spills which may occur.

1.2 Properties of defective HSV-1 vectors

Defective HSV-1 vectors have multiple attractive features for genetic intervention in the brain (2) which are based on specific properties of HSV-1:

(a) A capability to infect quiescent cells, specifically post-mitotic neurones and glia.
(b) An unusually wide host range; HSV-1 is capable of infecting multiple cell types such as fibroblasts, macrophages, and neural cells in many mammals including humans, non-human primates, and rodents (22).
(c) An HSV-1 genome can be stably maintained in a latent state in neurones (23).
(d) In the latent state, HSV-1 is quiescent. Expression of genes from the virus is limited to a latency associated transcript(s), HSV-1 DNA is not replicated, progeny virus are not produced, and electrophysiological properties of the infected neurones do not appear to be changed (23).

Upon injection of wild-type HSV-1 into the brain, the lytic cycle supports the production of progeny virus which spread the infection through the brain and kill the animal, often within several days. In contrast, injection of temperature sensitive (ts) mutants of HSV-1 into the brain results in a latent infection (24). These ts mutants contain a single base change in a gene required for productive virus growth, resulting in a single amino acid substitution in an essential HSV-1 protein. The resulting mutant protein can function properly at 31 °C but not at 37–39 °C, the normal temperature of the brain. Consequently, the lytic cycle can proceed in cultured cells at 31 °C, but not in the brain. Furthermore, because these temperature sensitive (ts) mutants do not grow in the brain, the infection is restricted to cells around the injection site and neurones which project to the injection site (24). These HSV-1 ts mutants, (and in subsequent studies deletion mutants in the same genes) were used to develop the HSV-1 vector system.

During the initial stages of developing the vector system, HSV-1 strain 17 ts K (25) was used as the helper virus to package HSV-1 vectors into virus particles (6). Because ts mutants revert to wild-type (wt) HSV-1 at a finite frequency (25), the usefulness of this packaging system was limited. However, deletion mutants of HSV-1 revert to wt only at very low frequencies. In subsequent work, therefore, we developed a deletion mutant packaging system for defective HSV-1 vectors (8). A cell line which contains the deleted gene in its genome can support growth of (complements) a deletion mutant. To package HSV-1 vector DNA into HSV-1 particles using a deletion mutant as the helper virus, a cell line harbouring the deleted gene is transfected with vector DNA and subsequently superinfected with the appropriate deletion mutant virus.

HSV-1 strain 17 D30EBA (26) was the first deletion mutant used as helper virus. Both ts K and D30EBA harbour a mutation in the immediate early (IE) 3 gene in HSV-1 strain 17 and consequently both mutants have the same phenotype. This deletion mutant packaging system was an improvement over packaging using ts K as the helper virus and established that defective HSV-1 vectors could be packaged using a deletion mutant of HSV-1. This packaging system has supported several experiments on neuronal physiology with HSV-1 vectors, both in culture and in the brain. However, when D30EBA is grown on M64A cells (27), it reverts to wt HSV-1 at a frequency of approximately 1×10^{-5}. Consequently, this particular packaging system still has limitations and can not support use in humans. Recently we have begun using deletion mutants in either the IE 2 (KOS strain 5dl1.2) (29, 30) or IE 3 (KOS strain d120) (31) genes which have a lower reversion frequency to wt HSV-1 of $\leqslant 10^{-7}$. Goals for additional improvements to the packaging system (see below) include a further reduction in the reversion frequency and reducing the acute cytopathic effects associated with virus infection. These potential improvements would aid in experiments and might eventually allow use of HSV-1 vectors in humans.

1.3 pHSVlac, the prototype defective HSV-1 vector

pHSVlac was the first vector we developed and was used to characterize the properties of the vector system (6). This vector is still used in most packaging procedures as a positive control. pHSVlac contains three types of genetic elements:

(a) A transcription unit consisting of the HSV-1 IE 4/5 promoter, an intervening sequence (intron) from HSV-1 immediately after the IE 4/5 promoter, the *E. coli lacZ* gene, and the SV40 early region polyadenylation site. The IE 4/5 promoter is a constitutive promoter which functions in multiple cell types and the *lacZ* gene encodes a β-galactosidase, providing a routine assay for expression from the vector.

(b) Two genetic elements from HSV-1 which are sufficient to support packaging of a vector into HSV-1 particles; HSV-1 ori_S, which supports replication of the vector DNA, and the HSV-1 **a** sequence which contains the packaging site.

(c) Bacterial plasmid sequences (from pBR322) that allow selection and growth of the vector in *E. coli*.

In other vectors, the *lacZ* gene is replaced with other genes and/or the IE 4/5 promoter is replaced with other promoters (see *Figure 1*).

1.4 Experimental design

The design of an experiment using HSV-1 vectors in the adult mammalian brain is delineated by the three major variables which at present are amenable to manipulation by the experimenter (1, 2). These are:

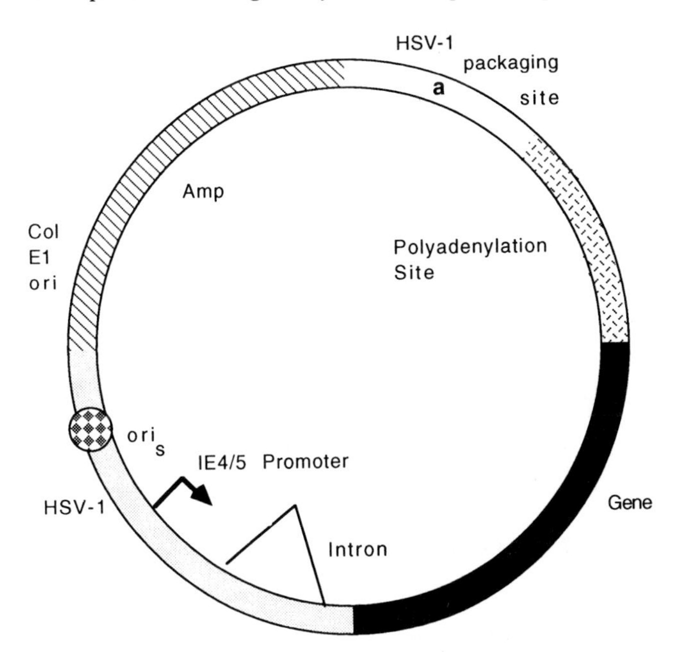

Figure 1. The structure of a defective HSV-1 vector. A defective HSV-1 vector contains three kinds of genetic elements. (1) Two sequences from the HSV-1 genome are sufficient to support packaging of vector DNA into HSV-1 particles; an HSV-1 origin of DNA replication, ori_s (circle filled with diamonds), supports replication of the vector DNA and the HSV-1 **a** sequence (clear segment) contains the packaging site which directs insertion of vector DNA into a HSV-1 particle. (2) Each vector contains a transcription unit which is composed of a promoter (the HSV-1 IE 4/5 promoter is shown by an arrow but can be replaced with other promoters), an intron or intervening sequence (the HSV-1 encoded intervening sequence following the HSV-1 IE 4/5 promoter is shown by a triangle although other introns have been used), a gene to be transcribed (black segment), and a poly-adenylation site (segment filled with short lines, usually the SV40 early region poly-adenylation site although other polyadenylation sites have been used). (3) Sequences from pBR322 (cross-hatched segment) allow the vector to be grown in *E. coli*; these sequences include the Col E1 ori, an origin of DNA replication, and the amp^r gene (Amp) which encodes ampicillin resistance.

- the location and number of the infected cells
- the promoter which directs transcription of the recombinant gene in the vector
- the recombinant gene

The location and class(es) of cells infected is determined by the choice of the site of injection, the number and location of the neurones which project to the injection site, the amount of virus in the inoculum, and the extent of diffusion of virus particles through the extracellular space prior to infection.

The promoter contained in the vector is the second variable subject to control by the experimenter. pHSVlac, the prototype vector, contains a constitutive promoter, the HSV-1 IE 4/5 promoter, which functions in many cell types. The human neurofilament L promoter has been used to restrict expression of recombinant gene from a HSV-1 vector to neurones (19). Expression may be further limited to a particular class of neurones; for example, we have used the tyrosine hydroxylase promoter to limit expression to catecholaminergic neurones (20) and other investigators have used the preproenkephalin promoter to limit expression to enkephalinergic neurones (21).

The recombinant gene that is expressed is the third variable that is amenable to manipulation by the experimenter. Almost any gene can be inserted into a HSV-1 vector; for example, expression of the catalytic domain of the yeast adenylate cyclase results in elevated intracellular cAMP levels and protein kinase A activity which directs a long-term increase in neurotransmitter release (16). Furthermore, by fusion of the gene to subcellular targeting sequences, the expressed protein might be directed to a particular subcellular compartment of a neurone.

2. Growth of helper virus

Helper viruses usually carry a deletion in an essential HSV-1 gene and are grown on a complementing cell line. Our laboratory has successfully used three different deletion mutants as helper virus each of which harbours a single deletion in either the IE 2 or the IE 3 gene (see *Table 1*). In the future it is likely that new helper viruses harbouring deletions in multiple HSV-1 genes will be introduced; they will be used in an analogous fashion with the appropriate complementing cell line.

Cell lines are grown in Dulbecco's modified essential medium (DMEM) supplemented with penicillin/streptomycin, 4 mM glutamine (from 50 × stock), and 10% foetal bovine serum (FBS) at 37°C in the presence of 5% CO_2. Cell lines are maintained in selection medium (M64A cells, HAT selection; RR1, E5, or 2.2 cells, G418 selection), and the cell lines are passaged one to three times in normal (non-selective) medium before use in experiments. At the time of virus infection the medium is replaced with fresh medium and the FBS is reduced from 10% to 5%. HSV-1 strain 17 can grow

Table 1 Deletion mutants and cell lines used for packaging defective HSV-1 vectors

Deletion virus	Deleted gene	Cell line	Parental cell line	Reversion frequency	Reference
Strain 17 D30EBA	IE 3	M64A (or RR1)	BHK tk$^-$	10^{-5} (RR1, $\leq 10^{-7}$)	26–28
KOS strain d120	IE 3	E5	VERO	$\leq 10^{-7}$	31
KOS strain 5dl1.2	IE 2	2.2	VERO	$\leq 10^{-7}$	29, 30

on BHK or VERO cells. KOS strain HSV-1 can grow on VERO cells but grows only poorly on BHK cells. To ensure the properties of a cell line remain constant, freeze multiple vials of a cell line at an early passage and then thaw a fresh vial once every four to six weeks. For a virus infection the cells should be almost confluent. In the following protocols we specify plating conditions and incubation times which work in our hands, but the investigator should verify these before growing virus.

To ensure that the virus stock is as genetically uniform as possible, a virus stock is grown from a single infectious virus particle (*Protocol 1*). This is accomplished by infecting a culture with a small amount of virus and then adding semi-solid medium. The semi-solid medium contains a high molecular weight polymer, either agarose or methocel, to retard the diffusion of virus particles through the culture medium. This ensures that the progeny virus infect cells adjacent to the cell they were produced in. Thus, a group of adjacent cells is infected by the progeny from a single virus and these cells die, resulting in a hole in the monolayer of cells which is termed a plaque. The virus in a plaque are then isolated and used to prepare a virus stock. The semi-solid medium should be added during the interval following virus absorption to the cells (which occurs rapidly in small volumes, e.g. one hour) but before production of the first progeny virus (which usually occurs six to eight hours after infection).

Protocol 1. Preparation of a plaque-purified virus stock

Equipment and reagents

- Laminar flow hood and incubator
- Tissue culture microscope
- 60 mm tissue culture plates
- VERO cells or BHK cells (see step 1 below)
- DMEM containing penicillin/streptomycin, 4 mM glutamine (from 50 × stock), and 10% FBS
- Virus stock

For agarose overlay method (step 4 below):

- DMEM, 5% FBS, 1% agarose. Prepare this by making 2% SeaPlaque low melting point agarose (tissue culture grade). To liquid (hot) 2% agarose, add an equal volume of 2 × DMEM and FBS to give DMEM, 5% FBS, 1% agarose final concentration. Place in a water-bath at 42°C to equilibrate. Remove just before required and use when the temperature is ≤40°C.
- DMEM, 5% FBS

For methocel overlay method (step 4 below):

- Methocel solution. Prepare this as follows. Add 3.6 g methocel (4000 centipoise; Fisher) to 100 ml distilled water in a 500 ml flask and autoclave. Shake at 4°C for two to three days, until the solution is uniform. Add 100 ml 2 × DMEM (without phenol red), penicillin/streptomycin, 4 ml 200 mM glutamine, and 10 ml FBS (5% FBS final concentration) to the methocel. Shake at room temperature until the solution is uniform (10–30 min). Store at 4°C.

Method

1. Seed 60 mm tissue culture plates with 5×10^5 VERO cells/plate, or 2×10^5 BHK cells/plate, in 5 ml DMEM, 10% FBS.

Protocol 1. *Continued*

2. Next day, remove 3 ml of the medium and add an amount of virus that will give well-separated plaques (about 50 plaque-forming units (p.f.u.)/plate).
3. Incubate at 37°C for 1.5–6 h.
4. Remove the medium and overlay with semi-solid medium, either agarose or methocel.
   (a) Agarose overlay method (for 60 mm plates). Remove the medium and add 3 ml DMEM, 5% FBS, 1% agarose solution. Leave to solidify before returning to the 37°C incubator. The following day add 2 ml DMEM, 5% FBS on top of the agarose. Repeat on the next day.
   (b) Methocel overlay method (for 60 mm plates). Add 5 ml of methocel solution to the plates. Incubate at 37°C for two to four days.
5. Two to four days later when plaques are visible, pick a single plaque by using a Pasteur pipette to stab through the semi-solid medium and into the cells, and then transfer the plug plus the cells to a 60 mm plate of cells plated as in step 1. It is advisable to pick several different plaques to ensure that at least one isolate has the appropriate phenotype for the virus.
6. After one or two days (depending on the amount of virus obtained from the plaque), the cells should show cytopathic effects (CPE). At this stage the cells should round up but remain stuck to the plate. Harvest the cells and prepare virus (see section 3).

3. Preparation of HSV-1 virus

The majority of HSV-1 particles (virions) remain inside the infected cells and HSV-1 virions have a lipid bilayer. Thus, a virus preparation procedure must create holes in a large lipid bilayer (the plasma membrane of the cell) without disrupting a small lipid bilayer (the coat of the virus). This is accomplished by using three freeze–thaw cycles and sonication (see *Protocol 2*). Detergents destroy HSV-1 virions and are therefore not useful for virus preparation.

Protocol 2. Harvesting HSV-1 virus

Equipment and reagents

- Dry ice–ethanol bath
- Cup sonicator (e.g. from Heat Systems Ultrasonics) with a pump for circulating ice water from an ice water-bath
- Screw-top 15 ml plastic centrifuge tubes
- Cell scraper (e.g. rubber policeman or plastic cell scraper)
- HSV-1-infected cell monolayers (from *Protocol 1*)

Method

1. Harvest the cells when the cells have rounded up and show CPE by scraping the cells off the plate and into the medium with a rubber policeman or plastic cell scraper.
2. Place 5 ml of the cell suspension into a 15 ml screw-top centrifuge tube.
3. Freeze–thaw three times using a dry ice–ethanol bath and a 37°C water-bath.
4. Sonicate in a cup sonicator cycling between on and off such that the sum of the sonication bursts is 2 min. The length of the on off cycle can vary from 5 sec to at least 30 sec. The cup in the sonicator should be cooled with circulating water from an ice water-bath (circulated using a submersible pump).
5. Remove the cellular debris by centrifugation at 1500 *g* for 5 min.
6. Transfer the supernatant to a new tube, freeze the resulting virus stock in a dry ice–ethanol bath, and store at −70°C.

4. Titring of virus

Titring is designed to determine the concentration of infectious particles of virus in a virus stock. Titring is performed by counting the number of plaques (determined using semi-solid medium) produced by a given volume of virus, so that the titre is expressed as p.f.u./ml. A virus harbouring a mutation in an essential gene is titred on the cell line which supports its growth (e.g. D30EBA is titred on M64A cells), and on the parent cell line (e.g. BHK cells are the parent cell line for M64A cells) to determine the concentration of wild-type virus in the virus stock. The procedure is described in *Protocol 3*.

Protocol 3. Virus titring

Equipment and reagents

- VERO or BHK cells
- Tissue culture plates (60 mm diameter)
- DMEM, 10% FBS
- DMEM, 5% FBS
- Virus stock to be titred

When using agarose overlay:

- DMEM, 5% FBS, 10% agarose (see *Protocol 1*)
- DMEM, 5% FBS
- 5% methanol, 10% acetic acid
- Crystal violet stain: prepare a solution containing 0.5% crystal violet and 0.2% sodium acetate, adjust the pH to 3.6 with acetic acid, and filter through Whatman 1MM paper.

When using methocel overlay:

- Methocel solution (see *Protocol 1*)
- TD solution. Prepare by mixing 16 g NaCl, 0.76 g KCl, 0.2 g Na_2HPO_4, 6 g Tris base, and water to 500 ml. Add 5 ml concentrated HCl. The pH should be pH 7.4. Add water to 2 litres final volume. Filter sterilize. Aliquot and store at room temperature.
- Neutral red solution. First prepare a 10 × neutral red stock by dissolving 1 g neutral red in 1 litre water. Boil, filter, and store at room temperature. Also prepare Mg/Ca solution (0.4 g $MgCl_2$, 0.4 g $CaCl_2$ in 200 ml H_2O) and sterilize. To prepare neutral red solution, mix 10 ml of 10 × neutral red stock, 95 ml TD solution, and 5 ml Mg/Ca solution.

Protocol 3. *Continued*

Method

1. Plate cells in 60 mm tissue culture plates and grow for one to three days to confluence in DMEM 10% FBS. Prepare one plate for each dilution of virus stock to be titred (see step 2). For VERO cells plate 1×10^6 cells/60 mm dish. For BHK cells, plate 3×10^5 cells/60 mm dish.[a]
2. Prepare tenfold serial dilutions of the virus stock to be titred as follows. Add 1 ml DMEM, 5% FBS to each of a series of snap-cap tubes (usually six to eight dilutions). Add 100 μl virus stock to tube 1. Mix the tube gently, but thoroughly, by tapping on the side. Remove 100 μl from tube 1 and add to tube 2. Repeat this process with the other tubes to create a dilution series.
3. Infect the cells by one of the following methods. Both methods (a) and (b) are standard virological methods and work well for these purposes. It is somewhat easier to prepare the agarose overlay than the methocel overlay, and the agarose method allows feeding of the cells during the procedure. However, the choice is largely arbitrary and both methods are used in our laboratory.
   (a) Remove 3 ml of medium from the plate, add 100 μl of the virus dilution to the plate. Allow the virus to adsorb for at least 90 min, but for less than 6 h.
   (b) Remove the medium from plates and add 100 μl of each dilution to successive plates. Keep the plates at room temperature for 1 h and tilt plates every 15 min to distribute the virus evenly. If there is a problem with the cells drying out during the virus absorption period then add 0.1–1.0 ml DMEM, 5% FBS to each plate in addition to the virus inoculum.
4. Add semi-solid medium, either agarose or methocel, as described in *Protocol 1*, step 4.
5. Two to four days later, process each plate as follows so that plaques are visible for counting.
   (a) If agarose overlay was used in step 4, remove the medium and fix the cells with 3 ml of 5% methanol, 10% acetic acid for at least 15 min. Remove the methanol/acetic acid solution, and add 1 ml crystal violet stain. Remove the stain (which can be reused), wash the cells with 1 ml water, and air dry. Plaques should be visible to the naked eye as holes.
   (b) If a methocel overlay was used in step 4, add 5 ml of neutral red solution to each plate and incubate at 37°C for 4–12 h. Neutral red is actively taken up by living cells which then appear red; the plaques will be visible as clear holes on a red background.

6. Count the number of plaques, and using the dilution factor of the virus stock tested, calculate the virus titre in p.f.u./ml.

[a] BHK cells grow rapidly and pile up on top of each other, especially in the middle of the well, so do not plate the cells at a higher density.

5. Growth of large amounts of virus from seed stock

Large amounts of virus are grown from the seed stock for use in the packaging procedure as the helper virus. The virus is inoculated at a low multiplicity of infection (m.o.i.) because at high m.o.i. spontaneously arising mutations will be complemented by the virus and these mutations will contaminate the virus stock. A m.o.i. of about 0.01 is typically used. The procedure is described in *Protocol 4*.

Protocol 4. Large scale preparation of virus

Equipment and reagents

- VERO or BHK cells (see step 1 below)
- 100 mm tissue culture plates
- DMEM, 10% FBS
- DMEM, 5% FBS
- Virus stock
- For other reagents, see *Protocols 2* and *3*

Method

1. Seed four 100 mm plates with 1×10^6 VERO cells/plate, or 4×10^5 BHK cells/plate, in 10 ml DMEM 10% FBS and grow for two days or until confluent.
2. Change the medium to 10 ml DMEM 5% FBS and add 50 μl of seed stock virus.[a]
3. When cells show CPE, harvest them and prepare the virus as described in *Protocol 2*.
4. Titre the virus (*Protocol 3*).
5. Divide the virus stock into aliquots and store at −70°C.

[a] Expect $0.5–2 \times 10^7$ cells/plate and infect the cells at a m.o.i. of 0.1 or lower. At high m.o.i., spontaneously arising defective viruses are able to grow and interfere with virus replication, resulting in lower p.f.u./ml.

6. Packaging procedure

6.1 Packaging using LipofectAMINE transfection

For packaging, vector DNA is delivered into the host cells which are then infected with helper virus. The resulting virus stock is then serially passaged several times both to increase the ratio of vector to helper and to increase the amount of virus (*Protocol 5*). The vector is titred by infecting cells and assaying for expression of the encoded gene product. With vectors that express *E. coli* β-galactosidase, this is performed using PC12 cells and the X-Gal procedure (see section 7). Vectors that express other gene products can be assayed using an antibody against the relevant gene product and immunocytochemical procedures for that antibody. The titre of the helper virus and the titre of the wild-type virus are determined by plaque assays (see *Protocol 3*). We have successfully used this procedure with 5dl1.2 virus (29) and 2–2 cells (30) and with d120 virus and E5 cells (31). A good virus stock should contain a ratio of vector to helper virus of $\geqslant 1$ and the ratio of helper virus to wild-type virus of $\geqslant 10^6$. In our hands, larger vectors (15 kb) will yield somewhat lower titres than smaller vectors (5–8 kb).

Protocol 5. Packaging using LipofectAMINE transfection

Equipment and reagents

- VERO cells
- DMEM, 10% FBS
- DNA for transfection (purified by two bandings by CsCl centrifugation or using QIAGEN (QIAGEN Inc.)
- OPTI-MEM (Gibco-BRL)
- LipofectAMINE (Gibco-BRL)
- Control vector (e.g. pHSVlac)
- PBS: prepare a 10 × PBS stock by mixing 1 g KH_2PO_4, 10.8 g $Na_2HPO_4.7H_2O$, 1 g KCl, 40 g NaCl, and water to 500 ml final volume (the pH should be approx. 7.0). Filter sterilize. Dilute tenfold before use.
- DMEM, 5% FBS

Method

1. Two days before transfection, trypsinize the stock cells and plate at 3×10^5/60 mm dish in 5 ml DMEM, 10% FBS. The growth state of the cells is important for ensuring good infectivity and packaging. As a guide-line, do not trypsinize the cells two days in succession.
2. Dilute 2 μg DNA with 100 μl OPTI-MEM in a microcentrifuge tube. Mix 12 μl LipofectAMINE with 100 μl OPTI-MEM in another tube and then add to the 100 μl DNA in the first tube. Leave the solution for 20 min at room temperature to allow liposomes to form. As a control for the transfection procedure, include a tube with a vector that expresses β-galactosidase (e.g. pHSVlac) and a tube without DNA, and one day later assay expression of β-galactosidase by immunocytochemical staining with X-Gal (*Protocol 7*). To monitor the packaging efficiency during the subsequent steps, include a second tube with a vector that expresses β-galactosidase and use the X-Gal assay (*Protocol 7*) to monitor this vector at each step.
3. Remove the medium from the plates and wash each plate once with 2 ml OPTI-MEM.
4. Remove the wash and replace with 2 ml fresh OPTI-MEM.

5. Add 800 μl OPTI-MEM to the DNA/LipofectAMINE mix from step 2. Add this 1 ml to the cells dropwise to distribute the solution evenly over the entire plate. Incubate at 37°C for 5 h.
6. The next step is repeated washing to get rid of excess LipofectAMINE which if not removed will inactivate virus particles. Pre-warm PBS and DMEM, 10% FBS for at least 20 min at 37°C. Remove the medium from the plates, add 2 ml of the pre-warmed PBS, and swirl the plates to wash the cells. Discard this wash and repeat twice more using pre-warmed PBS. Then remove the final wash and add 5 ml of pre-warmed DMEM 10% FBS. Incubate the cells overnight at 37°C.
7. Allow the cells to recover for at least 20 h after the transfection (from the time of the last wash in step 6). Pre-warm DMEM, 5% FBS for at least 20 min at 37°C. Remove the medium from the plates and add 5 ml of pre-warmed DMEM, 5% FBS. Add 6×10^5 p.f.u. of virus (m.o.i. 0.2–0.6) and incubate overnight at 37°C.
8. Check the cells one day after infection; the cells should have rounded up but remain stuck to the plate. Harvest the cells and prepare the virus as described in *Protocol 2*. This is designated p0 (passage 0). Titre the virus stock for the vector (*Protocol 3*) on PC12 cells. Continue with the packaging protocol only if the titre is $\geq 1 \times 10^4$ infectious particles (IP)/ml.
9. First amplification (p1). Plate the cells at 4×10^5/60 mm dish in 5 DMEM 10% FBS. Two days later replace the medium with 4 ml DMEM 5% FBS and add 4 ml of the p0 preparation (from step 8). Next day check the cells and harvest the virus (*Protocol 2*) when the cells have rounded up. For each passage include an uninfected (mock) plate as a control.
10. Second amplification (p2). Plate the cells at 1×10^6/100 mm dish (two dishes per sample) in 10 ml DMEM, 10% FBS. Two days later replace the medium in each plate with 6 ml DMEM, 5% FBS and add 4 ml of the p1 preparation (from step 9). On the next day check the cells and harvest the virus (*Protocol 2*) when the cells have rounded up.
11. Third amplification (p3). Plate the cells at 1×10^6/100 mm dish (four dishes per sample) in 10 ml DMEM, 10% FBS. Two days later replace the medium in each plate with 6 ml DMEM, 5% FBS and add 4 ml of the p2 preparation (from step 10). On the next day check the cells and harvest the virus (*Protocol 2*) when the cells have rounded up.

6.2 Packaging using calcium phosphate transfection

We are now predominantly using the LipofectAMINE packaging procedure. However, we also describe here the calcium phosphate transfection procedure

(*Protocol 6*). We have successfully used this procedure with D30EBA virus and M64A cells. We include this procedure because it is useful, calcium phosphate is cheaper than LipofectAMINE, the calcium phosphate transfection procedure is used in many laboratories, and because we have published several studies on neuronal physiology using vectors packaged with this procedure.

Protocol 6. Packaging using calcium phosphate transfection

Equipment and reagents

- BHK cells
- 2 M $CaCl_2$, filter sterilize
- 2 × HBS: mix 5 g Hepes, 8 g NaCl, 0.37 g KCl, 0.1 g Na_2HPO_4 (anhydrous), 1 g glucose, and water to 500 ml final volume, adjust the pH to exactly 7.05 using 2 M NaOH, filter sterilize
- 1 mg/ml salmon sperm DNA (the DNA is sheared by passing it through a 22 gauge needle several times)
- DNA for transfection (see *Protocol 5*)
- Control vector (e.g. pHSVlac)
- DMEM, 10% FBS
- 1 × HBS, 15% glycerol
- D30EBA virus solution (0.5–2.0 × 10^6 p.f.u./ml). Prepare this by mixing 100–200 μl of D30EBA virus (5–10 × 10^6 p.f.u./ml) with DMEM, 5% FBS to give 1 ml total volume for each transfection plate. Calculate the total volume of virus that will be needed and prepare this before adding to the plates (see below).
- DMEM, 5% FBS
- Other reagents are described in *Protocol 2*

A. *Transfection with DNA/$CaPO_4$*

1. For transfections, plate 1.5 × 10^5 cells in a 60 mm tissue culture dish. Shake in a circular motion to evenly distribute the cells. Use for a transfection 16–24 h later.
2. Add 937 μl of 2 M $CaCl_2$ to 6.11 ml H_2O. Mix and equilibrate to room temperature.
3. For each packaging, add the following to a microcentrifuge tube: 0.25 ml 2 × HBS, 10 μl 1 mg/ml sheared salmon sperm DNA, and 1 μg DNA. Vortex the tube. As a control for the transfection procedure, include a tube with a vector that expresses β-galactosidase (e.g. pHSVlac) and a tube without DNA, and one day later assay expression of β-galactosidase by immunocytochemical staining with X-Gal (*Protocol 7*). To monitor the packaging efficiency during the subsequent steps, include a second tube with a vector that expresses β-galactosidase and use the X-Gal assay to monitor this vector at each step.
4. Add dropwise, 235 μl of diluted $CaCl_2$ to each tube containing vector DNA (and to the no DNA control), and vortex immediately. The quality of the DNA/$CaPO_4$ co-precipitate depends upon rapid mixing. Incubate for 25 min at room temperature. Start a timer at the time of adding the diluted $CaCl_2$ to the first tube.
5. Remove medium from the tissue culture plates and add the DNA/$CaPO_4$ co-precipitate from step 4. Incubate for 25 min at room temperature, tilting the plates every 5 min. Add the co-precipitate to the plates

in same sequence as the $CaCl_2$ was added to the tubes in step 4 so that each co-precipitate forms for 25 min before addition to a plate. To facilitate handling, remove the medium from four to six plates at a time for addition of the relevant DNA/$CaPO_4$ co-precipitates.

6. Add 5 ml DMEM, 10% FBS to each plate. Add the medium to the plates in same order as above so that each plate is exposed to the undiluted DNA/$CaPO_4$ co-precipitate for 25 min. Incubate the plates at 37 °C for 4 h.

B. *Glycerol shock*

1. Process four to six plates at a time. Remove the medium and add 1 ml 1 × HBS, 15% glycerol to each plate. Mix quickly.

2. Aspirate off the glycerol.

3. Wash the plates three times with pre-warmed 2 ml DMEM, 10% FBS using a circular swirling motion.

4. Add 5 ml DMEM, 10% FBS and incubate the plates at 37°C for 24 h.

C. *Superinfection*

1. Perform X-Gal staining (*Protocol 7*) on the pHSVlac and 'no DNA' controls to verify the efficiency of the transfection; proceed only if the transfection has worked.

2. Process four to six plates at a time. Remove the medium and add 1 ml D30EBA virus solution (0.5–2.0 × 10^6 p.f.u./ml) to each plate.[a] Add 1 ml DMEM, 5% FBS (without virus) to a control plate (mock).

3. Start a timer when all of the plates have received the D30EBA virus.

4. Incubate the plates at room temperature. Tilt the plates to distribute the virus every 15 min.

5. After 1 h at room temperature, add 4 ml DMEM, 5% FBS to each plate. Incubate the plates at 37°C for 24 h.

D. *Amplification*

1. Harvest the virus (*Protocol 2*) when the cells round up, usually after 16–24 h. This virus stock is designated p0 (passage 0).

2. Passage 1 (p1). Measure the volume of the p0 preparation from step 1 when adding it to the plate for p1. Add the virus to a 60 mm tissue culture plate and then add DMEM, 5% FBS to make the total volume 5–5.5 ml. Harvest the virus (*Protocol 2*) when the cells round up, usually after 24–36 h. Include a mock-infected plate with no virus as a negative control in each passage.

3. Passage 2 (p2). Measure the volume of virus in the p1 preparation from step 2 when adding it to the plate for p2. Add the virus to a 100 mm plate containing 6 ml DMEM, 5% FBS to make the total volume 10–10.5 ml

Protocol 6. *Continued*

total. Harvest the virus (*Protocol 2*) when the cells round up, usually after 24 h.

4. Passage 3 (p3). Measure the volume of virus in the p2 preparation from step 3 when adding it to the plates for p3. Add the virus to two 100 mm plates each containing 6 ml DMEM, 5% FBS to make the total volume 10–10.5 ml. Harvest the virus (*Protocol 2*) when the cells round up, usually 24 h.

[a] Virus solution: 100–200 μl of D30EBA virus (5–10 × 10^6 p.f.u./ml) with DMEM, 5% FBS to give 1 ml total volume for each plate. Mix in one tube the required amount of virus and DMEM, 5% FBS for the entire packaging before adding to plates.

7. Titring vectors that express β-galactosidase

To measure the titre of vector in a virus stock, cells are infected with the vector and one day later the number of cells expressing the recombinant gene product is determined and used to calculate the titre of the vector. For vectors expressing β-galactosidase the titring can be performed using PC12 cells and the X-Gal assay (*Protocol 7*). This X-Gal procedure works with most cell lines and cultured peripheral neurones. Some cultured neurones from the central nervous system have a high level of endogenous β-galactosidase so this procedure can not be used, but cultured CNS neurones can be stained with X-Gal using the procedure described in Emson *et al.* (33).

Protocol 7. X-Gal staining of transfected cells

Equipment and reagents

- Poly-D-lysine (PDL): prepare 1 mg/ml poly-D-lysine in water, filter sterilize, and store at −20°C (this is a 50 × stock—dilute to 20 μg/ml just before use)
- DMEM (or RPMI 1640) medium
- Virus stock
- PBS (see *Protocol 5*)
- Fe solution: mix the following reagents; 200 ml PBS, 0.332 g potassium ferricyanide, 0.424 g potassium ferrocyanide, 0.2 ml 1 M $MgCl_2$, 0.2 ml 20% Nonidet P-40, 0.2 ml 10% sodium deoxycholate (store at 4°C covered with aluminium foil since this reagent is light-sensitive)
- 50 mg/ml X-Gal in DMSO (store in aliquots at −20°C)

Reagent for fixation of cells
Either 0.1–1.5% glutaraldehyde
or 4% paraformaldehyde pH 7.0.
Prepare this by adding 20 g paraformaldehyde to 300 ml H_2O and heating to 55–60°C. Slowly add 1 M NaOH dropwise over approx. 10 min until the solution becomes clear. Cool the solution to room temperature. Use pH paper to check that the pH is 7.0–7.5 (add more NaOH if necessary). Add 100 ml of 0.5 M sodium phosphate buffer pH 7.0 and then water to a final volume of 500 ml. The final pH should be 7.0–7.5. Store at 4°C.

Method

1. Coat 24-well plates with PDL to enable PC12 cells to adhere properly. Use 0.5 ml of 20 μg/ml PDL for at least 5 min at room temperature.

Then aspirate the PDL completely before plating the cells (in addition, optionally, rinse once with 0.5 ml PBS).

2. Plate the PC12 cells at 3×10^5 well in DMEM (or RPMI 1640), 5% FBS, 10% horse serum (HS). Incubate at 37 °C.
3. The following day change the medium. Add the test virus (up to 100 μl of a crude virus preparation). Always include an uninfected culture in a well (mock infection) for a negative control to detect the background level of staining.
4. One day after infection, remove the medium and fix the cells with either 4% paraformaldehyde, pH 7.0 for 20–60 min at room temperature or in 0.1–1.5% glutaraldehyde for 5–15 min.
5. Wash the fixed cells three times with 0.5 ml PBS for 5 min each wash.
6. Warm (in a 37 °C water-bath) 0.5 ml of Fe solution for each well of cells to be stained. Add 50 mg/ml X-Gal stock to the warm Fe solution to a final concentration of 1 mg/ml. If the solutions are not warm, a precipitate will form.
7. Remove the final wash from the cells and add the X-Gal/Fe solution (0.5 ml per well).
8. Incubate the cells at 37 °C. Depending on the cell type and the promoter in the vector, the staining will develop in a few hours to overnight.
9. To stop the staining reaction, remove the X-Gal solution and wash the wells two or three times with PBS. The plates can be stored in PBS at room temperature for several weeks.

8. Purification of the virus

Virus is purified to remove cytotoxic contaminants (for example, cell membranes or proteins) that injure some types of cells such as central nervous system neurones. Virus is also concentrated during the purification procedure so that a small volume (1–3 μl) can be injected into tissues *in vivo* (e.g. into the rat brain). The purification is based on centrifugation through a discontinuous sucrose gradient. The virus particles sediment through 25–30% sucrose but do not pass through 60% sucrose. Instead they are retained as a band at the top of a 60% sucrose layer. Cellular contaminants do not localize here; most proteins and membranes can not enter 25–30% sucrose while nucleic acids will pass through 60% sucrose. The virus is then collected, diluted, pelleted by centrifugation, and resuspended in a small volume. Because virus particles are heat labile, all solutions must be chilled on ice and all procedures and centrifugations must be performed at 4°C. Two procedures are presented (*Protocols 8* and *9*). Both procedures work well and are used in our laboratory; we have not yet quantitatively compared the procedures to determine which gives a greater purification of HSV-1 virus particles.

Protocol 8. Purification of virus using a 10–60% discontinuous sucrose gradient

Equipment and reagents

- Ultracentrifuge and appropriate rotor (Beckman SW28 or equivalent)
- Large (40 ml capacity) ultracentrifuge tubes (e.g. Beckman SW28 40 ml tubes or equivalent)
- Small (20 ml capacity) ultracentrifuge tubes (e.g. Beckman SW28 small tubes or equivalent)
- PBS (see *Protocol 5*)
- Solutions of 60% (w/v) sucrose, 30% (w/v) sucrose, and 10% (w/v) sucrose in PBS. Filter sterilize (filter sterilization is used instead of autoclaving in order to maintain the desired concentration).
- Virus stock for purification

Method

1. Prepare a sucrose step gradient in large (40 ml) SW28 tubes or their equivalent. To do this, add the following sucrose solutions in order to the bottom of the centrifuge tube:
   - 7 ml 60% (w/v) sucrose
   - 6 ml 30% (w/v) sucrose
   - 3 ml 10% (w/v) sucrose
2. Gently load 20 ml of crude virus on to the gradient.
3. Centrifuge for 1 h at 125 000 *g* (e.g. Beckman SW28 rotor, 25 000 r.p.m.). The virus bands at the interface between the 30% and 60% sucrose layers.
4. Carefully remove the top layers of sucrose and then pipette off the virus which should be contained in about 2 ml.
5. Dilute the purified virus with PBS (and mix) to fill a tall thin ultracentrifuge tube (approx. 20 ml; e.g. Beckman SW28 tube) and then centrifuge at 125 000 *g* (Beckman SW28, 25 000 r.p.m.) for 1 h.
6. Carefully pipette off the supernatant, resuspend the pellet (which contains the virus) in 200 μl of 10% sucrose in PBS, and dispense this into 30 μl aliquots.

Protocol 9. Purification of virus using a 25–70% discontinuous sucrose gradient

Equipment and reagents

- Ultracentrifuge and appropriate rotors (Beckman SW28 and SW50.1 or equivalent)
- Small (20 ml capacity) ultracentrifuge tubes (e.g. Beckman SW28 small tubes or equivalent) and tubes for SW50.1 rotor (or equivalent)
- PBS
- Solutions of 10% (w/v) sucrose, 25% (w/v) sucrose, 60% (w/v) sucrose, 70% (w/v) sucrose, and saturated sucrose in PBS. Filter sterilize.
- Virus stock for purification

Method

1. Centrifuge 10.5 ml of crude virus preparation for 5 min at 5600 *g* to remove material. Re-centrifuge the supernatant (same conditions).
2. Resuspend the pellet from the first centrifugation in 2 ml PBS and centrifuge it again at 5600 *g* for 5 min.
3. Combine the two supernantants from steps 1 and 2. This is the clarified virus.
4. Gently layer the clarified virus on top of a four step discontinuous sucrose gradient consisting of 1.5 ml 25% (w/v) sucrose; 1.5 ml 60% (w/v) sucrose, 0.5 ml 70% (w/v) sucrose; and 0.5 ml saturated sucrose in a tall thin ultracentrifuge tube (approx. 20 ml, e.g. Beckman SW28 tube).
5. Centrifuge the sucrose gradient for 90 min at 125 000 *g* (Beckman SW28, 25 000 r.p.m.).
6. Collect 2 ml from the 25%–60% sucrose interface, mix this with 3 ml PBS, and centrifuge for 90 min at 52 500 *g* (e.g. Beckman SW50.1 rotor; 25 000 r.p.m.).
7. Carefully pipette off the supernatant. Resuspend the pellet (which contains the virus) in 200 μl of 10% sucrose in PBS, and dispense this into 30 μl aliquots.

9. Limitations of the present HSV vector system and potential solutions

Although this defective HSV-1 vector system can be used for genetic intervention in the brain, it is a developing technology and hence potential technological advances may improve the system. First, the current packaging systems use HSV-1 mutants which harbour a single deletion in an essential gene, resulting in a reversion frequency to wild-type HSV-1 of $\sim 10^{-5}$ to $\leqslant 10^{-7}$. Use of a helper virus harbouring deletions in two or more essential genes should further lower the reversion frequency to $< 10^{-10}$. Secondly, an HSV-1 particle contains specific proteins which have acute and temporary cytopathic effects. Deleting these non-essential HSV-1 genes from the helper virus may allow the experimenter to increase the amount of vector used, especially for *in vivo* experiments in which cells proximal to the injection site are vulnerable to infection with multiple HSV-1 particles. Thirdly, while cell type-specific promoters do confer cell type-specific expression on a recombinant gene, frequently there is a low level of expression in inappropriate cell types. This inappropriate expression could be due either to specific elements in the vector, particularly those proximal to the HSV-1 ori_S, and/or to specific genes in the helper virus. Thus, improvements to the helper virus and/or to the vector may also enhance cell type-specific expression. If at least some of these proposed improvements are successful, it is possible they will result in a defective HSV-1 vector system which can support a wide range of experiments in neuronal physiology and which might be used to perform human gene therapy.

Acknowledgements

We are grateful to Drs R. D. Everett, N. D. Stow, P. A. Schaffer, R. M. Sandri-Goldin, and P. A. Johnson for viruses and/or cell lines. We thank Drs C. Fraefel, J. A. Majzoub, W. J. Mitchell, C. L. Wilcox, and R. Neve for sharing unpublished data and/or their comments on the manuscript. We thank Cindy Moore for assistance in preparing this manuscript. This work was supported by AG10827 (NIH), the American Health Assistance Foundation, the Burroughs Wellcome Fund, and the National Parkinson Foundation (A.I.G.); MH09823 (NIH) and the French Foundation (D.H.); and the Pharmaceutical Manufacturers Foundation (S.S.).

References

1. Geller, A. I., During, M. J., and Neve, R. L. (1991). *Trends Neurosci.*, **14**, 428.
2. Freese, A., Geller, A. I., and Neve, R. L. (1990). *Biochem. Pharmacol.*, **40**, 2189.
3. O'Malley, K. and Geller, A. I. (1992). In *Neurological disorders: novel experimental and therapeutic strategies* (ed. L. Vecsei, A. Freese, K. J. Swartz, and M. F. Beal), pp. 223–48. Ellis Horwood Limited, Simon and Schuster, West Sussex, England.
4. Geller, A. I. and Federoff, H. J. (1991). In *Human gene transfer* (ed. O. Cohen-Haguenauer and M. Boiron), Vol. 219, pp. 63–00. John Libbey Eurotext Ltd., Paris.
5. Weiss, R., Tsnich, N., Varmus, H., and Coffin, J. (1985). *RNA tumour viruses*. Cold Spring Harbor Laboratory, NY.
6. Geller, A. I. and Breakefield, X. O. (1988). *Science*, **241**, 1667.
7. Geller, A. I. and Freese, A. (1990). *Proc. Natl Acad. Sci. USA*, **87**, 1149.
8. Geller, A. I., Keyomarski, K., Bryan, J., and Pardee, A. B. (1990). *Proc. Natl Acad. Sci. USA*, **87**, 8950.
9. Spaete, R. R. and Frenkel, N. (1982). *Cell*, **30**, 295.
10. Freese, A. and Geller, A. I. (1991). *Nucleic Acids Res.*, **19**, 7219.
11. Battleman, D. S., Geller, A. I., and Chao, M. V. (1993). *J. Neurosci.*, **13**, 941.
12. Federoff, H. J., Geschwind, M. D., Geller, A. I., and Kessler, J. A. (1992). *Proc. Natl Acad. Sci. USA*, **89**, 1636.
13. Geschwind, M. D., Kessler, J. A., Geller, A. I., and Federoff, H. J. (1994). *Mol. Brain Res.*, **24**, 327.
14. Ho, D. Y., Mocarski, E. S., and Sapolsky, R. M. (1993). *Proc. Natl Acad. Sci. USA*, **90**, 3655.
15. Bergold, P. J., Casaccia-Bonnefil, P., Xiu-Liu, Z., and Federoff, H. J. (1993). *Proc. Natl Acad. Sci. USA*, **90**, 6165.

16. Geller, A. I., During, M. J., Haycock, J. W., Freese, A., and Neve, R. L. (1993). *Proc. Natl Acad. Sci. USA*, **90**, 7603.
17. Geller, A. I., Freese, A., During, M. J., and O'Malley, K. L. (1995). *J. Neurochem.*, **64**, 487.
18. During, M. J., Naegele, J., O'Malley, K., and Geller, A. I. (1994). *Science*, **266**, 1399.
19. Federoff, H. J., Geller, A., and Lu, B. (1990). *Abstr. Soc. Neurosci.*, **16**, 353.
20. Oh, Y. J., Wong, S. C., Moffat, M., Ullrey, D., Geller, A. I., and O'Malley, K. L. (1992). *Abs. Soc. Neurosci.*, **18**, 578.14.
21. Kaplitt, M. G., Kwong, A. D., Kleopoulos, S. P., Mobbs, C. V., Rabkin, S. D., and Pfaff, D. W. (1994). *Proc. Natl Acad. Sci. USA*, **91**, 8979.
22. Spear, P. G. and Roizman, B. (1981). In *DNA tumor viruses* (ed. J. Tooze), pp. 615–71. Cold Spring Harbor Laboratory, NY.
23. Stevens, J. G. (1975). *Curr. Top. Microbiol. Immunol.*, **70**, 31.
24. Watson, K., Stevens, J. G., Cook, M. L., and Subak-Sharpe, J. H. (1980). *J. Gen. Virol.*, **49**, 149.
25. Davison, M. J., Preston, V. G., and McGeoch, D. J. (1984). *J. Gen. Virol.*, **65**, 859.
26. Patterson, T. and Everett, R. D. (1990). *J. Gen. Virol.*, **71**, 1775.
27. Davidson, I. and Stow, N. D. (1985). *Virology*, **141**, 77.
28. Johnson, P. A., Miyanohara, A., Levine, F., Cahill, T., and Friedmann, T. (1992). *J. Virol.*, **66**, 2952.
29. McCarthy, A. M., McMahan, L., and Schaffer, P. A. (1989). *J. Virol.*, **63**, 18.
30. Smith, I. L., Hardwicke, M. A., and Sandri-Goldin, R. M. (1992). *Virology*, **186**, 74.
31. DeLuca, N. A., McCarthy, A. M., and Schaeffer, P. A. (1985). *J. Virol.*, **56**, 558.
32. Maniatis, T., Fritsch, E. F., and Sambrook, J. (ed.) (1982). *Molecular cloning*. Cold Spring Harbor Laboratory Press, Cold Spring Harbor, NY.
33. Emson, P. C., Shoham, S., Feler, C., Buss, T., Price, J., and Wilson, C. J. (1990). *Exp. Brain Res.*, **79**, 427.

9

Adenovirus vectors

ROBERT D. GERARD and ROBERT S. MEIDELL

1. Introduction

The introduction of foreign genetic material into somatic cells in culture and in intact organisms is an important investigational technique and holds promise as a therapeutic tool. Among the available vectors for gene transfer, recombinant adenoviruses possess several important advantages for direct introduction of foreign genes into mammalian cells, including:

- simple techniques for generation of recombinant viral genomes
- ready propagation of recombinant virus to high titres
- broad species and tissue trophism
- a high efficiency of gene transfer

Stable transformation following infection of replicating mammalian cells in culture by recombinant adenovirus, however, is infrequent and ranges from 10^{-3} to 10^{-6} (1). Similarly, expression of foreign genes introduced into quiescent somatic cells in intact animals has thus far proven transient. At present, recombinant adenovirus vectors are primarily useful in achieving high efficiency, transient expression of foreign genes in mammalian cells, either *in vitro* or *in vivo* (2–7).

The methods described in this chapter have allowed us to generate over 200 different recombinant adenovirus vectors for a variety of purposes. For example, such vectors allow the specific and efficient introduction of genes into differentiated cells refractory to chemical transfection protocols (4, 6–8). They also permit the efficient transfer of foreign genes into animals for the study of metabolic processes (2, 3, 9). Ultimately, they may be useful for human gene therapy.

2. Biology of adenovirus

Human adenovirus is a non-enveloped, icosahedral, double-stranded DNA virus approximately 130 nm in size. Numerous serotypes of adenovirus have been identified, although subgroup C human adenoviruses, principally

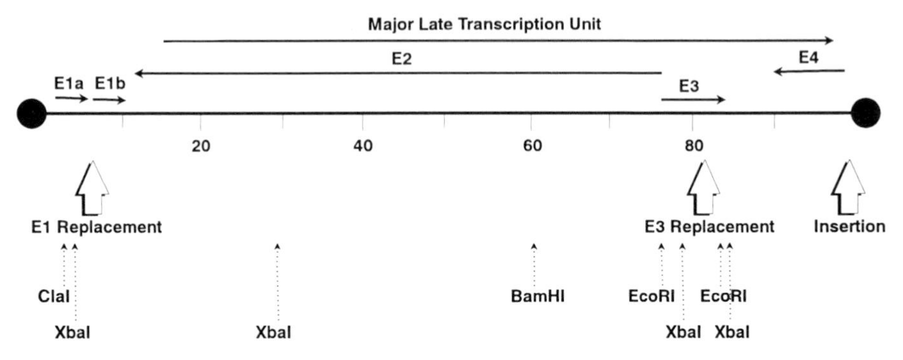

Figure 1. Map of the adenovirus 5 genome from 0–100 map units (mu). One map unit corresponds to 360 bp. The ends of the linear DNA containing the inverted terminal repeats are shown by filled circles. Large arrows indicate the positions at which foreign genes have been inserted into recombinant vectors. Transcription units are indicated by solid horizontal arrows and restriction enzyme sites are shown by dotted vertical arrows.

derivatives of adenovirus types 2 and 5, predominate as vectors for gene transfer. These adenoviruses promiscuously infect a broad range of mammalian cell types, presumably reflecting the ubiquitous expression of a specific, as yet unidentified, cell surface receptor. Cell attachment is mediated by the fibre protein component of the adenovirus capsid and the virus enters the cell via receptor-mediated endocytosis. Following escape from the lysosomal compartment, viral DNA replication occurs in the cell nucleus.

The linear adenoviral genome, illustrated schematically in *Figure 1*, is approximately 36 kb in length and encodes multiple overlapping transcriptional units. Expression of adenoviral genes occurs in two phases, early and late, defined with respect to the onset of replication of the adenoviral genome. The early genes are encoded in four distinct transcriptional units, E1–E4. All but the E3 transcriptional unit are required for lytic growth in cell culture. The late gene products include the principal capsid proteins hexon, penton, and fibre which are transcribed from a single (major late) promoter.

Early region 1 (E1) encodes transcription factors which interact with the host cell transcription apparatus and activate subsequent viral gene expression. The dependence of other viral gene expression, and ultimately replication of the viral genome, on E1 gene products is the basis for construction of conditionally replication-defective adenovirus vectors in which the endogenous E1 genes are replaced by a foreign gene of interest (10).

During lytic infection, the viral genome is replicated to several thousand copies per cell. Replicated genomes rapidly associate with 'core' proteins and are later packaged into incomplete capsids formed by self-assembly of the major capsid proteins. The size of the adenoviral genome which can be packaged into nascent capsids is limited to approximately 105% of native genome length (or approx. 38 kb), and is an important consideration in the construction of recombinant adenoviral vectors.

3. Generation of recombinant adenoviruses

Foreign genes have been inserted into recombinant adenoviral genomes as replacements for the E1 or E3 regions, or as insertions between the E4 region and the right end of the adenoviral genome (*Figure 1* and ref. 5). Vectors based on the insertion of foreign genetic material into the right end of the adenovirus genome or in place of E3 are replication-competent. In contrast, replacement of early region 1 results in a conditionally replication-defective vector which can only be propagated either in a cell line which supplies the missing functions *in trans* or in the presence of helper virus. While recombinant adenoviruses of each class have been employed as gene transfer vectors, replication-defective adenoviruses based on replacements of the early region 1, either with or without an intact early region 3, have been used most commonly.

A human embryonic kidney cell line termed 293 (11) that contains an integrated copy of the leftmost 12% of the adenovirus 5 genome complements the defect in E1 replacement vectors. This line is routinely employed to propagate vectors of this class in a helper-independent manner. True packaging cell lines for adenovirus are not available, nor are they likely to become available since the high level expression of capsid proteins needed to package adenovirus is toxic to cells.

3.1 Strategies available

Complete, recombinant adenovirus genomes are generated either by ligation of subgenomic fragments *in vitro* (12), or by homologous recombination *in vivo* (13) between overlapping fragments following co-transfection into the 293 host cell line. Recombinant virus construction has been facilitated by the development of phenotypically wild-type adenovirus variants from which specific restriction sites have been eliminated, such as Ad5*dl*309 (14), and by the development of bacterial plasmids carrying adenoviral genomic sequences.

The specific protocols outlined below and illustrated schematically in *Figures 2A* and *B* describe construction of E1 replacement adenoviruses. While the reagents used in construction of E3 replacement vectors differ (illustrated schematically in *Figure 2C*), the principles and general techniques are the same. The reader is referred to other papers on this subject for specific advice (10).

3.2 Construction of E1 replacement adenovirus vectors

The initial step in the construction of an E1 replacement recombinant adenovirus is insertion of the foreign gene into either the bacterial plasmid pAC

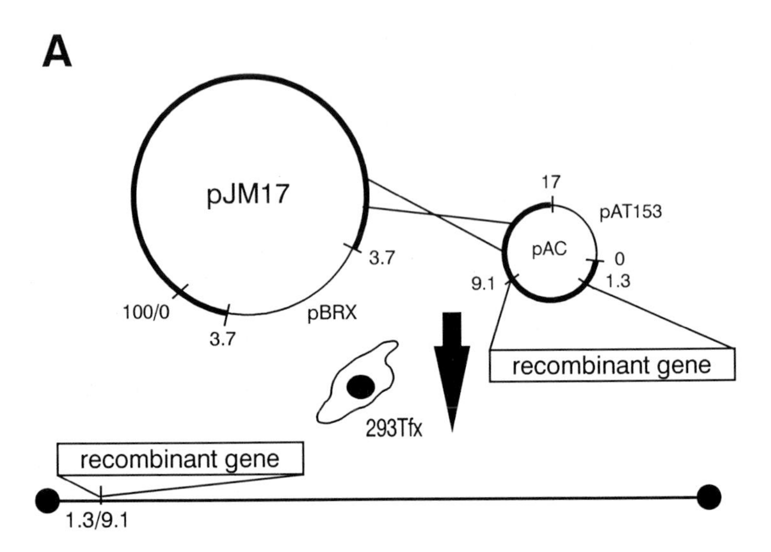
A
pJM17
17
pAT153
pAC
3.7
0
9.1
1.3
100/0
3.7
pBRX
recombinant gene
293Tfx
recombinant gene
1.3/9.1

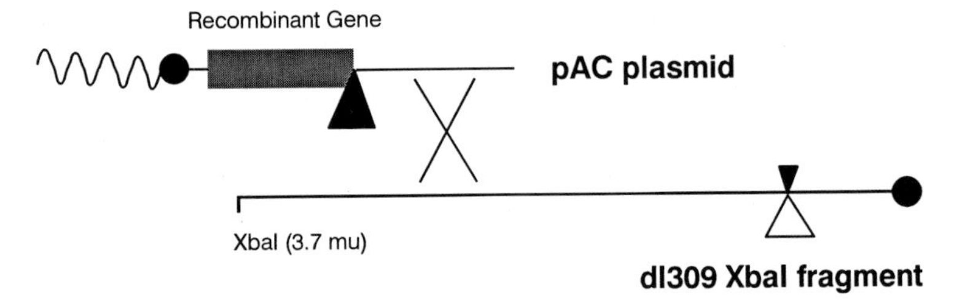
B
Recombinant Gene
pAC plasmid
XbaI (3.7 mu)
dl309 XbaI fragment

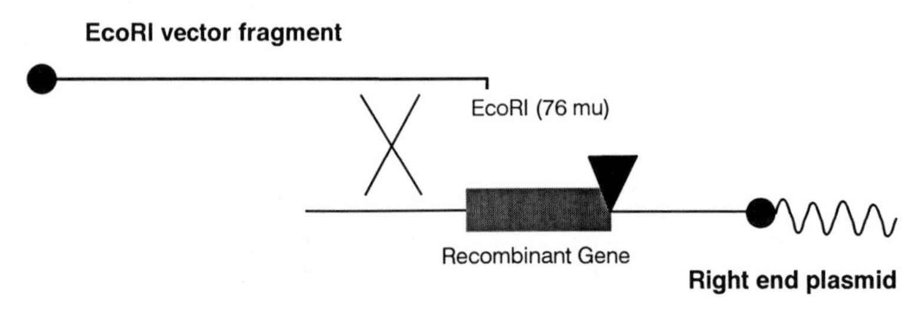
C
EcoRI vector fragment
EcoRI (76 mu)
Recombinant Gene
Right end plasmid

Figure 2. Diagrams indicating the various strategies used to construct adenovirus vectors by homologous recombination. The segment of viral DNA containing the origin of DNA replication and inverted terminal repeat is shown by a filled circle. Numbers refer to map units on the adenovirus genome. (A) The plasmid pJM17 is co-transfected into 293 cells with the pAC plasmid containing the foreign gene. In the diagrams of the plasmids, adenovirus DNA is indicated by thick lines and bacterial vector sequences are indicated by thinner lines. Recombination between the two DNAs results in the generation of a recombinant genome of packageable size in which the foreign gene replaces early region 1. (B) The pAC plasmid is co-transfected into 293 cells with a linear restriction fragment of adenovirus genomic DNA, such as the *Xba*I fragment of Ad5*dl*309 DNA shown. Recombination again yields a recombinant genome in which the foreign gene replaces early region 1. Although the pAC plasmid is shown as a linear molecule, both circular and linear DNA will generate a recombinant genome. Adenovirus DNA is indicated by straight lines and bacterial vector sequences are indicated by wavy lines. Deletions are indicated by filled triangles and insertions by open triangles. (C) A right end plasmid, such as pFGdX1 harbouring sequences from 60–100% of the viral genome (10) is co-transfected into 293 cells with a linear restriction fragment of adenovirus genomic DNA, such as the *Eco*RI fragment of Ad5 DNA shown. In this case, recombination yields a recombinant genome in which the foreign gene replaces early region 3. Adenovirus DNA is indicated by straight lines and bacterial vector sequences are indicated by wavy lines. The E3 deletion is indicated by a filled triangle.

---

(13) or one of a number of similar plasmids which contain adenovirus 5 sequences from the left end of the adenovirus genome including the origin of replication and the packaging signal. pAC contains nucleotides 1–453 and 3334–6231 of the adenovirus genome (Joe Alcorn, personal communication) and is shown in *Figure 3A*. The foreign gene is inserted into the site of the 453–3334 deletion in place of the native adenovirus early region 1 genes. To facilitate construction of recombinant viruses, several derivatives of pAC containing multiple cloning sites (e.g. pACRR.25 and pACRR.5) (6) and/or promoter and polyadenylation sequences (e.g. pACCMVpLpA) (7) have been produced in our laboratory.

(a) The parental pAC plasmid (*Figure 3A*) contains an *Xba*I restriction site inserted in place of the E1 deletion.
(b) pACRR was derived from pAC by digestion with *Eco*RI, filling the recessed ends using the Klenow fragment of DNA polymerase I, and ligation. pAC-RRHR was similarly derived from pACRR by sequential digestion with *Hind*III, treatment with the Klenow fragment of *E. coli* DNA polymerase, and ligation. pAC-ESHR was derived from pAC-RRHR by *Sal*I digestion, treatment with the Klenow fragment, and ligation.
(c) pACRR.5 and pACRR.25 (*Figure 3B*) were derived from pACRR by digestion with *Xba*I, filling the ends using Klenow fragment, and ligation of a Klenow-filled *Eco*RI–*Hind*III fragment from pUC19 that contains the multiple cloning site sequence to yield the two plasmids with the

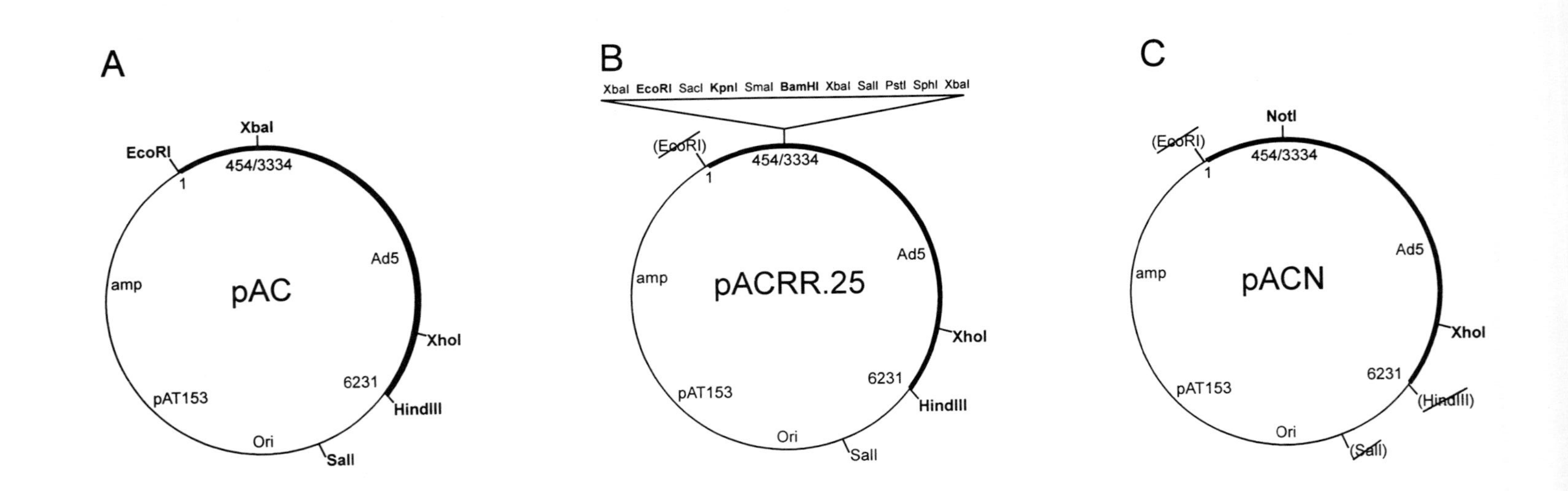
A
pAC
EcoRI
XbaI
1
454/3334
Ad5
XhoI
6231
HindIII
SalI
Ori
pAT153
amp
B
pACRR.25
XbaI EcoRI SacI KpnI SmaI BamHI XbaI SalI PstI SphI XbaI
(EcoRI)
1
454/3334
Ad5
XhoI
6231
HindIII
SalI
Ori
pAT153
amp
C
pACN
NotI
(EcoRI)
1
454/3334
Ad5
XhoI
6231
(HindIII)
(SalI)
Ori
pAT153
amp

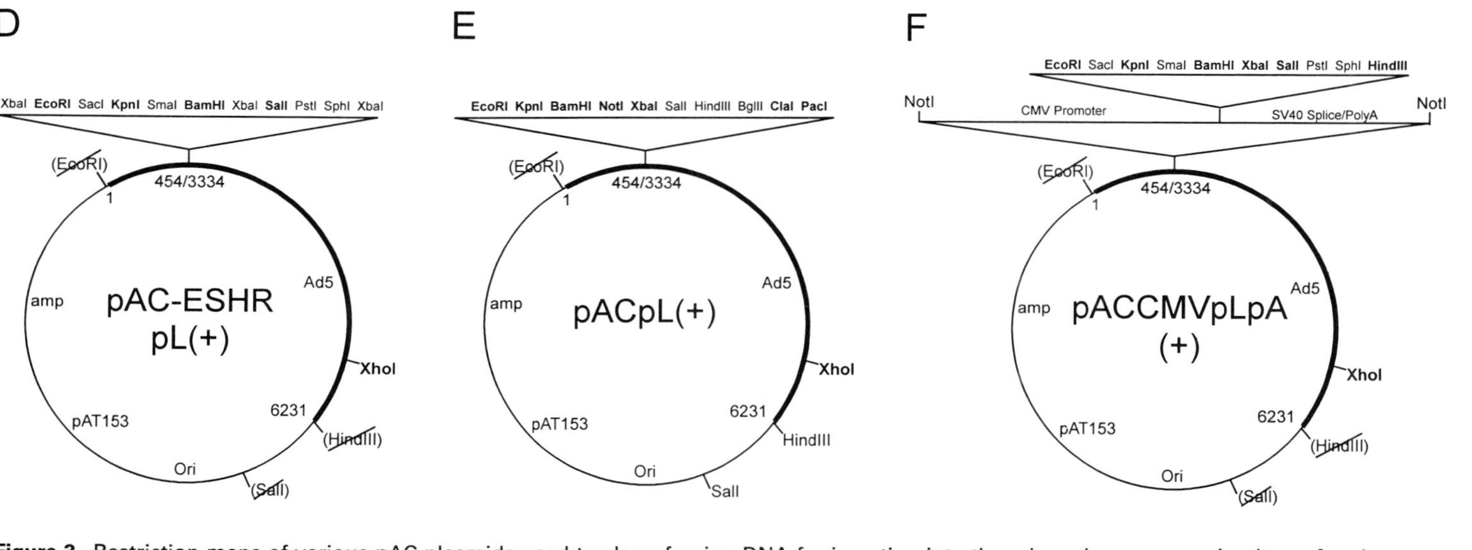

Figure 3. Restriction maps of various pAC plasmids used to clone foreign DNA for insertion into the adenovirus genome in place of early region 1. The parental plasmid, pAC, has adenovirus sequences (nucleotide positions 1–454 and 3334–6231) cloned into the bacterial vector pAT153 (Y. Gluzman, personal communication) between the *Eco*RI and *Hin*dIII sites. Unique restriction sites within the plasmids are shown in bold type. (A) The parental plasmid, pAC, has a unique *Xba*I site for cloning. The unique *Xho*I site within the sequences used for homologous recombination is also shown. (B) The derivative pACRR.25 contains the pUC19 polylinker in the plus orientation whereas pACRR.5 (not shown) is the minus orientation. The *Eco*RI site at the junction of the bacterial vector and adenovirus sequences has been eliminated. (C) In the derivative pACN, the *Eco*RI, *Sal*I, and *Hin*dIII sites have been eliminated and the *Xba*I site replaced by a *Not*I site. (D) pAC-ESHRpL(+) and its counterpart, pAC-ESHRpL(−) (not shown) are similar to pACRR.25 and pACRR.5 except that the extraneous *Sal*I and *Hin*dIII sites were first eliminated to generate additional unique sites within the polylinker. (E) pACpL(+), and its counterpart with the polylinker in the minus orientation, pACpL(−) (not shown), contains a different polylinker than pACRR.25 and pACRR.5. (F) The derivative pACCMVpLpA contains a 780 bp fragment of the human cytomegalovirus immediate early promoter flanking the pUC19 polylinker. The 470 bp fragment containing mRNA splicing and polyadenylation signals from SV40 were derived from pko-neo (1).

polylinker inserted in opposite orientations relative to the adenovirus sequences. The convention we use is that left-to-right is the + orientation (pACRR.25).

(d) pACN (*Figure 3C*) was derived from pAC-ESHR by *Xba*I digestion, filling the ends using Klenow fragment, and ligation of *Not*I linkers.

(e) pAC-ESHRpL(+) (*Figure 3D*) was derived from pAC-ESHR by insertion of a Klenow-blunted *Eco*RI–*Hin*dIII polylinker fragment from pUC19 into the blunted *Xba*I site. The *Xba*I sites at the ends of the polylinker were therefore restored.

(f) pACpL(+) (*Figure 3E*) was derived from pACRR by insertion of a synthetic polylinker sequence that contains the indicated restriction sites in place of the *Xba*I site.

(g) pACCMVpLpA(+) (*Figure 3F*) was designed to directly accept cDNA fragments for high level expression. It was derived from pACN by insertion of a *Not*I fragment containing the CMV immediate early promoter (15), pUC19 polylinker, and SV40 small t antigen splicing/polyadenylation signals from pko-neo (1). Two different orientation plasmids were obtained, designated + and −, depending on the orientation of the *Not*I fragment with respect to the adenovirus sequences (7).

3.2.1 Construction of recombinant adenovirus by *in vitro* ligation

A full-length, recombinant adenovirus genome can be reconstructed *in vitro* by ligation of the pAC plasmid, linearized to the right of the foreign gene by digestion at a unique restriction site, to a fragment of the linear adenovirus 5 genome encoding sequences to the right of this cleavage site (the right 91% of the Ad genome). Because the native adenovirus 5 genome lacks a unique restriction site between positions 3334 and 6231 to serve as a site for cleavage and re-ligation, this strategy is most readily employed using a variant of pAC [such as pACpL(+)] and a recombinant vector (AdpL+) derived from that plasmid that has been engineered to contain a unique restriction site at position 3334. Replication of the viral DNA resolves the viral DNA sequences from the plasmid vector through a process that may involve pairing of the inverted terminal repeats. Adenovirus construction *in vitro* is described in *Protocol 1* and transfection of the recombinant genome into 293 cells is described in *Protocol 3*.

Protocol 1. Adenovirus construction by direct ligation

Reagents

- An appropriate pAC plasmid (*Figure 3*) containing the foreign gene of interest
- CsCl purified adenovirus DNA
- Alkaline phosphatase
- Appropriate restriction enzyme
- T4 DNA ligase

- 10 mM Tris–HCl pH 7.4, 0.1 mM EDTA
- 10 mM Tris–HCl pH 7.4, 1 mM EDTA
- 10 × ligation buffer: 0.5 M Tris–HCl pH 7.8, 0.1 M $MgCl_2$, 0.1 M DTT, 10 mM ATP
- 1 mg/ml BSA
- Phenol buffered with 0.1 M Tris–HCl pH 7.4, 1 mM EDTA
- Chloroform/isoamyl alcohol (24:1)
- 3 M sodium acetate pH 6.5

Method

1. Digest the recombinant pAC plasmid with the appropriate restriction endonuclease [either *Cla*I or *Pac*I for pACpL(+)].
2. Digest the CsCl purified viral DNA with the same enzyme [*Cla*I or *Pac*I for Ad5pL(+)]. Dephosphorylate the viral DNA using alkaline phosphatase according to the supplier's recommended protocol.
3. Purify the restricted pAC plasmid DNA and adenovirus DNA as follows:
   (a) To each DNA, add an equal volume of phenol. Mix and then centrifuge at 13 000 *g* for 1 min to separate the phases. Recover the aqueous phase.
   (b) To the aqueous phase, add an equal volume of chloroform/isoamyl alcohol (24:1). Mix and then centrifuge as above to separate the phases. Recover the aqueous phase.
   (c) Add 3 M sodium acetate pH 6.5 to a final concentration of 0.3 M, then add 2.5 vol. ethanol. Leave on ice for 5 min.
   (d) Collect the precipitated DNA by centrifugation at 13 000 *g* for 5 min.
4. Redissolve each pellet in 10 mM Tris–HCl pH 7.4, 0.1 mM EDTA at approximately 1 μg/μl.
5. Set-up the following reaction mixture for ligation:[a]
   - adenoviral DNA (1 μg/μl) 1 μl
   - linearized pAC plasmid (1 μg/μl) 2 μl
   - 10 × ligation buffer 2 μl
   - 1 mg/ml BSA 2 μl
   - water 12 μl
   - T4 DNA ligase (100 U/μl) 1 μl

   Incubate overnight at 13°C.
6. Add 10 mM Tris–HCl pH 7.4, 1 mM EDTA to 200 μl final volume and extract the DNA with phenol, then with chloroform/isoamyl alcohol (24:1).
7. Ethanol precipitate the DNA and redissolve the pellet in 50 μl sterile 10 mM Tris–HCl pH 7.4, 0.1 mM EDTA.
8. Transfect the DNA into 293 cells as described in *Protocol 3*.

[a] Reaction volumes can be scaled up for the preparation of larger quantities of DNA.

3.2.2 Construction of recombinant adenovirus by homologous recombination

More commonly, a recombinant viral genome is generated by homologous recombination in 293 cells following co-transfection of the pAC plasmid and *either* a subgenomic fragment of adenovirus DNA *or* pJM17 (16).

1. *Using adenovirus DNA*. The adenovirus 5 subgenomic fragment used is generally the right 97% fragment of Ad5*dl*309 or the right 91% fragment of a recombinant adenovirus prepared by digestion of viral DNA with a restriction endonuclease which cleaves the viral genome at a unique site to the left of position 6231 (e.g. Ad5*dl*309 digested with *Xba*I and/or *Cla*I). The insert length which can be accommodated in an Ad5*dl*309-derived vector is about 2.0 kb (105% of wild-type genome length). Coupled with an El deletion such as that found in pAC, this length increases to 4.9 kb. The maximum insert length that can be accommodated in currently available vectors is provided by using double replacement vectors and is 7.0 kb. Such vectors are constructed using pAC and Ad5*dl*324 which contains a deletion of E3 between naturally occurring *Xba*I sites at 78.5 and 84.3 map units (17). Attempts to construct vectors containing inserts longer than these limits will probably result in the evolution and propagation of a virus deleted of some portion of the foreign DNA insert.

2. *Using pJM17*. pJM17 is a plasmid constructed by Frank Graham and colleagues (16). It is an infectious bacterial plasmid in which a circularized adenovirus 5 (variant *dl*309) genome is interrupted at the unique *Xba*I site at 3.7 map units by insertion of the plasmid pBRX. The plasmid vector encodes both amp^r and tet^r. We have found that while deletions of adenoviral sequences from the 40 kb plasmid occur frequently during propagation of transformed bacteria, frequent re-selection of clonally-derived bacterial stocks will minimize contamination of pJM17 preparations with deleted forms. During purification of the plasmid, care should be taken to avoid shearing.

In practice, the choice between these two alternative strategies for homologous recombination is largely a matter of convenience since each offers advantages and disadvantages. Exclusive use of plasmid DNAs will generally yield a very low background of non-recombinant plaques, but is inherently less efficient. Use of authentic viral DNA is generally more efficient, but will suffer from a higher background of non-recombinant viral clones that must be eliminated by screening. Simple screening methods and efficient transfection of 293 cells will benefit any approach.

Construction of recombinant adenovirus by homologous recombination in 293 cells is described in *Protocols 2* and *3*.

Protocol 2. Adenovirus construction by homologous recombination

Reagents

- An appropriate pAC plasmid (*Figure 3*) containing the foreign gene of interest
- CsCl purified adenovirus DNA (Ad5*dl*309 or Ad5*dl*324) or pJM17 DNA
- Appropriate restriction endonuclease
- Reagents for extraction with phenol and chloroform/isoamyl alcohol, and for ethanol precipitation (see *Protocol 1*)
- 10 mM Tris–HCl pH 7.4, 0.1 mM EDTA

Method

1. (a) If pJM17 is used, it may be purified using any common procedure for preparing bacterial plasmid DNA, taking care to avoid shearing (avoid vigorous vortexing). Dissolve the DNA in 10 mM Tris–HCl pH 7.4, 0.1 mM EDTA at 1 μg/μl.

   (b) If adenovirus DNA is used, digest Ad5*dl*309 DNA with *Xba*I and *Cla*I (or Ad5*dl*324 DNA with *Cla*I), and purify the linearized DNA by extraction with phenol, then chloroform/isoamyl alcohol (24:1), and finally ethanol precipitation (see *Protocol 1*, step 3). Dissolve the DNA in 10 mM Tris–HCl pH 7.4, 0.1 mM EDTA at 1 μg/μl.
2. For every 60 mm dish of cells to be transfected, mix 5 μg of the pAC plasmid and *either* 1 μg restriction enzyme digested adenovirus DNA *or* 2–5 μg of pJM17. Co-precipitate the DNAs with ethanol and redissolve in 50 μl of 10 mM Tris–HCl pH 7.4, 0.1 mM EDTA.
3. Transfect 293 cells with the DNA as described in *Protocol 3*.

3.3 Cell transfection

Irrespective of the strategy adopted to produce recombinant adenovirus (see section 3.2), transfection into 293 cells is carried out as described in *Protocol 3*. Of the available techniques for transfection of mammalian cells, we prefer to use calcium phosphate co-transfection. DEAE–dextran is quite toxic to 293 cells at concentrations normally employed to facilitate DNA transfection. While liposome-mediated transfection of DNA also works well for 293 cells, calcium phosphate co-precipitation is much less expensive and is a commonly used technique in many laboratories.

While, in general, we prefer the approach described in *Protocol 3*, modifications of this protocol also give acceptable results. Thus, following transfection,

Protocol 3. Transfection of DNA into 293 cells by calcium phosphate co-precipitation

Equipment and reagents

- Fetal bovine serum (FBS; Hyclone A1111L)
- Dulbecco's modified Eagle medium (DMEM) with 4.5 g/litre glucose (Gibco-BRL 11965–068)
- 293 cells (American Type Culture Collection, CRL 1573) cultured in DMEM supplemented with 10% fetal bovine serum (FBS)[a]
- 100 × penicillin/streptomycin (Gibco-BRL 15140–031)
- Hepes-buffered saline (HBS): 10 mM Hepes pH 7.15, 137 mM NaCl, 5 mM KCl, 1 mM $MgCl_2$, 0.7 mM Na_2HPO_4 (dissolve in autoclaved water, adjust the pH to 7.15, and then filter through a 0.22 μm filter unit)
- 2 M $CaCl_2$ (filtered through a 0.22 μm Millipore filter and sterilized by autoclaving)
- Purified adenovirus 2 DNA (Gibco-BRL 15270–010)
- 1000 × Fungizone (Gibco-BRL 15295–017)
- *Either* purified pAC plasmid DNA plus pJM17 DNA, *or* pAC plasmid plus *Xba*I/*Cla*I digested Ad5*dl*309 DNA (from *Protocol 2*, step 3), *or* purified ligation products (from *Protocol 1*, step 6)
- Carrier DNA[b]
- DMEM supplemented with 15–18% (v/v) sterile glycerol
- 2 × MEM (Gibco-BRL 11935–012)
- 1.3% Noble agar in water (Difco 0142–02)[c] (autoclave to sterilize, microwave to re-melt)
- 100 mm culture dishes
- 60 mm culture dishes
- Sterile Pasteur pipettes
- 45°C water-bath

Method

1. Propagate 293 cells in 100 mm tissue culture dishes in DMEM supplemented with 10% FBS. 24 h prior to transfection, split a confluent 100 mm dish to three 60 mm dishes. Best results are obtained with cells 95% confluent at the time of transfection.
2. The DNA for transfection is *either* purified pAC plasmid DNA plus pJM17 DNA, *or* pAC plasmid plus *Xba*I/*Cla*I digested Ad5*dl*309 DNA (from *Protocol 2,* step 3), *or* purified ligation products (from *Protocol 1,* step 6). Add the appropriate DNA to 0.5 ml HBS (per dish). Add purified carrier DNA, if necessary, to bring the final quantity of DNA to 20 μg. Mix well to achieve a homogeneous solution.
3. Add 32 μl of 2 M $CaCl_2$ and mix the solution rapidly by vigorous shaking. Allow the mixture to stand at room temperature for 5–10 min to form a fine co-precipitate of calcium phosphate and DNA.
4. Add the precipitate directly to the medium on the culture dishes and gently swirl to mix.
5. Incubate the dishes for 4–6 h at 37°C in a humidified atmosphere of 5% CO_2.
6. Aspirate the transfection medium from the plates and replace it with 4–5 ml DMEM supplemented with 15–18% glycerol for 2–3 min at room temperature.[d]
7. Following the glycerol shock, wash the cells gently with DMEM taking care not to disturb the cell monolayers. Incubate in 5 ml DMEM

supplemented with 10% FBS, antibiotics, and antimycotics overnight.[a]

8. The following day, overlay the monolayers with agar. To do this, re-melt Noble agar and cool it in a 45°C water-bath. Mix it 1:1 just prior to use with 2 × MEM supplemented with 4% FBS, 2 × antibiotics, and 2 × Fungizone that has been pre-warmed to 37°C. Then layer this solution (6 ml for 60 mm culture plates) gently on to the transfected monolayers by addition to the side of the dish. Swirl gently to mix any residual medium on the monolayers into the agar. Allow to gel at room temperature before moving the dishes.
9. Incubate the dishes at 37°C in a humidified atmosphere containing 5% CO_2 for 10–20 days. Monitor the dishes for the appearance of opaque plaques representing individual foci of lytic viral infection (see *Figure 4*). At intervals of approximately seven days, layer an additional 3 ml DMEM/Noble agar on to each transfected dish.
10. As a control for the efficiency of transfection, transfect a parallel dish with 100 ng of purified adenovirus 2 DNA. Typically, approximately 1000 recombinant viral plaques per microgram purified adenoviral DNA are obtained. Recombinant adenovirus plaques typically appear several days after plaques arising from transfection of wild-type adenovirus DNA.
11. Harvest recombinant viral clones from the dishes in step 10 by aspirating plugs of agar overlying well-separated plaques 1–2 mm in diameter using a sterile Pasteur pipette. Disperse each plug into 1 ml DMEM, 2% FBS and subject it to one freeze–thaw cycle. Use the resulting plaque lysate to infect a fresh monolayer of 293 cells for propagation of the virus.

[a] The growth medium may be supplemented with 50 μg/ml penicillin, 10 μg/ml streptomycin, and/or 25 μg/ml Fungizone, but not with gentamicin as this antibiotic kills 293 cells. Normally, the cells are not maintained in antibiotic-containing medium as this promotes poor sterile technique. Split the cells 1:5 at three to four day intervals. Since lower passage number cells tend to yield higher transfection efficiencies and to form plaques more quickly, a cell bank containing low passage number 293 cells should be set-up in liquid nitrogen at the earliest opportunity.

[b] Low molecular weight DNA works best as carrier, since high molecular weight DNA favours the formation of large aggregates. High molecular weight HeLa cell DNA can be conveniently sheared by repeated passage through a 28 gauge needle or by sonication. Alternatively, plasmid DNA (e.g. pUC19) can be used as carrier.

[c] Some batches of Noble agar are not suitable for this purpose since they do not gel properly, although this is an infrequent problem.

[d] This concentration of glycerol and the duration of the glycerol shock have been found to increase the transfection efficiency.

the monolayers can be re-incubated in DMEM, 2% FBS until more than 90% of the cells show evidence of infection. The cells can then be lysed by one freeze–thaw cycle and the viral lysate collected and used to infect fresh monolayers of 293 cells at an appropriate dilution. The newly infected cells are then overlaid with DMEM/Noble agar, and individual plaques isolated as described in *Protocol 3*, step 11. Following infection, plaques typically appear within approximately seven to ten days. Some investigators have found plaques more readily identifiable using this variation. This is particularly true if the transfection procedure results in significant disruption of the 293 cell monolayer, since clumps of cells can be confused with recombinant viral plaques. As another variation, some investigators prefer to perform transfections in 100 mm culture dishes rather than 60 mm dishes. With medium volumes scaled appropriately, we observe no difference in efficiency.

3.4 Characterization of adenovirus recombinants

Potential recombinant clones generated by the procedures described above must be checked to ensure that they contain the foreign gene of interest and that no rearrangements have occurred. This is most readily carried out by purifying recombinant adenovirus DNA as described in *Protocol 4* followed by PCR analysis or restriction enzyme mapping (see refs 18, 19).

Protocol 4. Characterization of recombinant adenoviral DNA

Equipment and reagents

- Hirt buffer: 0.5% SDS, 10 mM EDTA pH 8.0[a]
- 2 mg/ml Proteinase K (Fisher BP1700–100) in water (stored at −20°C)
- 293 cells grown on 60 mm culture dishes (see *Protocol 3*)
- Plaque lysate (see *Protocol 3*, step 12)
- DMEM, 2% FBS with penicillin/streptomycin (see *Protocol 3*)
- 5 M NaCl
- Reagents for extraction of DNA with phenol and chloroform/isoamyl alcohol (24:1), and for ethanol precipitation (see *Protocol 1*)
- Polypropylene centrifuge tubes
- TE buffer: 10 mM Tris–HCl pH 7.4, 1 mM EDTA
- TE buffer containing 10 μg/ml RNase A

Method

1. Grow 293 cells to confluence in 60 mm tissue culture dishes. For each dish, aspirate the culture medium and infect the monolayers by adding 1 ml of plaque lysate (from *Protocol 3*, step 12) directly to the culture dish.
2. Incubate the cultures at 37°C and rock at 15 min intervals for 1 h.
3. Add an additional 4 ml DMEM, 2% FBS with penicillin/streptomycin and re-incubate the dishes at 37°C until > 90% of the cells demonstrate infection (typically five to seven days).
4. When extensive cell death is observed, remove the medium and save this as a source of virus (P_1 stock). Lyse the adherent cells by the

addition of 2 ml Hirt buffer supplemented with 20 μg/ml Proteinase K directly to the culture dish. Incubate the plates at 37°C for 1 h to digest covalently attached terminal protein from the viral DNA.

5. Add 0.25 vol. of 5 M NaCl to the cell lysate and collect the viscous solution using a wide-bore pipette and transfer it to a polypropylene centrifuge tube. After mixing by gentle inversion, place the tube on ice for 4–16 h.
6. Pellet the high molecular weight cellular DNA by centrifugation at 13000 *g* for 20 min at 4°C. Recover the supernatant (containing viral DNA) and sequentially extract this with phenol and chloroform/isoamyl alcohol (24:1), then precipitate the nucleic acids with ethanol (see *Protocol 1*, step 12).
7. Dissolve the DNA pellet in 1 ml TE buffer containing 10 μg/ml RNase A and incubate at 37°C for 30 min.
8. Purify the DNA by extraction with phenol, chloroform/isoamyl alcohol, (24:1) and then precipitate it with ethanol.
9. Redissolve the DNA pellet in 100 μl TE buffer and analyse it for the presence of the foreign gene by restriction endonuclease mapping and Southern blotting (see refs 18, 19), or by PCR analysis.

[a] See ref. 20.

4. Propagation of recombinant adenoviruses

Once recombinant adenoviruses have been obtained and verified for the proper DNA structure, they are usually tested for functional expression of the foreign gene by any convenient assay (18, 19). Adenovirus can be propagated to very high titre on 293 cell monolayers as described in *Protocol 5* for small scale growth of virus stocks and in *Protocol 6* for large scale growth and purification of virions. Following preparation of the crude virus stock or the purified virion particles, the titre of the preparation is determined by plaque assay on 293 cells (see *Protocol 7*). This procedure also serves as a cloning step and allows for the isolation of a pure population of virus. In addition, we have found that repeated plaque assay and re-cloning of the large plaques that first appear on the dish tends to select for a population of recombinant adenovirus that grows both more rapidly and to higher titre. Frequent re-cloning also will eliminate any virus that has rearranged, although we have not observed any rearrangement of the genome with passage of the virus providing the packaging limit of adenovirus has not been exceeded.

Protocol 5. Small scale propagation of recombinant adenovirus stocks

Reagents

- 293 cells cultured in 100 mm tissue culture plates in DMEM, 10% FBS
- Plaque lysate (prepared as described in *Protocol 3*, step 11)
- DMEM, 2% FBS

Method

1. Infect fresh monolayers of 293 cells in 100 mm dishes with 1 ml per dish of plaque lysate (see *Protocol 3*, step 11). Incubate the cells at 37°C until >90% of the cells show evidence of infection.
2. Lyse the cells by one freeze–thaw cycle.
3. Collect the conditioned medium/viral lysate and clarify it by centrifugation at 13 000 *g* for 10 min at 4°C. This viral supernatant (derived by infection of cells with a plaque lysate) is a primary (P_1) stock. Store it in aliquots at −20°C until use.
4. Secondary (P_2) stocks of recombinant virus can be generated in a similar manner after infection of fresh monolayers of 293 cells with primary virus stock (from step 3) using a multiplicity of infection (plaque-forming unit (p.f.u.)/cell ratio) of 0.001–0.01. In practice, this is usually about 1 ml of a 1:100 or 1:1000 dilution of the virus stock.

Protocol 6. Large scale preparation of purified recombinant adenovirus

Equipment and reagents

- 293 cells cultured in 150 mm tissue culture plates in DMEM supplemented with 10% FBS
- P_1 or P_2 virus stock (see *Protocol 5*)
- DMEM, 2% FBS
- 10% Nonidet P-40 (NP-40; US Biochemical Corp. 19628) in sterile water
- 250 ml Nalgene centrifuge bottles
- Sorvall RC-5B centrifuge with GSA and SS-34 rotors (or equivalent)
- 20% (w/v) polyethylene glycol (PEG) 8000 (Fisher BP233–1), 2.5 M NaCl (filter through a 0.45 μm filter and autoclave)
- CsCl (density 1.10 g/ml) in 20 mM Tris–HCl pH 8.0 (filter through a 0.45 μm filter)
- CsCl (density 1.40 g/ml) in 20 mM Tris–HCl pH 8.0 (filter through a 0.45 μm filter)
- CsCl (density 1.30 g/ml) in 20 mM Tris–HCl pH 8.0 (filter through a 0.45 μm filter)
- Sterile 14 ml polypropylene tubes
- Sorvall TH641 ultracentrifuge rotor, tubes, and ultracentrifuge (or equivalent)
- Isotonic saline buffer: 10 mM Tris–HCl pH 7.4, 137 mM NaCl, 5 mM KCl, 1 mM $MgCl_2$ (filter and sterilize by autoclaving)
- Sepharose CL-4B (Pharmacia LKB Biotechnology 17–0150–01) equilibrated with isotonic saline buffer and packed into an appropriate size column for chromatography (e.g. 10–30 ml for the volumes in the protocol described below)
- 10 mg/ml BSA in autoclaved water (filter sterilize through a 0.45 μm filter)
- Sterile 6 ml polystyrene tubes
- Nalgene screw-top cryotubes, 2 ml size

Method

1. Culture the 293 cells in 150 mm tissue culture plates in DMEM, 10% FBS to near 100% confluency.[a]
2. Aspirate the medium and infect the dishes by adding P_1 or P_2 virus stock diluted into 2.5 ml DMEM, 2% FBS. A multiplicity of infection (m.o.i.) of 0.01 is typically used. Rock the plates at 15 min intervals for 1 h.
3. Re-feed the dishes with 15 ml DMEM, 2% FBS, and incubate at 37°C until $>$ 90% of the cells are infected (typically about five to seven days).
4. Lyse the infected monolayers by adding 10% Nonidet P-40 to the culture medium to a final concentration of 0.5%.
5. Collect the virus-containing lysate in 250 ml Nalgene centrifuge bottles. Centrifuge at 12 000 *g* for 10 min at 4°C in a Sorvall GSA rotor (or equivalent) to remove cellular debris.
6. Transfer the supernatant in 160 ml aliquots to fresh sterile Nalgene centrifuge bottles and add 80 ml (0.5 vol.) 20% PEG 8000, 2.5 M NaCl to each bottle. Incubate on ice for 1 h to precipitate the virus particles.
7. Collect the precipitated virus by centrifugation at 12 000 *g* for 20 min at 4°C.
8. Resuspend the viral pellet in 5 ml (per 250 ml centrifuge bottle) of CsCl (density 1.10 g/ml) in 20 mM Tris–HCl pH 8.0. Transfer the virus to a 14 ml polypropylene centrifuge tube and pellet residual debris by centrifugation for 5 min at 8000 *g* at 4°C in a Sorvall SS-34 rotor (or equivalent).
9. Prepare CsCl step gradients in TH641 ultracentrifuge tubes (or equivalent) by adding 2.0 ml CsCl (density=1.40 g/ml) in 20 mM Tris–HCl pH 8.0 to the bottom of the tube. Carefully layer 3.0 ml CsCl (density=1.30 g/ml) in 20 mM Tris–HCl pH 8.0 over the previous layer without disturbing the interface. Mark the location of the interface on the side of the tube.
10. Carefully layer the virus from step 8 over the CsCl step gradient using 5 ml virus per tube.
11. Centrifuge in TH641 rotor (or equivalent) at 22 000 r.p.m. (60 000 *g*) for 2 h at 20°C.
12. After centrifugation, the recombinant viral particles appear as a white opalescent band at the interface between the 1.30/1.40 CsCl layers. Aspirate off the top layers carefully without disturbing the virus band.
13. Harvest the virus using a sterile Pasteur pipette and transfer it to a sterile polypropylene tube.

Protocol 6. *Continued*

14. Apply the harvested virus to the Sepharose CL-4B column and elute in isotonic saline buffer. The virus elutes at the void volume of the column, and is readily visible by its opalescence in the effluent. Collect the entire viral volume in sterile 6 ml polystyrene tubes.
15. Add sterile 10 mg/ml BSA to a final concentration of 0.1 mg/ml to stabilize the virus upon freezing.
16. Aliquot the purified virus in 2 ml Nalgene screw-top cryovials and snap freeze in liquid nitrogen.
17. Store the virus aliquots at −70°C; the infectious titres are stable for months when stored in this manner. Alternatively, purified virus can be stored on ice in the cold room. Under these conditions it is stable for at least one month.
18. Freeze a small aliquot of purified virus (10–20 μl) separately. Use this to determine the titre of the purified virus preparation (see *Protocol 7*) after storage at −70°C followed by one freeze–thaw cycle to reflect the subsequent handling of the entire preparation.

[a] We typically prepare virus from 20–60 plates simultaneously, although the procedure can be scaled to fit individual requirements.

Protocol 7. Plaque assay

Equipment and reagents

- Virus stock to be titred
- DMEM, 2% FBS
- 1.3% Noble agar
- 2 × MEM, 4% FBS
- Sterile 6 ml polystyrene tubes
- 293 cells cultured in 100 mm dishes in DMEM supplemented with 10% FBS
- 60 mm tissue culture dishes

Method

1. On the day before the procedure, split each 100 mm plate of 293 cells to three 60 mm dishes. The number of dishes prepared should be six per virus stock to be titred (two for each dilution to be assayed). Care should be taken to prevent clumps of cells in the monolayer by vigorously dispersing the cell suspension. The dishes should be just confluent when used.
2. On the day of the procedure, thaw the virus stock to be titred. Prepare serial 10- or 100-fold dilutions of the virus stocks to be titred in DMEM, 2% FBS and penicillin/streptomycin in 6 ml polystyrene culture tubes in a final volume of 2.0 ml.

3. Aspirate the growth medium from the monolayers of 293 cells.
4. Overlay two 60 mm plates of 293 cells with 0.5 ml of each prepared virus dilution. Record on each dish details of the virus and the dilution of stock.[a]
5. Incubate the cells at 37 °C for 1 h, rocking at 15 min intervals.
6. Melt 1.3% Noble agar in a microwave and place in a 45 °C water-bath for it to equilibrate. Place 2 × MEM supplemented with 4% FBS in a 37 °C water-bath. Immediately before using, add an equal volume of the molten 1.3% Noble agar to the 2 × MEM.
7. Aspirate infectious virus medium from the cells.
8. Overlay the cells with 6.0 ml of the Noble agar/MEM, 2% FBS. This must be done very gently so as not to disturb the cell monolayer; 293 cells are not strongly adherent, so care is important. Add the agar preparation slowly to the side of the dish. Rock or swirl the dish very gently to cover the cells and mix in any remaining trace of liquid DMEM. Allow the overlaid cultures to stand at room temperature until the agar solidifies.
9. Incubate the infected plates at 37 °C. Plaques should begin to be visible within about seven days.
10. Count the plaques. Do this as follows:
    (a) Observe the plates against a dark background, or against fluorescent lighting, whichever makes the plaques most readily visible (see *Figure 4*). Small plaques will appear as white pin-points on the relatively clear background. As the plaques grow, they will appear as larger white spots on the cell lawn and later will form white rings. Cell clumps look macroscopically like plaques at first, but will not grow. Microscopically, plaques are comprised of rounded up cell areas on the monolayer, whereas cell clumps look just as their name suggests as a very dense, raised area of cells.
    (b) Mark the location of visible plaques with a felt-tipped pen.
    (c) Count the plaques over a period of several days until the number remains constant.
11. To determine the titre, multiply the number of plaques counted by the dilution factor. Average the results of the serial dilutions.

[a] For example, 1 μl of stock, diluted to 1 ml equals 1:1000 dilution. Infecting with 0.5 ml of a 1:1000 dilution results in a final dilution factor of 1:2000. For large scale purified virus preparations, assaying dilutions of 10^{-8}, 10^{-9}, and 10^{-10} generally provides an appropriate range.

Figure 4. Photograph of a dish of 293 cells infected with a recombinant adenovirus vector and overlayed with Noble agar for plaque assay as described in *Protocol 7*. Small plaques appear as white dots. Larger plaques appear as white rings.

5. Isolation of virion DNA

Isolation of virion DNA for construction of recombinant adenovirus can be achieved as described in *Protocol 8*.

Protocol 8. Isolation of virion DNA

Equipment and reagents

- Equipment and reagents for propagation of adenovirus and CsCl centrifugation (see *Protocol 6*)
- Hirt buffer: 0.5% SDS, 10 mM EDTA pH 8.0
- 10 mg/ml Proteinase K
- Reagents for extraction with phenol and chloroform/isoamyl alcohol, and ethanol precipitation (see *Protocol 1*)
- 10 mM Tris–HCl pH 7.4, 0.1 mM EDTA

Method

1. Propagate virus and purify viral particles by discontinuous CsCl density gradient centrifugation as described in *Protocol 6*.
2. Following centrifugation, dilute the virus at least fivefold with Hirt buffer.[a] Incubate at 50°C for 10 min to disrupt the viral particles.
3. Add Proteinase K to 20 μg/ml final concentration. Incubate at 37°C for 1 h.

4. Extract the DNA twice with phenol, then extract with chloroform/isoamyl alcohol (24:1), and precipitate with ethanol.
5. Pellet the DNA by centrifugation at 13000 *g* for 5 min. Dissolve it in 10 mM Tris–HCl pH 7.4, 0.1 mM EDTA.

[a] Desalting of the virus over Sepharose CL-4B is not necessary for the preparation of viral DNA.

6. Biosafety considerations

Recombinant viruses described herein are very efficient vectors for the introduction of foreign genetic material into mammalian cells. Recombinant viruses of the E3 replacement class are additionally replication-competent, and capable of infecting humans and other permissive species. All recombinant adenoviruses should therefore be handled as biohazardous materials. Applicable guidelines for their handling must be followed meticulously and appropriate biocontainment precautions must be observed. Standard decontamination procedures, including autoclaving or exposure to bleach, are sufficient to inactivate adenovirus. Insertion of foreign genes encoding potentially toxic or hazardous products into adenovirus vectors should be avoided.

Acknowledgements

The authors wish to express their gratitude to colleagues, fellows, and students at U.T. Southwestern for many helpful discussions, products of which have been incorporated without attribution. Robert Munford, Randall Moreadith, and Catherine Hedrick were kind enough to read the manuscript. Studies of adenovirus-mediated gene transfer in our laboratories are supported in part by funds from the NHLBI Specialized Center of Research in Ischemic Heart Disease (HL17669), a Grant-In-Aid from the American Heart Association (R.S.M.), and Texas Higher Education Coordinating Board Advanced Technology Program Grants Nos. 3660020 (R.S.M.) and 3660016 (R.D.G.). R.D.G. is the recipient of an American Heart Association-Genentech Established Investigator Award.

References

1. van Doren, K., Hanahan, D., and Gluzman, Y. (1984). *J. Virol.*, **50**, 606.
2. Herz, J. and Gerard, R. D. (1993). *Proc. Natl Acad. Sci. USA*, **90**, 2812.
3. Ishibashi, S., Brown, M. S., Goldstein, J. L., Gerard, R. D., Hammer, R. E., and Herz, J. (1993). *J. Clin. Invest.*, **92**, 883.
4. Berkner, K. L. (1988). *BioTechniques*, **6**, 616.
5. Gerard, R. D. and Meidell, R. S. (1993). *Trends Cardiovasc. Med.*, **3**, 171.

6. Alcorn, J. L., Gao, E., Chen, Q., Smith, M. E., Gerard, R. D., and Mendelson, C. R. (1993). *Mol. Endocrinol.*, **7**, 1072.
7. Gómez-Foix, A. M., Coats, W. S., Baqué, S., Alam, T., Gerard, R. D., and Newgard, C. B. (1992). *J. Biol. Chem.*, **267**, 25129.
8. McPhaul, M. J., Despleyere, J.-P., Allman, D. R., and Gerard, R. D. (1993). *J. Biol. Chem.*, **268**, 26063.
9. Becker, T. C., Noel, R. J., Coats, W. S., Gómez-Foix, A. M., Alam, T., Gerard, R. D., *et al.* (1994). In *Methods in cell biology: protein expression in animal cells* (ed. M. Roth), pp. 161–189. Academic Press, San Diego.
10. Graham, F. L. and Prevec, L. (1991). In *Methods in molecular biology, Vol. 7: gene transfer and expression protocols* (ed. E. J. Murray), pp. 109–128. Humana Press, Clifton, NJ.
11. Graham, F. L., Smiley, J., Russell, W. C., and Nairn, R. (1977). *J. Gen. Virol.*, **36**, 59.
12. Stow, N. (1981). *J. Virol.*, **37**, 171.
13. Gluzman, Y., Reichl, H., and Solnick, D. (1982). In *Eukaryotic viral vectors* (ed. Y. Gluzman), pp. 187–192. Cold Spring Harbor Laboratory Press, Cold Spring Harbor, NY.
14. Jones, N. and Shenk, T. (1979). *Cell*, **17**, 683.
15. Thomsen, D. R., Stenberg, R. M., Goins, W. F., and Stinski, M. F. (1984). *Proc. Natl Acad. Sci. USA*, **81**, 659.
16. McGrory, W. J., Bautista, D. S., and Graham, F. L. (1988). *Virology*, **163**, 614.
17. Thimmappaya, B., Weinberger, C., Schneider, R. J., and Shenk, T. (1982). *Cell*, **31**, 543.
18. Sambrook, J., Fritsch, E. F., and Maniatis, T. (ed.) (1989). *Molecular cloning: a laboratory manual*. Cold Spring Harbor Laboratory Press, Cold Spring Harbor, NY.
19. Glover, D. M. and Hames, B. D. (ed.) (1995). *DNA cloning 1: a practical approach. Core techniques*. Oxford University Press, Oxford and New York.
20. Hirt, B. (1967). *J. Mol. Biol.*, **26**, 365.

A1

List of suppliers

American Type Culture Collection, 12301, Parklawn Drive, Rockville, MD 20852–1776, USA.

Amersham

Amersham Corporation, 2636 South Clearbrook Drive, Arlington Heights, IL 60005, USA.

Amersham International plc., Lincoln Place, Green End, Aylesbury, Buckinghamshire HP20 2TP, UK.

Anderman and Co. Ltd., 145 London Road, Kingston-Upon-Thames, Surrey KT17 7NH, UK.

Arnold R. Horwell, 73 Maygrove Road, West Hampstead, London NW6 2BP, UK.

Bayer, Leverkusen, Germany.

Beckman Instruments

Beckman Instruments UK Ltd., Progress Road, Sands Industrial Estate, High Wycombe, Buckinghamshire HP12 4JL, UK.

Beckman Instruments Inc., PO Box 3100, 2500 Harbor Boulevard, Fullerton, CA 92634, USA.

Becton Dickinson

Becton Dickinson and Co., 2 Bridgewater Lane, Lincoln Park, NJ 07035, USA.

Becton Dickinson UK Ltd., Between Towns Road, Cowley, Oxford OX4 3LY, UK

Bio 101 Inc., PO Box 2284, La Jolla, CA 92038–2284, USA; c/o **Stratech Scientific Ltd.**, 61–63 Dudley Street, Luton, Bedfordshire LU2 0HP, UK.

Biometra Ltd., PO Box 167, Maidstone, Kent ME14 2AT, UK.

Bio-Rad Laboratories

Bio-Rad Laboratories Ltd., Bio-Rad House, Maylands Avenue, Hemel Hempstead HP2 7TD, UK.

Bio-Rad Laboratories, Division Headquarters, 3300 Regatta Boulevard, Richmond, CA 94804, USA.

Biotechnologies International, 61300 L'Aigle-France, Capital Social 500 000F, No. Siren 342246 840 00017, France.

Boehringer Mannheim

Boehringer Mannheim Ltd., Bell Lane, Lewes, East Sussex BN17 1LG, UK.

Boehringer Mannheim Corporation, Biochemical Products, 9115 Hague Road, PO Box 504, Indianapolis, IN 46250–0414, USA.
Boehringer Mannheim Biochemicals GmbH, Sandhofer Str. 116, Postfach 310120 D-6800 Mannheim 31, Germany.

British Drug Houses (BDH) Ltd., Poole, Dorset, UK.

BTX, 11199 A Sorrento Valley Road, San Diego, CA 92121, USA.

Carl Zeiss Jena, Mikroscopie Division, PO Box 1380, D-7082, Oberkochen, Germany.

Clark Electromedical Instruments, PO Box 8, Pangbourne, Reading RG8 7HU, UK.

Costar, One Alwife Center, Cambridge, MA 02140, USA.

Dako
Dako Corporation, 6392 Via Read, Carpinteria, CA 93013, USA.
Dako Ltd., 22 The Arcade, The Octagon, High Wycombe, Buckinghamshire HP11 2HT, UK.

David Kopf Instruments, 7324 Elmo Street, PO Box 636, Tujunga, CA 9104A–0636, USA.

Difco Laboratories
Difco Laboratories, PO Box 331058, Detroit, MI 48232–7058, USA.
Difco Laboratories Ltd., PO Box 14B, Central Avenue, West Molesey, Surrey KT8 2SE, UK.

DuPont
DuPont De Nemours, International SA., NEN Products, Pumpwerkstrasse 15, 8105 Regensdorf, Switzerland;
Dupont (UK) Ltd., Industrial Products Division, Wedgwood Way, Stevenage, Hertfordshire SG1 4QN, UK.
DuPont Co. (Biotechnology Systems Division), PO Box 80024, Wilmington, DE 19880–0024, USA.

Electron Microscopy Sciences, Fort Washington, PA 19034, USA.

European Collection of Animal Cell Culture (ECACC), Division of Biologics, PHLS Centre for Applied Microbiology and Research, Porton Down, Salisbury, Wiltshire SP4 0JG, UK.

Falcon (Falcon is a registered trademark of Becton Dickinson and Co.).

Fisher Scientific (Distributors for Nalgene and Corex brand tubes) (Headquarters), 711 Forbes Avenue, Pittsburgh, PA 15219–4785, USA.

Flow Laboratories Ltd., Woodcock Hill, Harefield Road, Rickmansworth, Hertfordshire WD3 1PQ, UK.

Fluka
Fluka-Chemie AG., CH-9470, Buchs, Switzerland.
Fluka Chemicals Ltd., The Old Brickyard, New Road, Gillingham, Dorset SP8 4JL, UK.

Gibco
Gibco-BRL (Life Technologies), Trident House, Renfrew Road, Paisley PA3 4EF, UK

Gibco-BRL (Life Technologies Inc.), 3175 Staler Road, Grand Island, NY 14072–0068, USA.
Harvard Apparatus, 22 Pleasant Street, South Natick, MA 01760, USA.
Heat Systems Ultrasonics, 1938 New Highway, Farmingdale, NY 11735, USA.
Hitachi, 6 Kanda-Surugadai 4-chome, Chiyoda-ku, Tokyo 101, Japan.
Hoffmann-La Roche Ltd., Basel, Switzerland.
Hybaid
Hybaid Ltd., 111–113 Waldegrave Road, Teddington, Middlesex TW11 8LL, UK.
Hybaid, National Labnet Corporation, PO Box 841, Woodbridge, NJ 07095, USA.
HyClone Laboratories, 1725 South HyClone Road, Logan, UT 84321, USA.
International Biotechnologies Inc., 25 Science Park, New Haven, Connecticut 06535, USA.
Invitrogen Corporation, 3985 B Sorrento Valley Building, San Diego, CA 92121, USA; c/o **British Biotechnology Products Ltd.,** 4–10 The Quadrant, Barton Lane, Abingdon, OX14 3YS, UK.
Janke and Kunkel GmbH, D-7813 Staufen, Germany.
Janssen Pharmaceuticals, Beerse, Belgium.
JRH Bioscience
JRH Biosciences, 13804 W. 107th Street, Lenexa, Kansas 66215, USA.
JRH Biosciences, Hophurst Lane, Crawley Down, Sussex RH10 4FF, UK.
Kodak:Eastman Fine Chemicals, 343 State Street, Rochester, NY, USA.
Leica
Leica Microskopie Umb Systems GmbH, D-6330 Wetzlar, Germany.
Leica UK Ltd., Davy Avenue Knowlhill, Milton Keynes MK5 8LB, UK.
Leitz, Davy Avenue, Knowlhill, Milton Keynes MK5 8LB, UK.
Life Technologies Inc., 8451 Helgerman Court, Gaithersburg, MN 20877, USA.
Lumac bv, PO Box 31101, 6370 AC Landgraaf, The Netherlands.
3M Company, St. Paul, MN 55144–1000, USA.
Merck
Merck Inc., 5 Skyline Drive, Nawthorne, NY 10532, USA;
Merck, Frankfurter Strasse 250, Postfach 4119, D-64293 Darmstadt, Germany.
Millipore
Millipore, The Boulevard, Blackmoor Lane, Watford, Hertfordshire WD1 8YW, UK;
Millipore Corp./Research, PO Box 255, 80 Ashby Road, Bedford, MA 01730, USA.
Nalge Company, Box 20365, Rochester, NY 14602–0365, USA.
Narishige Co. Ltd., Minamikarauyama 4-chome, Setagaya-ku, Tokyo, Japan.
New England Biolabs (NBL), 32 Tozer Road, Beverley, MA 01915–5510, USA; c/o **CP Labs Ltd.,** PO Box 22, Bishops Stortford, Hertfordshire CM23 3DH, UK.

Nikon Corporation, Fuji Building, 2–3 Marunouchi 3-chome, Chiyoda-ku, Tokyo, Japan.

Orme Scientific, PO Box 3, Stakehill Industrial Park, Middleton, Manchester M24 2RH, UK.

Perkin-Elmer

Perkin Elmer-Cetus, 761 Main Avenue, Norwalk, CT 0689, USA;

Perkin-Elmer Ltd., Beaconsfield, Buckinghamshire HP9 1QA, UK.

Perkin-Elmer Ltd., Post Office Lane, Beaconsfield, Buckinghamshire HP9 1QA, UK.

Pharmacia

Pharmacia Biotech Europe, Procordia EuroCentre, Rue de la Fuse-e 62, B-1130 Brussels, Belgium.

Pharmacia Biosystems Ltd., Davy Avenue, Knowlhill, Milton Keynes MK5 8PH, UK;

Pierce

Pierce, 3747 North Meridan Road, PO Box 117, Rockford, IL 61105, USA; c/o **Life Science Labs Ltd.,** Sedgewick Road, Luton, Bedfordshire LU4 9DT, UK;

Pierce, Pierce Europe BV, Box 1512, 3260 BA Ous-Beijerland, The Nederlands.

Planer Products Ltd., Windmill Road, Sunbury on Thames, Middlesex TW16 7HD, UK.

Promega

Promega Corporation, 2800 Woods Hollow Road, Madison, WI 53711–5399, USA.

Promega Ltd., Delta House, Enterprise Road, Chilworth Research Centre, Southampton SO1 4NS, UK.

Qiagen Inc., 9259 Eton Avenue, Chatsworth, CA 91311, USA; c/o **Hybaid,** 111–113 Waldegrave Road, Teddington, Middlesex TW11 8LL, UK.

Schleicher and Schuell

Schleicher and Schuell Inc., Keene, NH 03431A, USA;

Schleicher and Schuell Inc., D-3354 Dassel, Germany;

c/o *Schleicher and Schuell Inc.,* Anderman and Company Ltd., UK.

Shandon Scientific Ltd., Chadwick Road, Astmoor, Runcorn, Cheshire WA7 1PR, UK.

Sigma Chemical Company

Sigma Chemical Company, Fancy Road, Poole, Dorset BH17 7NH, UK;

Sigma Chemical Company, 3050 Spruce Street, PO Box 14508, St. Louis, MO 63178–9916, USA.

Sorvall DuPont Company, Biotechnology Division, PO Box 80022, Wilmington, DE 19880–0022, USA.

Stoelting, Wood Dale, Illinois, USA.

Stratagene
Stratagene Ltd., Unit 140, Cambridge Innovation Centre, Milton Road, Cambridge CB4 4FG, UK;
Stratagene Ltd., 11011 North Torrey Pines Road, La Jolla, CA 92037, USA.

Tel-Test Inc., PO Box 1421, 1511 County Road 129, Friendswood, TX 77546, USA.

Thomas Scientific, Swedesboro, NJ, USA.

Turner Designs, 845 W. Maude Avenue, Sunnyvale, CA 94086, USA.

United States Biochemical, PO Box 22400, Cleveland, OH 44122, USA.

VWR Scientific (distributor for Neubauer Counting Chamber/haemocytometer), PO Box 1002, 600C Corporate Court, South Plainfield, NJ 07080, USA.

Wellcome Reagents, Langley Court, Beckenham, Kent BR3 3BS, UK.

Index